Stability and Periodic Solutions of Ordinary and Functional Differential Equations

T. A. BURTON

Department of Mathematics
Southern Illinois University
Carbondale, Illinois

DOVER PUBLICATIONS, INC.
Mineola, New York

Bibliographical Note

This Dover edition, first published in 2005, is an unabridged, slightly corrected version of the edition published by Academic Press, Inc., Orlando, Florida, 1985.

Library of Congress Cataloging-in-Publication Data

Burton, T. A. (Theodore Allen), 1935–
    Stability and periodic solutions of ordinary and functional differential equations / T.A. Burton.
    p. cm.
    Originally published: Orlando, Fla. : Academic Press, 1985.
    Includes bibliographical references and index.
    ISBN 0-486-44254-3 (pbk.)
    1. Differential equations—Numerical solutions. 2. Functional differential equations—Numerical solutions. I. Title.

QA372.B885 2005
518'.63—dc22

          2004065114

Manufactured in the United States of America
Dover Publications, Inc., 31 East 2nd Street, Mineola, N.Y. 11501

To Professor Taro Yoshizawa

Stability and Periodic Solutions of Ordinary and Functional Differential Equations

Contents

Preface

This book contains a discussion of a broad class of differential equations. It was motivated by a series of results obtained during the past three years which bring into a single perspective the qualitative theory of ordinary differential equations, Volterra integrodifferential equations, and functional differential equations with both bounded and infinite delay. It was written to appeal to two audiences.

The book contains a basic introduction to ordinary differential equations suitable for a first graduate course; in fact, this is the material developed by the author over the past twenty-one years for such classes. We make very gentle transitions from that material to counterparts in the theory of Volterra and in functional differential equations with both finite and infinite delay. This makes the material very accessible and, we believe, ideal for a second semester graduate course in differential equations.

The book is, however, written primarily for mathematicians, engineers, physicists, and other scientists who investigate and apply a wide variety of differential equations. The accent throughout the book is on the unity and cohesiveness of the subject. Thus, we seek to show the applicability of techniques of ordinary differential equations to functional differential equations and to expose problems concerning the more general equations.

Chapter 0 contains an extensive overview of the book; it describes what the book does and how it accomplishes its goals. Chapter 1 discusses as a unit the structure of the solution space and the stability and periodic properties of linear ordinary and Volterra differential equations. Chapter 2 contains an extensive collection of applied problems using many types of differential equations. Chapter 3 discusses the background for and application to differential equations of the fixed-point theorems of Banach, Brouwer, Browder, Horn, Schauder, and Tychonov, including the asymptotic fixed-point theorems. Chapter 4 provides a unified presentation of the basic stability and periodicity theory for nonlinear ordinary and functional differential equations. Two main themes are developed: first, we show how the major results of Liapunov's direct method for ordinary differential equations can be advanced to functional differential equations with

ix

finite or infinite delay, focusing on very recent results; next, we show how certain boundedness properties of solutions of either ordinary or functional differential equations can be put into the framework of Horn's fixed-point theorem yielding periodic solutions.

I am indebted to many for assistance with the book: to Professors J. Haddock, C. Langenhop, W. Mahfoud, A. Somolinos, and S. Zhang for reading and criticizing portions of the manuscript; to the graduate students who studied from the manuscript and made suggestions, and particularly to David Dwiggins, Roger Hering, and Shou Wang; to the editors of Academic Press for their interest in and help with the project; to my wife, Fredda, for drawing the figures; and a very special indebtedness to Linda Macak for typing the manuscript.

Concerning the present Dover edition, it is a retyped version of the first printing and corrects several misprints, but is otherwise unchanged. It was not possible to make the pages correspond exactly. Thus, if a reader encounters a reference to a particular result from the first printing, it may be necessary to look for the result in nearby pages of this printing. I send a special thanks to Charles and Linda Gibson for typing the Latex copy and to the editor, John Grafton, for his interest in including the book in the Dover series.

T. A. Burton
Northwest Research Institute
Port Angeles, WA 98362
taburton@olypen.com
November 2004

Chapter 0

An Overview

To a reader who is unacquainted with differential equations, an overview such as this can be quite meaningless. Thus, for such readers, we state that Section 1.1 of Chapter 1, together with Section 4.1 of Chapter 4, will provide a basic introduction to theory of ordinary differential equations and constitutes a very standard and self-contained one-semester beginning graduate course. This is essentially the material used for such courses taught by the author during the past 21 years. And the reader without background in differential equations may do well to ignore the rest of this overview until those two sections are well in hand.

Most authors choose to treat separately the subjects of ordinary differential equations, Volterra equations, equations with bounded delay, and functional differential equations with infinite delay. Much too frequently the reader encounters these as entirely different subjects requiring a different frame of mind, different notation, different vocabulary, different techniques, and having different goals. [An exception to this is found in the excellent introductory book by Driver (1977) who presents the elementary theory of differential and finite delay equations in a unified manner.] When one moves from ordinary differential equations to functional differential equations it seems that one moves into a different world. One of the goals here is to show that many aspects of these worlds are so similar that one may move freely through them so that the distinctions between them are scarcely noticed.

But this is not merely a book on point of view. During the last three years there have been very significant advances in stability theory of various types of functional differential equations so that they can be brought into line with classical ordinary differential equations. There have also been dramatic advances in periodic theory. One may now discern unifying themes

1

through very different types of problems which we attempt to expose in this book.

Thus, the second group of readers we address consists of those mathematicians, engineers, physicists, and other scientists who are interested in the broad theory of differential equations both from the point of view of applications and further investigations. Much of the material in this book is taken from preprints of papers soon to be published in research journals. It provides a synthesis of recent results.

And the third group of readers we address consists of those who have completed an introductory graduate course in ordinary differential equations and are interested in broadening their base, as opposed to exploring fewer concepts at great depth such as is accomplished by a study of the excellent book of Hartman (1964). We recommend a second course of study as follows:

(a) A quick reading of Chapter 2 allows one to focus on a broad range of applications pointing out the need for a variety of types of differential equations.

(b) Follow (a) with a careful study of Sections 1.2, 1.3, and 1.4, a quick look at Sections 1.4.1 and 1.5, and a careful study of Section 1.6.

(c) Chapter 3 contains fundamental fixed point theory and existence theorems.

(d) Finish with the balance of Chapter 4.

The remainder of this overview is aimed at the seasoned investigator of some area of ordinary or functional differential equations. It should be helpful for any reader to frequently review it in order to maintain an overall perspective. For this is not a collection of results discovered about the subject, but rather it is an exposition of the cohesion of a very large area of study. Thus, in no manner is this a comprehensive treatment. With each topic we give references for full discussion which the reader can consult for something more than an introduction.

0.1   Survey of Chapter 1

Section 1.1 presents the standard introductory theory concerning stability and structure of solution spaces of linear ordinary differential equations. The following properties emerge as central. It is interesting to see the easy transition from the elementary to the quite general theory.

(a) For the system

$$(0.1.1) \qquad\qquad x' = Ax$$

with A an $n \times n$ constant matrix, the solution through (t_0, x_0) can be written as

(0.1.2) $$x(t, t_0, x_0) = e^{A(t-t_0)}x_0;$$

and every solution of

(0.1.3) $$x' = Ax + f(t)$$

can be expressed by the variation of parameters (VP) formula

(0.1.4) $$x(t, t_0, x_0) = e^{A(t-t_0)}x_0 + \int_{t_0}^{t} e^{A(t-s)}f(s)\, ds.$$

(b) For the system

(0.1.5) $$x' = B(t)x$$

with B continuous, $n \times n$, and $B(t + T) = B(t)$ for all t and some $T > 0$, there is a periodic matrix $P(t)$ and a constant matrix J such that a solution of (0.1.5) through $(0, x_0)$ can be expressed as

(0.1.6) $$x(t, 0, x_0) = P(t)e^{Jt}x_0;$$

and every solution of

(0.1.7) $$x' = B(t)x + f(t)$$

through $(0, x_0)$ can be expressed by the VP formula

(0.1.8) $$x(t, 0, x_0) = P(t)e^{Jt}x_0 + \int_{0}^{t} P(t)e^{J(t-s)}P^{-1}(s)f(s)\, ds.$$

(c) For the system

(0.1.9) $$x' = C(t)x$$

with C continuous and $n \times n$, there is an $n \times n$ matrix $Z(t)$ satisfying (0.1.9), and a solution of (0.1.9) through (t_0, x_0) is written

(0.1.10) $$x(t, t_0, x_0) = Z(t)Z^{-1}(t_0)x_0;$$

and a solution of

(0.1.11) $$x' = C(t)x + f(t)$$

is expressed by

(0.1.12) $$x(t, t_0, x_0) = Z(t)Z^{-1}(t_0)x_0 + \int_{t_0}^{t} Z(t)Z^{-1}(s)f(s)\, ds.$$

Moreover, if x_p is any solution of (0.1.11), then any solution of (0.1.11) can be expressed as

$$(0.1.13) \qquad x(t, t_0, x_0) = Z(t)Z^{-1}(t_0)[x_0 - x_p(t_0)] + x_p(t).$$

In Section 1.2 several periodic results are discussed. It is noted that if all solutions of (0.1.1) tend to zero as $t \to \infty$ and if $f(t+T) = f(t)$ for all t and some $T > 0$, then all solutions of (0.1.3) approach

$$(0.1.14) \qquad x(t) = \int_{-\infty}^{t} e^{A(t-s)} f(s) \, ds,$$

a T-periodic solution of (0.1.3). And, in the same way, if all solutions of (0.1.5) tend to zero as $t \to \infty$, then all solutions of (0.1.7) approach

$$(0.1.15) \qquad x(t) = \int_{-\infty}^{t} P(t)e^{J(t-s)} P^{-1}(s)f(s) \, ds,$$

a T-periodic solution of (0.1.7).

In Section 1.3 we examine a variety of Volterra integrodifferential equations and note a complete correspondence with the preceding.

(d) For the system

$$(0.1.16) \qquad x' = Ax + \int_{0}^{t} D(t-s)x(s) \, ds$$

with A constant and $n \times n$ and with D continuous and $n \times n$, there is an $n \times n$ matrix $Z(t)$ satisfying (0.1.16) on $[0, \infty)$ such that each solution of (0.1.16) through $(0, x_0)$ can be expressed as

$$(0.1.17) \qquad x(t, 0, x_0) = Z(t)x_0;$$

and a solution of

$$(0.1.18) \qquad x' = Ax + \int_{0}^{t} D(t-s)x(s) \, ds + f(t)$$

though $(0, x_0)$ is expressed by the VP formula

$$(0.1.19) \qquad x(t, 0, x_0) = Z(t)x_0 + \int_{0}^{t} Z(t-s)f(s) \, ds.$$

(e) For the system

$$(0.1.20) \qquad x' = B(t)x + \int_{0}^{t} E(t, s)x(s) \, ds$$

with B and E continuous, there is an $n \times n$ matrix $Z(t)$ satisfying (0.1.20) and each solution through $(0, x_0)$ can be expressed as

$$(0.1.21) \qquad x(t, 0, x_0) = Z(t)x_0;$$

and if $x_p(t)$ is any solution of

$$(0.1.22) \qquad x' = B(t)x + \int_0^t E(t, s)x(s)\, ds + f(t),$$

then every solution through $(0, x_0)$ can be written as

$$(0.1.23) \qquad x(t, 0, x_0) = Z(t)[x_0 - x_p(0)] + x_p(t).$$

Equation (0.1.22) also has a VP formula given by (0.1.26), below.

In Section 1.4 we note for (0.1.18) that if D and Z are $L^1[0, \infty]$ and if $f(t + T) = f(t)$, then every solution of (0.1.18) converges to

$$(0.1.24) \qquad x(t) = \int_{-\infty}^t Z(t - s)f(s)\, ds,$$

a T-periodic solution of the *limiting equation*

$$x' = Ax + \int_{-\infty}^t D(t - s)x(s)\, ds + f(t).$$

Considerable stability theory is introduced to verify $Z \in L^1[0, \infty)$. There is also a subsection on the existence of periodic solutions when $Z \notin L^1[0, \infty)$.

In Section 1.5 we consider periodic theory for (0.1.22) with $B(t + T) = B(t)$, $E(t + T, s + T) = E(t, s)$, and $f(t + T) = f(t)$. There is an $n \times n$ matrix $R(t, s)$ satisfying

$$(0.1.25) \qquad \begin{aligned} \partial R(t, s)/\partial s &= -R(t, s)B(s) - \int_s^t R(t, u)E(u, s)\, du, \\ R(t, t) &= I, \end{aligned}$$

so that a solution of (0.1.22) through $(0, x_0)$ can be expressed by the VP formula

$$(0.1.26) \qquad x(t, 0, x_0) = R(t, 0)x_0 + \int_0^t R(t, s)f(s)\, ds.$$

Moreover,

$$(0.1.27) \qquad R(t + T, s + T) = R(t, s).$$

If $R(t,0) \to 0$ as $t \to \infty$ and if both $\int_0^t |R(t,s)|\,ds$ and $\int_0^t |E(t,s)|\,ds$ are bounded, then each solution of (0.1.22) approaches

$$(0.1.28) \qquad\qquad x(t) = \int_{-\infty}^t R(t,s)f(s)\,ds,$$

a T-periodic solution of the limiting equation

$$(0.1.29) \qquad x' = B(t)x + \int_{-\infty}^t E(t,s)x(s)\,ds + f(t).$$

The classical result of Perron for ordinary differential equations holds for Volterra equations also and allows one to verify that $\int_0^t |R(t,s)|ds$ is bounded.

Remark 0.1.1. The solution spaces of the various equations are seen to be similar because of the variation of parameters formulas and especially by (0.1.13) and (0.1.23). But the property which brings unity to the periodic theory for these linear systems is isolated in (0.1.27); this property is valid for all kernels of the VP formulas in the periodic case. And in each of those cases we are able to consider $x(t+nT, 0, x_0)$ in the VP formula, change the variable of integration, and let $n \to \infty$ obtaining the integral representation for the periodic solution, all because of (0.1.27).

Remark 0.1.2. In each case in which we conclude that a periodic solution exists there is a fundamental boundedness principle which can be deduced from the VP formula. It turns out that solutions are uniform bounded and uniform ultimate bounded for bound B, terms defined later. And this boundedness property yields periodic solutions for linear and nonlinear equations as well as for equations with finite and infinite delay. It is this boundedness property that unites the entire book by means of asymptotic fixed-point theorems. While the VP formula shows boundedness for linear systems, Liapunov's direct method is well suited to the nonlinear study. And it turns out that as we use the Liapunov theory to extend the periodic theory from the nonlinear ordinary differential equations to the delay equations, then the Liapunov theory itself extends to the delay equations in a very unified and cohesive fashion.

Section 1.6 consists of an introduction to Liapunov's direct method for Volterra equations and construction of Liapunov functionals.

0.2   Survey of Chapter 2

In this chapter we look at a number of concrete applications ranging from ordinary differential equations models to models requiring functional differential equations with infinite delays. These include mechanical and electrical systems, biological problems, arms race models, and problems from economic theory.

0.3 Survey of Chapter 3

Many of the interesting problems in differential equations involve the use of fixed-point theory. [Cronin (1964) has a classical treatment of their uses.] And the fixed-point theorems frequently call for compact sets in the initial condition spaces which may be subsets of R^n, subsets of continuous functions on closed bounded intervals into R^n, or subsets of continuous functions from $(-\infty, 0]$ into R^n. Thus, one searches for a topology which will provide many compact sets, but is yet strong enough that solutions depend continuously on initial conditions in the new topology.

In Section 3.1 we discuss compactness in metric spaces and a variety of examples. We also introduce the rudiments of Banach space theory. Section 3.2 introduces contraction mappings, fixed-point theorems, asymptotic fixed-point theorems, and error bounds. Section 3.3 applies the contraction mapping theory to obtain solutions of linear ordinary and Volterra differential equations.

In Section 3.4 we discuss retracts, Brouwer's fixed-point theorem, and the sequence of results leading to the two common forms of Schauder's fixed-point theorem. We then discuss the asymptotic fixed-point theorems of Browder and Horn, as well as the Schauder-Tychonov theorem for locally convex spaces. These are the main results needed for existence of solutions and existence of periodic solutions of nonlinear ordinary and functional differential equations. Section 3.5 contains existence theorems for nonlinear ordinary differential equations as well as functional differential equations with bounded or infinite delays.

0.4 Survey of Chapter 4

This chapter contains three main themes; limit sets, periodicity, and stability. Section 4.1 is the basic introductory theory of nonlinear ordinary differential equations which, together with Section 1.1, constitutes a standard first graduate course in the subject. We introduce limit sets and arrive at the Poincaré-Bendixson theorem. This theorem concerns simple second-order problems, but its conclusion represents universal good advice: If we are interested in the behavior of bounded solutions, then we should concentrate on periodic solutions and equilibrium points.

Theorem 0.4.1. Poincaré-Bendixson *Consider the scalar equations*

(0.4.1) $$x' = P(x, y), \qquad y' = Q(x, y)$$

with P and Q locally Lipschitz in (x, y). If there is a solution ϕ of (0.4.1) which is bounded for $t \geq 0$, then either

(a) *ϕ is periodic,*

(b) *ϕ approaches a periodic solution, or*

(c) *ϕ gets close to an equilibrium point infinitely often.*

We are now ready to enlarge on Remark 0.1.2. Examine (0.1.4) with f bounded and all characteristic roots of Λ having negative real parts. It follows readily that

$$|e^{At}| \leq K e^{-\alpha t}$$

for $t \geq 0$ and some positive constants K and α. One then easily shows that the following definition holds for (0.1.3) using (0.1.4). In fact, the definition holds for all those linear systems for which we conclude that a periodic solution exists.

Definition 0.4.1. *Solutions of the system*

(0.4.2) $$x' = F(t, x)$$

are uniform bounded if for each $B_1 > 0$ there exists $B_2 > 0$ such that $[t_0 \in R, \ |x_0| \leq B_1, \ t \geq t_0]$ imply that $|x(t, t_0, x_0)| < B_2$. Solutions of (0.4.2) are uniform ultimate bounded for bound B if for each $B_3 > 0$ there exists $L > 0$ such that $[t_0 \in R, \ |x_0| \leq B_3, \ t \geq t_0 + L]$ imply that $|x(t, t_0, x_0)| < B$.

Theorem 0.4.2. *In (0.4.2) let F be continuous, locally Lipschitz in x, and satisfy $F(t + T, x) = F(t, x)$ for all t and some $T > 0$. If solutions of (0.4.2) are uniform bounded and uniform ultimate bounded for bound B, then (0.4.2) has a T-periodic solution.*

The conditions actually implied in Theorem 0.4.2 can be enumerated as follows:

(i) For each (t_0, x_0) there is a unique solution $x(t, t_0, x_0)$ of (0.4.2) defined on $[t_0, \infty)$.

(ii) The solution $\mathrm{x}(t, t_0, x_0)$ is continuous in x_0.

(iii) Solutions of (0.4.2) are uniform bounded and uniform ultimate bounded for bound B.

(iv) For each $\gamma > 0$ there is an $L > 0$ such that $[t_0 \in R, |x_0| < \gamma, t \geq t_0]$ imply that $|x'(t, t_0, x_0)| < L$.

(v) If $x(t)$ is a solution of $(0.4.2)$, so is $x(t + T)$.

Remark 0.4.1. One of the main themes of this book is that if (i)–(v) are properly extended to functional differential equations of either finite or infinite delay, then the equation has a T-periodic solution. Moreover, there is a single proof that works for all of these equations. Thus, these five properties of a general family of maps may well imply the existence of a periodic map.

Recall that the variation of parameters formulas led to periodic solutions of linear systems and also led to the uniform boundedness properties which, in turn, implied the existence of periodic solutions of nonlinear equations. But the nonlinear equations do not have VP formulas.

Thus, we will carry on the work (begun with VP formulas) using Liapunov's direct method which has proved to be very fruitful on nonlinear equations. Recall that the Poincaré-Bendixson theorem advised us to study periodic solutions and equilibrium points. We want to study what happens to solutions which get close to equilibrium points. If x_0 is an equilibrium point we can translate it to $x = 0$ and study the zero solution of $(0.4.2)$.

Definition 0.4.2. *The zero solution of* $(0.4.2)$ *is*

(a) stable *if for each* $\varepsilon > 0$ *and* $t_0 > 0$ *there exists* $\delta > 0$ *such that* $[|x_0| < \delta, \ t \geq t_0]$ *imply that* $|x(t, t_0, x_0)| < \varepsilon$,

(b) uniformly stable *if it is stable and if* δ *is independent of* t_0,

(c) asymptotically stable *if it is stable and for each* $t_0 \in R$ *there is an* $\eta > 0$ *such that* $|x_0| < \eta$ *implies that* $x(t, t_0, x_0) \to 0$ *as* $t \to \infty$, *and is*

(d) uniformly asymptotically stable *if it is uniformly stable and if there is an* $\eta > 0$ *with the following property: For each* $\mu > 0$ *there exists* $S > 0$ *such that* $[t_0 \in R, \ |x_0| < \eta, \ t \geq t_0 + S]$ *imply that* $|x(t, t_0, x_0)| < \mu$.

Properties (a), (b), and (d) have been characterized by Liapunov functions. The following theorem gives a streamlined summary of the main properties derived from Liapunov functions, denoted here by V. The point of Liapunov theory is that, while we do not know the solutions of $(0.4.2)$, we do know their derivatives; thus, if $V(t, x)$ is a scalar function and $x(t)$ is an (unknown) solution of $(0.4.2)$, we can compute

$$dV(t, x(t))/dt = \operatorname{grad} V(t, x) \cdot F(t, x) + \partial V/\partial t$$

and frequently determine if $V(t, x(t))$ is nonincreasing.

We denote this derivative of V by

$$V'_{(0.4.2)}(t, x).$$

Also, a symbol $W_i(r)$ will denote a continuous increasing scalar function satisfying $W_i(0) = 0$, $W_i(r) > 0$ if $r > 0$, and $W_i(r) \to \infty$ as $r \to \infty$.

Theorem 0.4.3. *Let D be an open neighborhood of $x = 0$ in R^n and let $V : (-\infty, \infty) \times D \to [0, \infty)$ be a differentiable function.*

(a) *If*

$$V(t, 0) = 0, \qquad W_1(|x|) \leq V(t, x), \qquad and \qquad V'_{(0.4.2)}(t, x) \leq 0,$$

then the zero solution of (0.4.2) is stable.

(b) *If*

$$W_1(|x|) \leq V(t, x) \leq W_2(|x|) \qquad and \qquad V'_{(0.4.2)}(t, x) \leq 0,$$

then $x = 0$ is uniformly stable.

(c) *If $F(t, x)$ is bounded for $|x|$ bounded and if*

$$V(t, 0) = 0, \quad W_1(|x|) \leq V(t, x), \quad and \quad V'_{(0.4.2)}(t, x) \leq -W_3(|x|),$$

then the zero solution of (0.4.2) is asymptotically stable.

(d) *If*

$$W_1(|x|) \leq V(t, x) \leq W_2(|x|) \quad and \quad V'_{(0.4.2)}(t, x) \leq -W_3(|x|),$$

then $x = 0$ is uniformly asymptotically stable.

(e) *If $D = R^n$ and if there is an $M > 0$ with*

$$W_1(|x|) \leq V(t, x) \leq W_2(|x|) \quad and \quad V'_{(0.4.2)}(t, x) \leq -W_3(|x|) + M,$$

then solutions of (0.4.2) are uniform bounded and uniform ultimate bounded for bound B.

Remark 0.4.2. In (e) it is not required that $F(t, 0) = 0$. In particular, if $F(t + T, x) = F(t, x)$ and if (e) holds with F Lipschitz in x, then from Theorem 0.4.2 we see that (0.4.2) has a T-periodic solution.

Remark 0.4.3. We wish to pick up a unifying theme once more. The VP formulas yielded periodic solutions and the uniform boundedness. The boundedness yielded periodic solutions of nonlinear systems (both with and without delays). Liapunov functions yield boundedness for nonlinear ordinary differential equations. We want to move Theorem 0.4.3 forward to apply to delay equations, thereby facilitating the study of limit sets and obtaining periodic solutions.

We remark also that now we could go back to (0.1.3) and obtain uniform ultimate boundedness when f is bounded and all characteristic roots of A have negative real parts (without the VP formula) using a Liapunov function; thus, in view of Theorem 0.4.3(e) and Remark 0.4.2 a Liapunov function can be the central unifying concept. For when f is bounded and all characteristic roots of A have negative real parts then there is a positive definite symmetric matrix B with $A^T B + BA = -I$. Then $V(x) = x^T Bx$ satisfies

$$V'_{(0.1.3)}(t, x) \leq -\frac{1}{2} x^T x + M$$

for some $M > 0$. By Theorem 0.4.3(e), solutions of (0.1.3) are uniform bounded and uniform ultimate bounded for bound B. When $f(t) \equiv 0$, this V satisfies all parts of Theorem 0.4.3.

Section 4.1 also considers a number of other results concerning the location of limit sets for (0.4.2). We now give a bit more detail on the advance of Theorems 0.4.2 and 0.4.3 to delay equations.

In Section 4.2 we introduce functional differential equations with fixed bounded delay,

(0.4.3) $$x' = F(t, x_t)$$

where x_t is the segment of $x(s)$ for $t - \alpha \leq s \leq t$, α a positive constant. To specify a solution we need a $t_0 \in R$ and a continuous function $\phi :$ $[t_0 - \alpha, t_0] \to R^n$, called an initial function; we then obtain a solution $x(t, t_0, \phi)$ defined on $[t_0 - \alpha, t_0 + \beta)$ with $x(t, t_0, \phi) = \phi(t)$ on $[t_0 - \alpha, t_0]$ and $x(t, t_0, \phi)$ satisfies (0.4.3) on $[t_0, t_0 + \beta)$. Definitions of boundedness and stability are patterned after those for ordinary differential equations. We say, for example, that solutions of (0.4.3) are uniform bounded if for each $B_1 > 0$ there is a $B_2 > 0$ such that $[t_0 \in R, |\phi(t)| \leq B_1$ on $[t_0 - \alpha, t_0]$, $t \geq t_0]$ imply that $|x(t, t_0, \phi)| < B_2$. We simply have a function ϕ from a Banach space as the initial condition instead of x_0 from the Banach space R^n. Then Theorem 0.4.2 is extended as follows.

Theorem 0.4.4. *Let F be continuous and locally Lipschitz in x_t and suppose $x(t+T)$ is a solution of (0.4.3) whenever $x(t)$ is a solution. If solutions of (0.4.3) are uniform bounded and uniform ultimate bounded for bound B, then (0.4.3) has a T-periodic solution.*

In extending Theorem 0.4.3 to (0.4.3) investigators have been puzzled over the types of upper bounds to use on V. If we use variants of an L^2-norm, then there is perfect unity between (0.4.2) and (0.4.3). Here, if $\phi : [t - \alpha, t] \to R^n$, then $\|\phi\|$ denotes the supremum norm, while $\||\phi\||$ denotes the L^2-norm.

Theorem 0.4.5. *Let $V(t, x_t)$ be a differentiable scalar function defined for $-\infty < t < \infty$ and x a continuous function into R^n with $|x(t)| < D$, $D \leq \infty$.*

(a) *If*

$$V(t, 0) = 0, \ W_1(|x(t)|) \leq V(t, x_t), \ and \ V'_{(0.4.3)}(t, x_t) \leq 0,$$

then the zero solution of (0.4.3) is stable.

(b) *If*

$$W_1(|x(t)|) \leq V(t, x_t) \leq W_2(\|x_t\|) \ and \ V'_{(0.4.3)}(t, x_t) \leq 0,$$

then $x = 0$ is uniformly stable.

(c) *If $F(t, x_t)$ is bounded for $\|x_t\|$ bounded and*

$$V(t, 0) = 0, \ W_1(|x(t)|) \leq V(t, x_t), \ and \ V'_{(0.4.3)}(t, x_t) \leq -W_3(|x(t)|),$$

then $x = 0$ is asymptotically stable.

(d) *If*

$$W_1(|x(t)|) \leq V(t, x_t) \leq W_2(|x(t)|) + W_3(\|x_t\|)$$

and

$$V'_{(0.4.3)}(t, x_t) \leq -W_4(|x(t)|),$$

then $x = 0$ is uniformly asymptotically stable.

(e) *If $D = \infty$ and if there is an $M > 0$ with*

$$W_1(|x(t)|) \leq V(t, x_t) \leq W_2(|x(t)|) + W_3\left(\int_{t-\alpha}^{t} W_4(|x(s)|) \, ds\right)$$

and

$$V'_{(0.4.3)}(t, x_t) \leq -W_4(|x(t)|) + M,$$

then solutions of (0.4.3) are uniform bounded and uniform ultimate bounded for bound B.

A large collection of Liapunov functionals and other limit set results are also given.

In Section 4.3 we discuss the system

$$(0.4.4) \qquad\qquad x' = h(t, x) + \int_{-\infty}^{t} q(t, s, x(s)) \, ds$$

and extend the concepts of uniform boundedness and uniform ultimate boundedness to this infinite delay case. To specify a solution we need a continuous initial function $\phi : (-\infty, 0] \to R^n$ and then obtain a solution $x(t, 0, \phi)$ satisfying (0.4.4) for $t \geq 0$ with $x(t, 0, \phi) = \phi(t)$ for $t \leq 0$.

Three interesting things happen. We quickly see that if we extend boundedness definitions as we did before for finite delay equations, we cannot define maps that will give fixed points. First, we must let our initial functions be unbounded. We derive a function $g : (-\infty, 0] \to [1, \infty)$ which is continuous, decreasing, and $g(r) \to \infty$ as $r \to -\infty$; our initial functions ϕ must satisfy $|\phi(t)| \leq \gamma g(t)$ for some $\gamma > 0$. Next, we must find compact subsets of initial functions so we use this g as a weight in the Banach space $(X, |\cdot|_g)$ of continuous initial functions for which

$$|\phi|_g = \sup_{-\infty < t \leq 0} |\phi(t)/g(t)|$$

exists; this makes "large" compact sets plentiful. And, finally, our solutions $x(t, 0, \phi)$ must be continuous in ϕ with the weighted norm.

It is now easy to extend our boundedness definitions to (0.4.4). For example, solutions of (0.4.4) are g-uniform bounded if for each $B_1 > 0$ there is a $B_2 > 0$ such that $[\phi \in X, |\phi|_g \leq B_1, t \geq 0]$ imply that $|x(t, 0, \phi)| < B_2$. Theorems 0.4.2 and 0.4.4 are now extended exactly according to properties (i)–(v) following Theorem 0.4.2.

Theorem 0.4.6. *Let the following conditions hold for* (0.4.4).

(i) *For each $\phi \in X$ there is a unique solution $x(t, 0, \phi)$ defined on $[0, \infty)$.*

(ii) *The solution $x(t, 0, \phi)$ is continuous in ϕ in the norm $|\cdot|_g$.*

(iii) *Solutions are g-uniform bounded and g-uniform ultimate bounded for bound B.*

(iv) *For each $\gamma > 0$ there is an $L > 0$ such that $[\phi \in X, |\phi|_g \leq \gamma, t \geq 0]$ imply that $|x'(t, 0, \phi)| \leq L$.*

(v) *If $x(t)$ is a solution of* (0.4.4), *so is $x(t + T)$.*

Then (0.4.4) *has a T-periodic solution.*

In Section 4.4 we consider a general system of functional differential equations

$$(0.4.5) \qquad x' = F(t, x(s); \ \alpha \leq s \leq t), \qquad t \geq 0,$$

where $\alpha \geq -\infty$. Definitions of stability and boundedness are extended from $x' = F(t, x)$ and $x' = F(t, x_t)$ using the supremum norm $\|\cdot\|$ on the initial functions $\phi : [\alpha, t_0] \to R^n$. The Liapunov theory of Theorems 0.4.3 and 0.4.4 is extended to (0.4.5) by using a variant of a weighted L^2-norm as an

upper bound on the Liapunov functional. In this context we also denote
the segment of a function $x(s)$ for $\alpha \leq s \leq t$ by x_t; a Liapunov functional
for (0.4.5) is denoted by $V(t, x_t)$.

Theorem 0.4.7. *Let $V(t, x_t)$ be a differentiable scalar functional defined
when $x : [\alpha, t] \to R^n$ is continuous and bounded by some $D \leq \infty$.*

(a) *If*
$$W_1(|x(t)|) \leq V(t, x_t), \qquad V(t, 0) = 0,$$

   and
$$V'_{(0.4.5)}(t, x_t) \leq 0,$$

   then $x = 0$ is stable.

(b) *If*
$$W_1(|x(t)|) \leq V(t, x_t) \leq W_2(\|x_t\|), \qquad V'_{(0.4.5)}(t, x_t) \leq 0,$$

   then $x = 0$ is uniformly stable.

(c) *Let $F(t, x_t)$ be bounded whenever $x : [\alpha, \infty] \to R^n$ is bounded. If*
$$W_1(|x(t)|) \leq V(t, x_t), \qquad V(t, 0) = 0,$$

   and
$$V'_{(0.4.5)}(t, x_t) \leq -W_3(|x(t)|),$$

   then $x = 0$ is asymptotically stable.

(d) *If there is a bounded continuous $\Phi : [0, \infty) \to [0, \infty)$ which is $L^1[0, \infty)$
   with $\Phi(t) \to 0$ as $t \to \infty$ such that*
$$W_1(|x(t)|) \leq V(t, x_t) \leq W_2(|x(t)|) + W_3\left(\int_\alpha^t \Phi(t - s)W_4(|x(s)|)\, ds \right)$$

   and
$$V'_{(0.4.5)}(t, x_t) \leq -W_5(|x(t)|),$$

   then $x = 0$ is uniformly asymptotically stable.

(e) *Let $M > 0$, $D = \infty$, Φ be as in (d) with $\Phi'(t) \leq 0$. If*
$$W_1(|x(t)|) \leq V(t, x_t) \leq W_2(|x(t)|) + W_3\left(\int_\alpha^t \Phi(t - s)W_4(|x(s)|)\, ds \right)$$

   and
$$V'_{(0.4.5)}(t, x_t) \leq -W_4(|x(t)|) + M,$$

   *then solutions of (0.4.5) are uniform bounded and uniform ultimate
   bounded for bound B.*

There are, of course, numerous other results and examples concerning limit sets.

Let us say a word about notation. In this book we look at seven essentially different types of differential equations and we do not want different notation for each. For example, the symbol x_t is used in three different ways in different parts of the book; we have just seen it used in two of those ways. We are careful in each section to advise the reader of the meaning of the symbols. Thus, the reader needs to always look at the context.

Chapter 1

Linear Differential and Integrodifferential Equations

1.0　The General Setting

This chapter contains

(a) a basic introduction to linear ordinary differential equations,

(b) a basic introduction to linear Volterra integrodifferential equations,

(c) the theory of existence of periodic solutions of both types of equations based on boundedness and stability.

While it provides an introduction to these subjects, it does not pretend to be an exhaustive review of either subject; rather, it is a streamlined treatment which leads the reader quickly and clearly into the periodic theory and forms the proper foundation for the subsequent nonlinear work.

The theme of the chapter is the cohesiveness of the theory. In so many ways the structure of the solution spaces of systems

$$x' = A(t)x + p(t)$$

and

$$x' = A(t)x + \int_0^t C(t, s)x(s) \, ds + p(t)$$

are completely indistinguishable. The proofs are identical and the variation of parameters formulae are frequently identical. Moreover, from the variation of parameters formulas there emerge beautiful formulas for periodic solutions.

The present section is a general introduction to notation concerning both linear and nonlinear ordinary differential equations. Section 1.1 is a lengthy account of the basic theory of linear ordinary differential equations. Section 1.2 is a brief treatment of the existence of periodic solutions of

$$x' = A(t)x + p(t).$$

Section 1.3 introduces the basic theory of linear Volterra equations and contains a subsection on equations of convolution type. Section 1.4 extends the periodic theory of Section 1.2 to Volterra equations of convolution type. Section 1.5 extends the periodic material to Volterra equations not of convolution type. Finally, Section 1.6 deals with stability and boundedness needed to implement the periodic results.

Throughout the following pages we will consider real systems of differential equations with a real independent variable t, usually interpreted as "time."

We use the following fairly standard notation: $[a, b] = \{t | a \leq t \leq b\}$, $(a, b) = \{t | a < t < b\}$, and $[a, b) = \{t | a \leq t < b\}$.

If U is a set in R^n and p is a point in R^n, then $\bar{U}$ is the closure of U, U^c is the complement of U, $S(U, \varepsilon)$ is an ε-neighborhood of U, and $d(p, U)$ denotes the distance from p to U.

If A is a matrix of functions of t, then $A' = dA/dt$ denotes the matrix A with each of its elements differentiated. The symbol $\int_a^t A(s)\, ds$ has the corresponding meaning.

Unless otherwise stated, if $x \in R^n$, then $|x|$ denotes the Euclidean length of x. If A is an $n \times n$ matrix, $A = (a_{ij})$, then

$$|A| = \left[\sum_{i,j=1}^{n} a_{ij}^2 \right]^{1/2}.$$

Additional norms are given in Sections 1.1 and 1.2.

Let D be an open subset of R^n, (a, b) an open interval on the t-axis, and $f : (a, b) \times D \to R^n$. Then

(1.0.1) $$x' = f(t, x)$$

is a system of first-order differential equations. Thus, as x and f are n-vectors, (1.0.1) represents n equations:

$$x_1' = f_1(t, x_1, \ldots, x_n),$$
$$x_2' = f_2(t, x_1, \ldots, x_n),$$
$$\vdots$$
$$x_n' = f_n(t, x_1, \ldots, x_n).$$

Definition 1.0.1. *A function* $\phi : (c, d) \to D$ *with* $(c, d) \subset (a, b)$ *is a solution of (1.0.1) if* ϕ' *is continuous on* (c, d) *and if* $t \in (c, d)$ *then* $\phi'(t) = f(t, \phi(t))$.

Example 1.0.1. Let $n = 1$, $(a, b) = (-\infty, \infty)$, $D = R$, and $x' = x^2$. Then one may verify that $\phi(t) = -1/t$ is a solution on $(-\infty, 0)$ and also on $(0, \infty)$. Thus, (c, d) could be chosen as any subinterval of either $(-\infty, 0)$ or $(0, \infty)$. We say that the solution has finite escape time because $|x(t)| \to \infty$ as $|t| \to 0$.

There is a simple geometrical interpretation of (1.0.1) and a solution ϕ. If $\phi(t)$ is graphed in R^n treating t as a parameter, then a directed curve is swept out. At each point $\phi(t_1)$ with $c < t_1 < d$ the vector $f(t_1, \phi(t_1))$ is the tangent vector to the curve (Fig. 1.1).

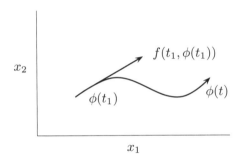

Fig. 1.1

Our discussion will be concerned primarily with the existence and properties of solutions of (1.0.1) through specified points $(t_0, x_0) \in (a, b) \times D$.

Definition 1.0.2. *For a given system (1.0.1) and a given point* $(t_0, x_0) \in (a, b) \times D$ *then*

(IVP) $$x' = f(t, x), \qquad x(t_0) = x_0$$

denotes the initial value problem for (1.0.1).

Definition 1.0.3. *A function* $\phi : (c, d) \to D$ *is a solution of (IVP) if* ϕ *is a solution of (1.0.1) and if* $t_0 \in (c, d)$ *with* $\phi(t_0) = x_0$. *We denote it by* $\phi(t, t_0, x_0)$ *or by* $x(t, t_0, x_0)$.

Example 1.0.2. Let $x' = x^2$, as in Example 1.0.1, and consider the IVP with $x(-1) = 1$. Then an appropriate ϕ is defined by $\phi(t) = -1/t$ for $-\infty < t < 0$.

Definition 1.0.4. *An IVP has a unique solution just in case each pair of solutions ϕ_1 and ϕ_2 on any common interval (c,d) with $\phi_1(t_0) = \phi_2(t_0) = x_0$ satisfy $\phi_1(t) = \phi_2(t)$ on (c,d).*

Example 1.0.3. Let $n = 1$,

$$x' = \begin{cases} 0 & \text{if } -\infty < x \leq 0, \\ x^{1/3} & \text{if } x > 0, \end{cases}$$

and let $x(0) = 0$ be the initial condition. Clearly, $\phi_1(t) = 0$ on $(-\infty, \infty)$ is one solution of the IVP, while

$$\phi_2(t) = \begin{cases} 0 & \text{if } t \leq 0, \\ (\frac{2}{3}t)^{3/2} & \text{if } t > 0 \end{cases}$$

is another solution on the same interval.

The following two results are among the most basic in differential equations. A proof in the linear case is given in Section 1.1 and in the general case in Chapter 3.

Theorem 1.0.1. *Let f be continuous on $(a,b) \times D$ and let $(t_0, x_0) \in (a,b) \times D$. Then (IVP) has a solution.*

Definition 1.0.5. *A function $f : (a,b) \times D \to R^n$ satisfies a local Lipschitz condition with respect to x at a point (t_0, x_0) if there is a neighborhood N of (t_0, x_0) and a constant K such that (t, x_1) and $(t, x_2) \in N$ imply $|f(t, x_1) - f(t, x_2)| \leq K|x_1 - x_2|$.*

Theorem 1.0.2. *Let f be continuous on $(a,b) \times D$ and let $(t_0, x_0) \in (a,b) \times D$. If f satisfies a local Lipschitz condition with respect to x at (t_0, x_0), then (IVP) has a unique solution.*

Exercise 1.0.1. Show that the function f in Example 1.0.3 does not satisfy a local Lipschitz condition at (0,0).

Exercise 1.0.2. A function $f(t, x)$ is linear in x if $f(t, x) = A(t, x) + b(t)$ where A is an $n \times n$ matrix of functions and $b(t)$ is an n-column vector function. Show that if f is linear in x and if each element of A is continuous for $a < t < b$, then f is locally Lipschitz at any point $(t_0, x_0) \in (a,b) \times R^n$.

Exercise 1.0.3. Show that if $n = 1$, if f is continuous, and if f has a bounded partial derivative with respect to x in a neighborhood of (t_0, x_0), then f satisfies a local Lipschitz condition with respect to x. Can you extend the result to arbitrary n? Is $f(x) = |x|$ a counterexample to the converse?

We remark that solutions are defined on open intervals $(c, d) \subset (a, b)$ only for the existence of right- and left-hand derivatives. If, for example, $f : [a, b) \times D \to R^n$ and f is continuous then Theorem 1.0.1 will be easily extended to produce a solution on $[a, d)$ for some $d > a$; the solution simply will not have a derivative from the left at $x = a$. In this manner, it will be understood that solutions on closed or half-closed intervals are permitted with one-sided derivatives being all that is required at closed end points.

We next show that certain scalar differential equations of order greater than one may be reduced to a system of first-order equations. If y is a scalar, then $y'' = g(t, y, y')$ denotes a second-order scalar differential equation, where g is defined on some suitable set. It is understood, without great formality, that ϕ is a solution if it satisfies the equation. We convert this to a system of two first-order equations as follows: define $y = x_1$ and then let $x_1' = x_2$ so that $y'' = x_1'' = x_2' = g(t, y, y') = g(t, x_1, x_2)$. Thus, we have the system

$$x_1' = x_2,$$
$$x_2' = g(t, x_1, x_2)$$

which we express as

$$x' = f(t, x)$$

with the understanding then that $x = (x_1, x_2)^T$ and $f = (x_2, g)^T$. If $\phi(t)$ is a solution of $y'' = g(t, y, y')$, then $(\phi(t), \phi'(t))^T$ is a solution of $x' = f(t, x)$. On the other hand, if $(\phi(t), \psi(t))^T$ is a solution of $x' = f(t, x)$ then $\phi(t)$ is a solution of $y'' = g(t, y, y')$. Thus, we say the system is equivalent to the single equation.

In exactly the same manner, the nth-order equation

$$y^{(n)} = g(t, y, y', \ldots, y^{(n-1)})$$

is equivalent to the system

$$x_1' = x_2,$$
$$x_2' = x_3.$$

$$\vdots$$

$$x_n' = g(t, x_1, \ldots, x_n).$$

One may find many equivalent systems for a given higher-order scalar equation. For example, the well-known Liénard equation

$$y'' + h(y)y' + r(y) = 0$$

has an equivalent system

$$x_1' = x_2 - \int_0^{x_1} h(s)\, ds,$$
$$x_2' = -r(x_1)$$

which has proved to be very useful in analyzing behavior of solutions.

By Theorem 1.0.1 we see that under very mild conditions indeed (IVP) has a solution. There are certainly many equations of the form of (IVP) which can be explicitly solved and that is the scope of most elementary courses in differential equations. However, most equations of interest cannot be solved and so one must resort to other methods. A good discussion of the problem of unsolvable equations may be found in Kaplansky (1957). In particular, it is shown that the simple equation $y'' + ty = 0$ cannot be solved in anything resembling closed form.

Not only are we faced with the impossibility of solving an equation, but often the solution of a solvable equation is so complex and so formidable that it is nearly impossible to extract from it the desired information. Moreover, since the initial condition is often known only approximately, one wishes to make a fairly concise statement concerning the behavior of all solutions starting near a given point (t_0, x_0). Finally, one may be anxious to learn the behavior of solutions for arbitrarily large t.

These difficulties and requirements often preclude effective use of approximate and computational devices. One is then led to the area of qualitative theory and, in particular, stability theory. These terms mean many things to many authors, and one author has isolated over 17,000 different types of stability. Our presentation here will relate in large measure to the following two types.

Definition 1.0.6. *Let $f : (a, \infty) \times R^n \to R^n$. Equation (1.0.1) is* Lagrange stable *if for each $(t_0, x_0) \in (a, \infty) \times R^n$, then each solution $\phi(t, t_0, x_0)$ of the IVP is defined on $[t_0, \infty)$ and is bounded.*

Definition 1.0.7. *Let $f : (a, \infty) \times D \to R^n$ and let ϕ be a solution of (1.0.1) defined on $[t_0, \infty)$. We say that ϕ is* Liapunov stable *if for each $t_1 \geq t_0$ and each $\varepsilon > 0$ there exists $\delta > 0$ such that $|\phi(t_1, t_0, x_0) - x_1| < \delta$ and $t \geq t_1$ imply $|\phi(t, t_0, x_0) - \phi(t, t_1, x_1)| < \varepsilon$.*

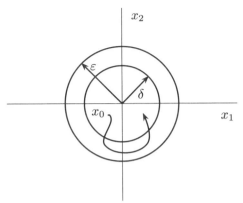

Fig. 1.2

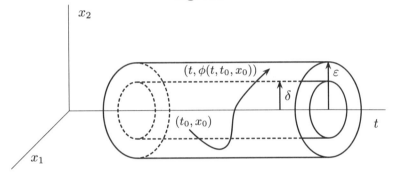

Fig. 1.3

Frequently, there is a point $p \in D$ with $f(t, p) = 0$ for all t so that $\phi(t) = p$ is a solution of (1.0.1), and it is found by inspection. It is called an *equilibrium solution*.

Definition 1.0.8. *Let $f : (a, \infty) \times D \to R^n$ and let $f(t, p) = 0$ for all t and some $p \in D$. We say that $\phi(t) = p$ is* Liapunov stable *if for each $t_0 > a$ and each $\varepsilon > 0$, there exists $\delta > 0$ such that $t \geq t_0$ and $|x_0 - p| < \delta$ imply $|p - \phi(t, t_0, x_0)| < \varepsilon$.*

For $p = 0$ this concept is portrayed in Fig. 1.2 when (1.0.1) is time independent, while Fig. 1.3 gives the time-dependent picture.

Definitions 1.0.6 and 1.0.7 lead us along quite divergent paths, both of which we will explore in some detail in the following sections. However, for a linear equation of the form $x' = A(t)x$, then Definitions 1.0.6 and 1.0.8 are, in fact, equivalent.

Our definitions clearly specify that we are working with a real equation, in a real domain, and with real solutions. However, the form of the equation may frequently allow the domain D to be part of the complex plane. And it may happen that complex solutions present themselves in an entirely natural way. In such cases it is often possible to extract real solutions from complex ones.

In the next example we attempt to present a concrete realization of most of the concepts presented to this point.

Example 1.0.4. Consider the equation $y'' + y = 0$ having complex solution $\phi(t) = \exp i(t - t_0)$. Note that both the real and imaginary parts also satisfy the equation. An equivalent system is

$$x_1' = x_2,$$
$$x_2' = -x_1$$

which may be expressed as

$$x' = Ax \qquad \text{with} \qquad A = \begin{pmatrix} 0 & 1 \\ -1 & 0 \end{pmatrix}$$

and thus with $f(t, x) = Ax$ a Lipschitz condition is satisfied. Two solutions of the system are $(\cos(t - t_0), -\sin(t - t_0))^T$ and $(\sin(t - t_0), \cos(t - t_0))^T$. One easily verifies that linear combinations of solutions are solutions. Also, if (t_0, x_0) is a given initial condition with $x_0 = (x_{01}, x_{02})^T$, then

$$\begin{pmatrix} \cos(t - t_0) & \sin(t - t_0) \\ -\sin(t - t_0) & \cos(t - t_0) \end{pmatrix} \begin{pmatrix} x_{01} \\ x_{02} \end{pmatrix}$$

is a solution satisfying the initial condition. As the IVP has unique solutions, this is the solution. Note that in Definition 1.0.8 we may take $p = 0$. Moreover, a computation shows that solutions of our equation have constant Euclidean length; hence, the zero solution is Liapunov stable and we may take $\delta = \varepsilon$.

1.1 Linear Ordinary Differential Equations

1.1.1 Homogeneous Systems

Let A be an $n \times n$ matrix of real, continuous functions on a real t-interval (a, b). Then

(1.1.1) $$x' = A(t)x$$

is called a linear homogeneous system of first-order differential equations. If $A = (a_{ij})$, then (1.1.1) can also be expressed as

$$(1.1.1a) \qquad x_i' = \sum_{j=1}^{n} a_{ij}(t)x_j \qquad \text{for} \qquad 1 \le i \le n.$$

The following result seems to have been first proved by Gronwall, although it was subsequently discovered by many others. It usually goes under the name of Gronwall's inequality, Reid's inequality, or Bellman's lemma. The result is continually being extended and has wide application. We use it here to obtain a simple uniqueness result for the IVP.

Theorem 1.1.1. Gronwall *Let k be a nonnegative constant and let f and g be continuous functions mapping an interval $[c, d]$ into $[0, \infty)$ with*

$$(1.1.2) \qquad f(t) \le k + \int_c^t f(s)g(s)\,ds \qquad \text{for} \qquad c \le t \le d.$$

Then

$$(1.1.3) \qquad f(t) \le k \exp \int_c^t g(s)\,ds \qquad \text{for} \qquad c \le t \le d.$$

Proof. We prove the result for $k > 0$ and then let $k \to 0$ through positive values for the result when $k = 0$. Multiply both sides of (1.1.2) by $g(t)$ and then divide both sides by the right side of (1.1.2). Then integrate both sides from c to t to obtain

$$\int_c^t \left\{ [f(u)g(u)] \bigg/ \left[k + \int_c^u f(s)g(s)\,ds \right] \right\} du \le \int_c^t g(u)\,du.$$

The integrand on the left contains the derivative of the denominator in the numerator and so this immediately yields

$$\ln \left\{ \left[k + \int_c^t f(s)g(s)\,ds \right] \bigg/ k \right\} \le \int_c^t g(u)\,du$$

or

$$\left[k + \int_c^t f(s)g(s)\,ds \right] \bigg/ k \le \exp \int_c^t g(u)\,du.$$

If we multiply by k and apply (1.1.2) again, then we have (1.1.3), as desired.

It is now convenient to introduce certain matrix norms and inequalities.

Definition 1.1.1. *If A is an $n \times n$ matrix, $A = (a_{ij})$, and if x is a column vector, then $\|A\| = \sum_{i,j=1}^{n} |a_{ij}|$ and $\|x\| = \sum_{i=1}^{n} |x_i|$.*

Exercise 1.1.1. Let A and B be $n \times n$ matrices, α a scalar, and x an n-column vector. Show that the following properties hold both as stated and with norm replaced by Euclidean length as defined in Section 1.0:

(1) $\|A\| \geq 0$ and $\|A\| = 0$ iff $A = 0$.

(2) $\|\alpha A\| = |\alpha| \, \|A\|$.

(3) $\|A + B\| \leq \|A\| + \|B\|$ and $\|x_1 + x_2\| \leq \|x_1\| + \|x_2\|$.

(4) $\|AB\| \leq \|A\| \, \|B\|$.

(5) $\|Ax\| \leq \|A\| \, \|x\|$.

The reader may also note that the definition of Liapunov stability (Def. 1.0.7) may be equivalently stated with Euclidean length replaced by norm.

Convergence of sequences of vectors is defined component by component, or by norm, or by distance function. For example, a sequence of vectors $\{x_k\}$ has limit x if $\|x_k - x\| \to 0$ or $|x_k - x| \to 0$ as $k \to \infty$.

Theorem 1.1.2. *Let $(c, d) \subset (a, b)$, with (c, d) a finite interval, and let $A(t)$ be continuous and bounded on (c, d). If $(t_0, x_0) \in (c, d) \times R^n$, then the IVP*

$$x' = A(t)x, \qquad x(t_0) = x_0$$

has a unique solution defined on (c, d).

Proof. Note that $x(t, t_0, x_0)$ is a solution of the IVP if and only if

$$(1.1.4) \qquad x(t, t_0, x_0) = x_0 + \int_{t_0}^{t} A(s)x(s, t_0, x_0)\, ds \qquad \text{for} \qquad t \in (c, d).$$

For brevity we write $x(t) = x(t, t_0, x_0)$ and define a sequence $\{x_n(t)\}$ on (c, d) inductively as

$$x_0(t) = x_0,$$

$$(1.1.5) \qquad x_1(t) = x_0 + \int_{t_0}^{t} A(s)x_0\, ds,$$

$$x_{k+1}(t) = x_0 + \int_{t_0}^{t} A(s)x_k(s)\, ds.$$

As A is continuous on (c, d) and $t_0 \in (c, d)$, the sequence is well defined and each $x_k(t)$ is differentiable. We will show that $\{x_k(t)\}$ converges uniformly on (c, d) and, as each $x_k(t)$ is continuous, we may pass the limit through the integral in (1.1.5) obtaining relation (1.1.4) for the limit function $x(t)$ (cf. Fulks, 1969, p. 417).

Using (1.1.5) twice and subtracting, we obtain

$$x_{k+1}(t) - x_k(t) = \int_{t_0}^{t} A(s)[x_k(s) - x_{k-1}(s)]ds.$$

Taking norms of both sides of the last equation and recalling that $\|A(t)\| \le m$ for some $m > 0$, we obtain

$$\|x_{k+1}(t) - x_k(t)\| \le \left| \int_{t_0}^{t} \|A(s)\| \|x_k(s) - x_{k-1}(s)\| ds \right|$$

$$\le m \left| \int_{t_0}^{t} \|x_k(s) - x_{k-1}(s)\| ds \right|.$$

Lemma. *For* $t \in (c, d)$ *we have*

$$(1.1.6) \qquad \|x_{k+1}(t) - x_k(t)\| \le \|x_0\| \frac{(m|t - t_0|)^{k+1}}{(k+1)!}.$$

Proof. By induction, if $k = 0$ then from (1.1.5) we have $x_1(t) = x_0 + \int_{t_0}^{t} A(s)x_0(s)\,ds$ so that

$$\|x_1(t) - x_0\| \le m \left| \int_{t_0}^{t} \|x_0(s)\| ds \right| \le m\|x_0\| |t - t_0|,$$

which verifies the statement. Assume that (1.1.6) holds for $k = p - 1$:

$$\|x_p(t) - x_{p-1}(t)\| \le \|x_0\| \frac{(m|t - t_0|)^p}{p!}.$$

Then

$$\|x_{p+1}(t) - x_p(t)\| \le m \left| \int_{t_0}^{t} \|x_p(s) - x_{p-1}(s)\| ds \right|$$

$$\le m \left| \int_{t_0}^{t} \|x_0\| \frac{(m|s - t_0|)^p}{p!} ds \right|$$

$$= m\|x_0\| \frac{m^p}{p!} \left| \int_{t_0}^{t} |s - t_0|^p ds \right|$$

$$= \|x_0\| \frac{m^{p+1}}{p!(p+1)} |t - t_0|^{p+1},$$

as desired for (1.1.6).

As (c, d) is a finite interval, the right side of (1.1.6) is bounded by

$$\|x_0\| \frac{[m(d-c)]^{k+1}}{(k+1)!} = M_k$$

and $\sum_{k=0}^{\infty} M_k$ converges by the ratio test. Thus, by the Weierstrass M-test the series

$$x_0(t) + (x_1(t) - x_0(t)) + (x_2(t) - x_1(t)) + \ldots$$

converges uniformly to some function $x(t)$. But the typical partial sum is $x_k(t)$ and hence $\{x_k(t)\}$ converges uniformly to $x(t)$. Thus, we take the limit through the integral in (1.1.5) obtaining the existence of a continuous function $x(t)$ satisfying

$$x(t) = x_0 + \int_{t_0}^{t} A(s)x(s)\, ds.$$

As the right side is differentiable, so is the left. The initial condition is satisfied and so the existence of a solution of the IVP is proved.

We now suppose that there are two solutions ϕ_1 and ϕ_2 of the IVP on (c, d). Then by (1.1.4) we have

$$\phi_i(t) = x_0 + \int_{t_0}^{t} A(s)\phi_i(s)\, ds, \qquad i = 1, 2.$$

Then

$$\phi_2(t) - \phi_1(t) = \int_{t_0}^{t} A(s)[\phi_2(s) - \phi_1(s)]\, ds$$

and so for $t \geq t_0$ we have

$$\|\phi_2(t) - \phi_1(t)\| \leq \int_{t_0}^{t} m\|\phi_2(s) - \phi_1(s)\|\, ds.$$

We now apply Gronwall's inequality to obtain $\|\phi_2(t) - \phi_1(t)\| = 0$ on $[t_0, d)$. A similar argument for $t \leq t_0$ completes the proof.

The result is fundamental and gives a simple criterion for existence and uniqueness. The technique of proof is called Picard's method of successive approximations and it has many other uses as well. The contraction mapping principal is based on it, which we use extensively in Chapter 3.

We next show that, under special conditions, one can obtain very exact information about the nature of solutions from the successive approximations.

Example 1.1.1. Let $x' = A(t)x$ where A is an $n \times n$ matrix of continuous functions on $(-\infty, \infty)$ with $A(t + T) = A(t)$ for all t and some $T > 0$, and suppose that $A(-t) = -A(t)$. Then all solutions are periodic of period T. We give the proof for $t_0 = 0$ and x_0 arbitrary. In (1.1.5) we note that for $k = 0$ we have

$$x_1(t) = x_0 + \int_0^t A(s)x_0 \, ds.$$

As the integrand is an odd periodic function, $x_1(t)$ is an even periodic function. By induction, if x_k is an even T-periodic function, then in

$$x_{k+1}(t) = x_0 + \int_0^t A(s)x_k(s) \, ds$$

the integrand is odd and T-periodic, so $x_{k+1}(t)$ is even and T-periodic. As $\{x_k(t)\}$ converges uniformly to $x(t, 0, x_0)$, it must also be even and T-periodic.

Remark 1.1.1. The successive approximations are constructive up to a point and, also, an error bound can be easily obtained. Such bounds are discussed in Chapter 3 in some detail.

It readily follows from Theorem 1.1.2 that for $A(t)$ continuous on (a, b), then for each $(t_0, x_0) \in (a, b) \times R^n$, the unique solution $x(t, t_0, x_0)$ exists on all of (a, b).

Theorem 1.1.3. *Let $A(t)$ be continuous on (a, b). The set of solutions of (1.1.1) on (a, b) forms an n-dimensional vector space over the reals:*

(a) $x(t) = 0$ *is a solution.*

(b) *If $x(t_1) = 0$ for some $t_1 \in (a, b)$, then $x(t) = 0$ on (a, b).*

(c) *If $x(t)$ is a solution and c is a scalar, then $cx(t)$ is a solution.*

(d) *If $x_1(t)$ and $x_2(t)x$ are solutions, so is $x_1(t) + x_2(t)$.*

(e) *There are exactly n linearly independent solutions.*

Proof. Parts (a)–(d) are clear. We must show that the dimension is n. First, we note that for a given $t_0 \in (a, b)$ the solutions $x_1(t), \ldots, x_n(t)$ in which $x_i(t_0) = (0, \ldots, 0, 1_i, 0, \ldots, 0)^T = e_i$ are linearly independent; for if not, then there are constants $c_1, \ldots, c_n$, not all zero, with $\sum_{i=1}^n c_i x_i(t) = 0$. But then $\sum_{i=1}^n c_i x_i(t_0) = (c_1, \ldots, c_n)^T = 0$, a contradiction. Now the dimension is at most n, for if $x(t)$ is any solution of (1.1.1), then we have $x(t_0) = (x_{10}, x_{20}, \ldots, x_{n0})^T$ and we write $x(t) = \sum_{i=1}^n x_{i0}x_i(t)$; thus, the $x_1(t), \ldots, x_n(t)$ span the space. This completes the proof.

Definition 1.1.2. *Any set of n linearly independent solutions of* (1.1.1) *is called a* fundamental system *of solutions or a* linear basis *for* (1.1.1).

Definition 1.1.3. *Given* $t_0 \in (a, b)$, *the* $n \times n$ *matrix* $Z(t, t_0)$ *whose columns are the* $x_i(t)$ *satisfying* $x_i(t_0) = e_i$ *(so that* $Z(t_0, t_0) = I$) *is called the* principal matrix solution (PMS) *of* (1.1.1).

Notice that the unique solution of (1.1.1) through (t_0, x_0) is $x(t, t_0, x_0) = Z(t, t_0)x_0$. Also, if $x_1(t), \ldots, x_n(t)$ is a linear basis for (1.1.1), then the matrix $H(t)$ whose columns are $x_1(t), \ldots, x_n(t)$, respectively, satisfies det $H(t) \neq 0$ (cf. Theorem 1.1.5); hence, $Z(t, t_0) = H(t)H^{-1}(t_0)$. In that connection, note that for any $n \times n$ matrix C, then $H(t)C$ has columns which are solutions of (1.1.1); however, $CH(t)$ normally does not share that property.

Theorem 1.1.4. *Let* $A(t)$ *be continuous on* $[t_0, \infty)$. *Then the zero solution of* (1.1.1) *is Liapunov stable if and only if* $Z(t, t_0)$ *is bounded.*

Proof. It is understood that Z is bounded if each of its elements is bounded. We first suppose that $Z(t, t_0)$ is bounded, say $|Z(t, t_0)| \leq M$ for $t_0 \leq t < \infty$. Let $t_1 \geq t_0$ and $\varepsilon > 0$ be given. We must find $\delta > 0$ such that $|x_0| < \delta$ and $t \geq t_1$ imply $|x(t, t_1, x_0)| < \varepsilon$. Now

$$
\begin{aligned}
|x(t, t_1, x_0)| &= |Z(t, t_0)Z^{-1}(t_1, t_0)x_0| \\
&\leq |Z(t, t_0)||Z^{-1}(t_1, t_0)||x_0| \\
&\leq Mm|x_0|
\end{aligned}
$$

where $|Z^{-1}(t_1, t_0)| = m$. Thus we take $\delta = \varepsilon/Mm$.

Exercise 1.1.2. Complete the proof of Theorem 1.1.4. That is, assume $x = 0$ is Liapunov stable and show that $Z(t, t_0)$ is bounded.

Exercise 1.1.3. Show that under the conditions of Example 1.1.1 we have $Z(t, t_0)$ bounded and hence $x = 0$ is Liapunov stable. Furthermore, show that δ may be chosen independent of t_1 in this case.

Recall that if A is an $n \times n$ matrix then

$$
\text{trace } A = \text{tr } A = \sum_{i=1}^{n} a_{ii}.
$$

The next result is known as Jacobi's identity or Abel's lemma.

Theorem 1.1.5. Jacobi-Abel *For $t_0 \in (a, b)$ and $A(t)$ continuous we have*

$$\det Z(t, t_0) = \exp \int_{t_0}^{t} \text{tr } A(s) \, ds \qquad for \qquad t \in (a, b).$$

Proof. Let $Z(t, t_0) = (\psi_{ij})$ and let $M(t) = \det Z(t, t_0)$. The derivative of an $n \times n$ determinant $M(t)$ is the sum of n determinants, say $M'(t) = M_1(t) + \cdots + M_n(t)$ in which $M_i(t)$ is obtained from $M(t)$ by differentiating the ith row and leaving the other elements unchanged. Consider

$$M_1(t) = \det \begin{bmatrix} \psi'_{11} & \cdots & \psi'_{1n} \\ \psi_{21} & \cdots & \psi_{2n} \\ & \vdots & \\ \psi_{n1} & \cdots & \psi_{nn} \end{bmatrix}$$

and note that the ψ_{1i}, for $i = 1, \ldots, n$, satisfy the first equation in the vector system (1.1.1) so $M_1(t)$ can be written as

$$\det \begin{bmatrix} \sum_{j=1}^{n} a_{1j}\psi_{j1} & \cdots & \sum_{j=1}^{n} a_{1j}\psi_{jn} \\ \psi_{21} & \cdots & \psi_{2n} \\ & \vdots & \\ \psi_{n1} & \cdots & \psi_{nn} \end{bmatrix}.$$

Now multiply the second, third, $\ldots$ row by $a_{12}, a_{13}, \ldots$, respectively, and subtract from the first row (which leaves the determinant unchanged) to get

$$\det \begin{bmatrix} a_{11}\psi_{11} & \cdots & a_{11}\psi_{1n} \\ \psi_{21} & \cdots & \psi_{2n} \\ & \vdots & \\ \psi_{n1} & \cdots & \psi_{nn} \end{bmatrix} = a_{11}M(t).$$

Perform similar operations on $M_2, \ldots, M_n$ to get

$$M'(t) = a_{11}M(t) + a_{22}M(t) + \cdots + a_{nn}M(t)$$
$$= (\text{tr } A)M(t)$$

which has one solution $M(t) = \exp \int_{t_0}^{t} \text{tr } A(s) \, ds$. As $M' = (\text{tr } A)M$ is a first-order linear equation, it has a unique solution for each initial condition. As $M(t_0) = \det Z(t_0, t_0) = 1$, the displayed solution is the desired one. This completes the proof.

Corollary. *Let A be continuous on $[t_0, \infty)$. Then the zero solution of (1.1.1) is Liapunov stable only if $\int_{t_0}^{t} \text{tr } A(s) \, ds$ is bounded above.*

Proof. We have $\det Z(t, t_0) = \exp \int_{t_0}^t \operatorname{tr} A(s)\, ds$. If $\int_{t_0}^t \operatorname{tr} A(s)\, ds$ is not bounded above, then $\det Z(t, t_0)$ is unbounded and, hence, some element of $Z(t, t_0)$ is unbounded. This means that $x = 0$ is unstable by Theorem 1.1.4.

On the other hand, we point out that for

$$A = \begin{pmatrix} -2 & 0 \\ 0 & 1 \end{pmatrix}$$

we have

$$Z(t, 0) = \begin{pmatrix} e^{-2t} & 0 \\ 0 & e^t \end{pmatrix} \quad \text{and} \quad \operatorname{tr} A = -1.$$

Thus, our corollary cannot be extended to ensure stability.

1.1.2 Nonhomogeneous Systems

The system

$$(1.1.7) \qquad x' = A(t)x + d(t)$$

in which A is an $n \times n$ matrix and d is an n-column vector with both A and d defined on (a, b) is called a system of n first-order linear nonhomogeneous differential equations. If A and d are continuous on (a, b), then the initial value problem for (1.1.7) will inherit existence from the existence of solutions of (1.1.1). It will inherit uniqueness from uniqueness of the zero solution of (1.1.1).

Theorem 1.1.6. *Let A and d be continuous on (a, b) and let $(t_0, x_0) \in (a, b) \times R^n$. Then there is a unique solution of the IVP*

$$x' = A(t)x + d(t), \qquad x(t_0) = x_0$$

and it is given by the variation of parameters formula

$$(1.1.8) \qquad x(t) = Z(t, t_0)\left[x_0 + \int_{t_0}^t Z^{-1}(s, t_0) d(s)\, ds\right]$$

where $Z(t, t_0)$ is the PMS of (1.1.1).

Proof. One may show by direct computation that (1.1.8) satisfies (1.1.7). Clearly, the initial condition is satisfied. We now show that (1.1.8) is the unique solution of the IVP. Let $\phi_1(t)$ and $\phi_2(t)$ be two solutions of the IVP on a common interval (c, e). We will show that $\psi(t) = \phi_1(t) - \phi_2(t)$ is a solution of (1.1.1) and, as $\psi(t_0) = 0$, it will then follow from the uniqueness of the zero solution of (1.1.1) that $\psi(t) = 0$ on (c, e). We have

$$\psi'(t) = A(t)\phi_1(t) + d(t) - A(t)\phi_2(t) - d(t)$$
$$= A(t)[\phi_1(t) - \phi_2(t)] = A(t)\psi(t).$$

This completes the proof.

Notice that the form of (1.1.8) agrees with that of a solution of a first-order scalar equation and hence it is fairly easy to remember. However, there is a simple way to arrive at (1.1.8) once existence of solutions has been established. Each solution of (1.1.1) can be expressed as $x(t) = Z(t, t_0)x_0$ and we are led to try for a solution of (1.1.7) of the form $x(t) = Z(t, t_0)r(t)$ where r is a differentiable vector function. Then

$$x'(t) = Z'(t, t_0)r(t) + Z(t, t_0)r'(t) = A(t)Z(t, t_0)r(t) + d(t)$$

is the required equation. We then have

$$A(t)Z(t, t_0)r(t) + Z(t, t_0)r'(t) = A(t)Z(t, t_0)r(t) + d(t)$$

or

$$Z(t, t_0)r'(t) = d(t).$$

Thus, we find that

$$r(t) = r(t_0) + \int_{t_0}^{t} Z^{-1}(s, t_0)d(s)\, ds$$

which solves the problem when $r(t_0) = x_0$.

The following fact was derived in the proof and is worth formally stating.

Corollary 1. *If ϕ_1 and ϕ_2 are two solutions of (1.1.7), then $\psi(t) = \phi_1(t) - \phi_2(t)$ is a solution of (1.1.1).*

As $\psi(t) = Z(t, t_0)c$ for an appropriate constant vector c, this result can be stated more to the point as follows.

Corollary 2. *If $Z(t, t_0)$ is the PMS of (1.1.1) and if ϕ is any solution of (1.1.7), then any other solution $x(t)$ of (1.1.7) can be expressed as*

$$(1.1.9) \qquad\qquad x(t) = Z(t, t_0)[x_0 - \phi(t_0)] + \phi(t).$$

The reader not acquainted with the mechanics of integrating matrix equations is advised to carry out the details in the following problem.

Exercise 1.1.4. Let

$$A = \begin{pmatrix} 0 & 1 \\ -1 & 0 \end{pmatrix},$$

$d(t) = (1, 0)^T$, and $x_0 = (0, 0)^T$. Find the solution $x(t, 0, x_0)$ of $x' = Ax + d(t)$. Verify that it is a solution. See Example 1.0.4 for $Z(t, 0)$.

1.1.3 Constant Coefficients, Autonomous Principles, and Complex Solutions

There are at least four very satisfactory and common methods of treatment of (1.1.1) when A is a matrix of constants: complex integration, Laplace transforms, characteristic solutions, and Jordan forms. The first three are preferable for actually displaying desired solutions, while the last is fundamental to a full understanding of the structure. All of the methods depend on being able to find the roots of a polynomial of degree n. In this section we present the last two methods. Let A be an $n \times n$ matrix of real constants and let

$$(1.1.10) \qquad x' = Ax.$$

It follows from Theorem 1.1.2 that for each $(t_0, x_0) \in (-\infty, \infty) \times R^n$, there is a unique solution $x(t, t_0, x_0)$ of the IVP and it is defined for $-\infty < t < \infty$.

By direct substitution into (1.1.10) one can see that if ϕ is any solution of (1.1.10), so is $\phi(t - t_0)$ for any t_0. Hence, if we have $\phi(t) = x(t, 0, x_0)$ and we wish to obtain $\psi(t) = x(t, t_0, x_0)$, then we let $\psi(t) = \phi(t - t_0)$ and we then have $\psi(t_0) = \phi(0) = x_0$. This is a fundamental principle which applies to all systems $x' = f(x)$ with right sides independent of t, called *autonomous equations*.

Autonomous Principle. *If $\phi(t) = x(t, 0, x_0)$ is a solution of*

$$(1.1.11) \qquad x' = f(x)$$

where $f : D \to R^n$ and $D \subset R^n$, and if for each $(t_0, x_0) \in (-\infty, \infty) \times D$ the IVP for (1.1.11) has a unique solution $\psi(t) = x(t, t_0, x_0)$, then $\psi(t) = \phi(t - t_0)$. If the IVP does not have a unique solution, then for each solution $\psi(t)$ satisfying $\psi(t_0) = x_0$, there is a solution ϕ with $\phi(0) = x_0$ and $\psi(t) = \phi(t - t_0)$.

Autonomous Convention. *We denote any solution of the IVP*

$$x' = f(x), \qquad x(0) = x_0$$

by $x(t, x_0)$ or $\phi(t, x_0)$.

Proposition 1.1.1. *If $x(t)$ is any solution of (1.1.10), then $x(t)$ has derivatives of all orders and each derivative is a solution.*

Proof. As $x(t)$ is a solution, we have $x'(t) = Ax(t)$ and, as the right side is differentiable, so is the left. Thus we have $x''(t) = Ax'(t) = A[Ax(t)]$, and this can be written as $[Ax(t)]' = A[Ax(t)]$, showing that $x'(t) = Ax(t)$ is a solution.

Now A is real and our basic assumption is that $x \in R^n$. However, it is possible to let $x \in C^n$ with C complex and to consider complex solutions with a real independent variable t. In fact, complex solutions do enter in an entirely natural way. Moreover, their exit is just a natural owing to the following two facts.

Proposition 1.1.2. *If $x(t) = u(t) + iv(t)$ is any complex solution of* (1.1.1) *with u and v real, then $u(t)$ and $v(t)$ are both solutions of* (1.1.1). *The converse is also true.*

Proof. As $x(t)$ is a solution of (1.1.1) we have $x'(t) = u'(t) + iv'(t) = A(t)[u(t) + iv(t)]$. The definition of equality of complex numbers then requires that $u'(t) = A(t)u(t)$ and $v'(t) = A(t)v(t)$. The converse is proved by noting that if $u(t)$ and $v(t)$ are real solutions, then $u(t) + iv(t)$ is a complex solution.

We emphasize the converse as follows.

Corollary. *If A is continuous on (a, b) and if $z_0 \in C^n$, then for each $t_0 \in (a, b)$ there is a unique complex solution $x(t, t_0, z_0)$ defined on (a, b).*

Proposition 1.1.3. *If $x(t)$ is a solution of* (1.1.1) *on an interval (c, d) with $t_0 \in (c, d)$ and if $x(t_0)$ is real, then $x(t)$ is real on (c, d).*

Proof. If $x(t)$ is complex, say $x(t) = u(t) + iv(t)$, then $x(t_0) = u(t_0) + iv(t_0)$. But $x(t_0)$ is real and so $v(t_0) = 0$. As $v(t)$ is a solution and as $v(t_0) = 0$, by the uniqueness theorem $v(t) = 0$ on (c, d).

It may frequently appear from the form of a solution that it is complex; however, Proposition 1.1.3 shows that if a solution of (1.1.1) is real at one point, then it remains real throughout its interval of definition.

Proposition 1.1.4. *If $x_1(t), \ldots, x_n(t)$ is any set of n linearly independent complex solutions of* (1.1.1)*, then a set of n linearly independent real solutions of* (1.1.1) *may be constructed from them.*

Proof. With complex solutions it is understood that the scalars in linear combinations are complex. Write $x_j(t) = u_j(t) + iv_j(t)$ and form the determinant whose columns are the x_j:

$$\det(u_1(t) + iv_1(t), \ldots, u_n(t) + iv_n(t))$$
$$= \det(u_1(t), x_2(t), \ldots, x_n(t)) + \det(iv_1(t), x_2(t), \ldots, x_n(t)).$$

Continue to decompose the determinant until a sum of determinants is obtained, each column of which is either $u_j(t)$ or $iv_j(t)$. Evaluate all these determinants at some $t_0 \in (a, b)$ and notice that $\det(x_1(t_0), \ldots, x_n(t_0)) \neq 0$ and so the sum of determinants is not zero. Thus, some one determinant in the sum is nonzero and, hence, its columns are linearly independent. Factor all of the i's out of that determinant. Its columns will still be linearly independent and the solutions through those constant vectors will be linearly independent.

We return now to the discussion in which A is constant.

Definition 1.1.4. *A characteristic vector of any real $n \times n$ constant matrix A is any real or complex vector $x \neq 0$ satisfying $Ax = \alpha x$ for some real or complex scalar α. Then α is a characteristic value of A and x is a characteristic vector belonging to α*

Now for A constant, $Ax = \alpha x$ implies $(A - \alpha I)x = 0$ which has a nontrivial solution x if and only if

$$(1.1.12) \qquad \det(A - \alpha I) = 0.$$

This is a polynomial of degree n in α, called the *characteristic polynomial*. By the fundamental theorem of algebra it has exactly n roots, not necessarily distinct.

Theorem 1.1.7. *If $x_1, \ldots, x_m$, are characteristic vectors of the constant matrix A belonging to distinct characteristic values $\alpha_1, \ldots \alpha_m$, respectively, then the x_i are linearly independent.*

Proof. The proof is by induction on m, the number of distinct characteristic roots of A. If $m = 1$, the result is true as $x_1 \neq 0$ by definition of a characteristic vector. Assume the theorem true for $m = k - 1$ and let $x_1, \ldots, x_k$ be the characteristic vectors belonging to $\alpha_1, \ldots, \alpha_k$. If there is a nontrivial relation $\sum_{i=1}^{k} c_i x_i = 0$, then upon left multiplication by A we obtain $\sum_{i=1}^{k} c_i A x_i = \sum_{i=1}^{k} c_i \alpha_i x_i = 0$. Now at least one of the $c_i \neq 0$, so by reordering if necessary we may assume $c_1 \neq 0$. Multiply $\sum_{i=1}^{k} c_i x_i = 0$, by α_k and subtract from the last equality obtaining $\sum_{i=1}^{k-1} c_i(\alpha_i - \alpha_k)x_i = 0$, a nontrivial linear relation since the α_i are distinct. This is a contradiction.

Definition 1.1.5. *Let A be constant and let α be a root of* $\det(A - \alpha I) = 0$. *A complex vector $x \neq 0$ is a vector associated with α of multiplicity m if for some nonnegative integer m we have* $(A - \alpha I)^m x = 0$, *but if h is a nonnegative integer less than m, then* $(A - \alpha I)^h x \neq 0$.

To this point our discussion has been quite self-contained. A proof of the next result is beyond the scope of this discussion and may be found in Finkbeiner (1960, p. 154).

Theorem 1.1.8. *Let A be constant and let $\alpha_1, \ldots, \alpha_k$ be the distinct roots of* $\det(A - \alpha I) = 0$ *with multiplicities $m_1, \ldots, m_k$, respectively (so $\sum_{i=1}^{k} m_i = n$). Then to each α_i there corresponds a set of exactly m_i vectors x_{ij}, for $j = 1, \ldots, m_i$ such that:*

(a) *x_{ij} is associated with α_i of multiplicity at most m_i for $j = 1, \ldots, m_i$.*

(b) *The set of n vectors x_{ij}, for $j = 1, \ldots, m_i$ and $i = 1, \ldots, k$, is linearly independent.*

Theorem 1.1.9. *Suppose that $x(t)$ is a solution of (1.1.10) for all t. If there exists t_0 such that $x(t_0)$ is a characteristic vector of A belonging to α (i.e., $Ax(t_0) = \alpha x(t_0)$) then $Ax(t) = \alpha x(t)$ for all t.*

Proof. As $x'(t)$ is a solution, so is $y(t) = x'(t) - \alpha x(t) = Ax(t) - \alpha x(t) = (A - \alpha I)x(t)$. As $y(t_0) = 0$, $y(t) = 0$ for all t, yielding the result.

Definition 1.1.6. *Any solution $x(t)$ satisfying Theorem 1.1.9 will be called a* characteristic solution.

Remark 1.1.2. The reader who is acquainted with the standard technique of solving an nth-order constant coefficient equation, say $L(y) = y^{(n)} + a_1 y^{(n-1)} + \cdots + a_n y = 0$, by letting $y = e^{\alpha t}$ may note the following correspondence. The characteristic polynomial obtained by substituting $y = e^{\alpha t}$ into $L(y) = 0$ is the same as that obtained by transforming $L(y) = 0$ to a system $x' = Ax$ and obtaining $\det(A - \alpha I) = 0$. Thus, one thinks of $y = e^{\alpha t}$ as a characteristic solution. For each root α_i, then $L(y) = 0$ has exactly one characteristic solution; however, there are systems $x' = Ax$ having more than one characteristic solution for each (or some) characteristic root. On the other hand, if $y = e^{\alpha t}$ is a solution of $L(y) = 0$ and α has multiplicity $m > 1$, then $te^{\alpha t}, t^2 e^{\alpha t}, \ldots, t^{m-1} e^{\alpha t}$ are all solutions of $L(y) = 0$; however, if α_1 is a root of $\det(A - \alpha I) = 0$ (for A arbitrary) of multiplicity $m > 1$ and if $x(t)$ is a characteristic solution, then the other solutions corresponding to α_1 are generally more complicated than $tx(t), t^2 x(t), \ldots, t^{m-1} x(t)$.

Theorem 1.1.10. *For each characteristic value α of A there is at least one characteristic solution $x(t)$ of (1.1.10) which may be written in the* *form* $x(t) = (x_1, \ldots, x_n)^T e^{\alpha t}$ *where* $(x_1, \ldots, x_n)^T$ *is a characteristic vector belonging to* α.

Proof. Let α be a characteristic root and let $(x_1, \ldots, x_n)^T = x_0$ be a characteristic vector belonging to α. There exists a solution $x(t)$ with $x(0) = x_0$ which, therefore, is a characteristic solution and so $x'(t) = Ax(t) = \alpha x(t)$ or $dx_i/dt = \alpha x_i(t)$. Thus, $x_i(t) = x_i(0)e^{\alpha t}$ and $x_i(0) = x_i$. This completes the proof.

Exercise 1.1.5. Convert the equation $y'' + 4y' + 3y = 0$ to a system, say $x' = Ax$: find all characteristic roots; find a characteristic vector belonging to each root; verify that the characteristic vectors are linearly independent; find all characteristic solutions; determine $Z(t, 0)$; and find a solution $x(t)$ with $x(0) = (1, 1)^T$.

Theorem 1.1.11. *Any set of characteristic solutions of* (1.1.10) *belonging to distinct characteristic roots is linearly independent.*

Proof. Let $x_1(t), \ldots, x_k(t)$ be the specified solutions. Then $x_1(0), \ldots, x_k(0)$ are linearly independent as they are characteristic vectors. If the $x_i(t)$ are not linearly independent, then there are constants $c_1, \ldots, c_k$, not all zero, with $c_1 x_1(t) + \cdots + c_k x_k(t) = 0$ for all t and so $c_1 x_1(0) + \cdots + c_k x_k(0) = 0$ is a nontrivial linear relation, a contradiction.

Let $D = d/dt$ so $D^k x = d^k x/dt^k$. Recall that if $x(t)$ is a solution of (1.1.10), then $Dx(t) = Ax(t)$ is also a solution and so $D^k x(t) = A^k x(t)$. Thus, if P is any polynomial with constant coefficients, then $P(D)x = P(A)x$ for any solution x of (1.1.10).

Theorem 1.1.12. *If* $x(t)$ *is a solution of* (1.1.10), *and if for some* t_0, $x(t_0)$ *is associated with a characteristic root* α *of* A *of multiplicity* m,

$$(A - \alpha I)^m x(t_0) = 0,$$

then

$$(A - \alpha I)^m x(t) = 0 \qquad for\ all\ t,$$

and

$$(D - \alpha)^m x(t) = 0 \qquad for\ all\ t.$$

Proof. As $P(D)x = P(A)x$, we have $(A - \alpha I)^m x(t) = (D - \alpha)^m x(t)$. Let $M(t) = (D - \alpha)^m x(t)$. Then M is a solution of (1.1.10). But $M(t_0) = 0$ and so $M(t) \equiv 0$.

Definition 1.1.6. *A primitive solution of* (1.1.10) *associated with the characteristic value* α *of multiplicity* m *is any nontrivial solution* $x(t)$ *of* (1.1.10) *for which* $(A - \alpha I)^m x(t) \equiv 0$.

Theorem 1.1.13. *If* $x(t)$ *is a primitive solution of* (1.1.10) *associated with* α *of multiplicity* m,
$$(A - \alpha I)^m x(t) \equiv 0,$$
then $x(t) = (P_1(t)e^{\alpha t}, P_2(t)e^{\alpha t}, \ldots, P_n(t)e^{\alpha t})^T$ *where the* P_i *are polynomials in* t *of degree* $\leq (m-1)$.

Proof. We have $(D-\alpha)^m x(t) \equiv 0$ which in scalar form is $(D-\alpha)^m x_i(t) \equiv 0$. Now if ϕ is any function with m derivatives, then $(D - \alpha)^m [e^{\alpha t}\phi(t)] = e^{\alpha t} D^m \phi(t)$. Thus,

$$(D - \alpha)^m [e^{\alpha t}(e^{-\alpha t}x_i(t))] = e^{\alpha t} D^m [e^{-\alpha t}x_i(t)] \equiv 0$$

or $D^m[e^{-\alpha t}x_i(t)] \equiv 0$. If we integrate this expression m times we get $e^{-\alpha t}x_i(t) = P_i(t)$ where $\deg P_i(t) \leq m - 1$.

Theorem 1.1.14. *A basis for* (1.1.10) *of the form*

$$x(t) = (P_1(t)e^{\alpha t}, \ldots, P_n(t)e^{\alpha t})^T$$

exists. If all the characteristic roots of A *are known, then this linear basis may be obtained by solving a finite number of linear algebraic equations.*

Proof. By Theorem 1.1.8 there are n linearly independent vectors x_{ij} which are associated with the characteristic values of A of certain multiplicities. By the existence theorem there are n solutions $x(t, x_{ij})$ and they are linearly independent as the x_{ij} are linearly independent.

To find this solution we proceed as follows. Let α be a root of A with multiplicity m. Assume a solution $x(t) = (c_1 e^{\alpha t}, \ldots, c_n e^{\alpha t})^T$ where the c_i are to be determined from the differential equation after canceling the $e^{\alpha t}$. Determine as many linearly independent constant vectors $(c_{1i}, \ldots, c_{ni})^T$ as possible.

Then assume a solution of the form

$$x(t) = [(c_1 + d_1 t)e^{\alpha t}, \ldots, (c_n + d_n t)e^{\alpha t}]^T.$$

(These c_i are not related to the previous c_i: in fact, one can see that if so, then $x(t) = [d_1 t e^{\alpha t}, \ldots, d_n t e^{\alpha t}]^T$ would also be a solution as it is the remainder after subtracting the previous solution. As that form satisfies $x(0) = 0$, then $x(t) \equiv 0$.) This will result in $2n$ equations when coefficients of t are equated after substituting this assumed solution into the differential equation. Again, determine as many linearly independent solutions of this form as possible.

Continue with assumed solutions of the form

$$x(t) = (P_1(t)e^{\alpha t}, \ldots, P_n(t)e^{\alpha t})^T$$

where the $P_i(t)$ have degree $\leq (m-1)$. We will obtain m linearly independent solutions in this way.

Complete the linear basis by treating each root α_i as above.

Exercise 1.1.6. Let

$$A_1 = \begin{pmatrix} 1 & 0 \\ 1 & 1 \end{pmatrix} \quad \text{and} \quad A_2 = \begin{pmatrix} 1 & 0 \\ 0 & 1 \end{pmatrix}.$$

Find linear bases for $x' = A_1 x$ and $x' = A_2 x$.

Exercise 1.1.7. Convert $y^{(4)} + y = 0$ to a system $x' = Ax$ and find a basis.

1.1.4 A Little Stability

In Section 1.0 we introduced the concept of Liapunov stability in Definitions 1.0.7 and 1.0.8. Theorem 1.1.4 showed that the zero solution of (1.1.1) is Liapunov stable if and only if $Z(t, t_0)$ is bounded. Example 1.1.1 showed that the periodic system $x' = A(t)x$ with $A(-t) = -A(t)$ had all solutions bounded (hence $x = 0$ is Liapunov stable), while Exercise 1.1.3 noted that δ is independent of t_0. All of this suggests certain extensions.

Definition 1.1.7. *Let $f : [t_0, \infty) \times D \to R^n$ be continuous and let $\phi : [t_0, \infty) \to R^n$ with ϕ a solution of*

(1.0.1) $$x' = f(t, x).$$

(a) *We say that ϕ is* uniformly stable *if ϕ is Liapunov stable and if δ is independent of $t_1 \geq t_0$.*

(b) *If ϕ is Liapunov stable and if for each $t_1 \geq t_0$ there exists $\eta > 0$ such that $|x_0 - \phi(t_1)| < \eta$ implies $|x(t, t_1, x_0) - \phi(t)| \to 0$ as $t \to \infty$, then ϕ is* asymptotically stable.

(c) *If $D = R^n$, if ϕ is Liapunov stable, and if for each $t_1 \geq t_0$ and each $x_0 \in R^n$ we have $\lim_{t \to \infty} |x(t, t_1, x_0) - \phi(t)| = 0$, then $x = \phi$ is* globally asymptotically stable.

Remark 1.1.3. From the autonomous principle stated in Section 1.1.3 of this chapter, if a constant solution of an autonomous system $x' = f(x)$ is Liapunov stable, then it is uniformly stable.

We saw in Theorem 1.1.4 that the zero solution of $x' = A(t)x$ is Liapunov stable if and only if $Z(t, t_0)$ is bounded. Our next result is a natural extension of that; however, *it is strictly for linear systems.*

Theorem 1.1.15. *Let $A(t)$ be continuous on $[t_0, \infty)$ and let $Z(t, t_0)$ be the* PMS *of (1.1.1). Then $x = 0$ is asymptotically stable if and only if $Z(t, t_0) \to 0$ as $t \to \infty$. Furthermore, $x = 0$ is asymptotically stable if and only if it is globally asymptotically stable.*

Proof. Let $x = 0$ be asymptotically stable. Then there exists $\eta > 0$ such that $|x_0| < \eta$ implies $|x(t, t_0, x_0)| \to 0$ as $t \to \infty$. Let $e_1, \ldots, e_n$ be the standard unit basis vectors for R^n. Then $|\eta e_i/2| < \eta$ so

$$|x(t, t_0, \eta e_i/2)| = |Z(t, t_0)e_i\eta/2| = |Z(t, t_0)e_i|\eta/2 \to 0$$

as $t \to \infty$ for $i = 1, \ldots, n$ but $Z(t, t_0)e_i$ is the ith column of Z and so $Z(t, t_0) \to 0$ as $t \to \infty$.

Now suppose that $Z(t, t_0) \to 0$ as $t \to \infty$. Then $Z(t, t_0)$ is bounded and so $x = 0$ is Liapunov stable. Let $t_1 \geq t_0$, let $\eta > 0$ be arbitrary, and let $|x_0| < \eta$. Then

$$|x(t, t_1, x_0)| = |Z(t, t_0)Z^{-1}(t_1, t_0)x_0| \leq |Z(t, t_0)|\,|Z^{-1}(t_1, t_0)|\,|x_0| \to 0$$

as $t \to \infty$, proving asymptotic stability.

To show that local asymptotic stability implies global asymptotic stability, note that each solution of (1.1.1) tends to zero if and only if $Z(t, t_0) \to 0$. This completes the proof.

Theorem 1.1.16. *Let $A(t)$ and $d(t)$ be continuous on $[t_0, \infty)$. The zero solution of*

$$(1.1.6) \qquad x' = A(t)x$$

is Liapunov stable if and only if every solution of

$$(1.1.7) \qquad x' = A(t)x + d(t)$$

is Liapunov stable.

Proof. Let ϕ be a solution of (1.1.7) on $[t_1, \infty)$ and let ψ be any other solution of (1.1.7) on $[t_1, \infty)$ for $t_1 \geq t_0$. Then $\phi(t) - \psi(t)$ is a solution of (1.1.1) and so it can be expressed as $\phi(t) - \psi(t) = Z(t, t_0)Z^{-1}(t_1, t_0)[\phi(t_1) - \psi(t_1)]$.

Now if ϕ is Liapunov stable, then for $\varepsilon = 1$, there exists $\delta > 0$ such that $|\phi(t_0) - x_0| < \delta$ and $t \geq t_0$ implies $|\phi(t) - x(t, t_0, x_0)| < 1$. Take

$\psi(t) = x(t, t_0, x_0)$ and choose $x_{01}, \ldots, x_{0n}$ such that $\phi(t_0) - x_{0i} = e_i \delta/2$ where $e_1, \ldots, e_n$ are the standard unit basis vectors. Then for each i we have $1 \geq |\phi(t) - x(t, t_0, x_{0i})| = |Z(t, t_0)e_i\delta/2|$ for $t \geq t_0$ and hence the matrix $Z(t, t_0)$ is bounded. Thus, $x = 0$ is Liapunov stable in (1.1.1).

Next suppose the zero solution of (1.1.6) is Liapunov stable (so $|Z(t, t_0)Z^{-1}(t_1, t_0)|$ is bounded for fixed $t_1 \geq t_0$) and let $t_1 \geq t_0$ and $\varepsilon > 0$ be given. We must conclude that

$$|\phi(t) - \psi(t)| = |Z(t, t_0)Z^{-1}(t_1, t_0)[\phi(t_1) - \psi(t_1)]| < \varepsilon$$

if $|\phi(t_1) - \psi(t_1)| < \delta$ for some $\delta > 0$. That this can be done is clear.

Remark 1.1.4. As it is less cumbersome to work with stability properties of constant solutions than with those of arbitrary solutions, it is of interest to note that they are equivalent. To see this, consider the system

$$(1.0.1) \qquad\qquad x' = f(t, x)$$

with $f : [t_0, \infty) \times D \to R^n$ and $\phi : [t_0, \infty) \to D$ a solution of (1.0.1). Let $y = x - \phi(t)$ so that

$$y' = x' - \phi'(t) = f(t, x) - \phi'(t)$$
$$= f(t, y + \phi(t)) - \phi'(t) \overset{\text{def}}{=} g(t, y)$$

has a solution $y = 0$. Thus

$$(1.0.1)' \qquad\qquad y' = g(t, y)$$

is equivalent to (1.0.1) and it has the zero solution.

Theorem 1.1.17. *Let $A(t)$ be continuous on $[t_0, \infty)$ and let the PMS of (1.1.1), $Z(t, t_0)$, be bounded. If there exists $K > 0$ with $\int_{t_0}^{t} \operatorname{tr} A(s)\,ds \geq -K$ for $t_0 \leq t < \infty$, then the zero solution of (1.1.1) is uniformly stable.*

Proof. We easily verify that the zero solution of (1.1.1) is uniformly stable if and only if there exists a constant M for which $|Z(t, t_0)Z^{-1}(t_1, t_0)| \leq M$ for $t_0 \leq t_1 \leq t < \infty$. As $Z(t, t_0)$ is bounded and as the elements of $Z^{-1}(t_1, t_0)$ are obtained from sums of products of elements of $Z(t_1, t_0)$ divided by $\det Z(t_1, t_0) = \exp \int_{t_0}^{t_1} \operatorname{tr} A(s)\,ds \geq \exp(-K)$ for some $K > 0$, we have $Z^{-1}(t_1, t_0)$ bounded. This completes the proof.

Exercise 1.1.8. By looking at a first-order linear equation, say $x' = -(1 + \cos t)x$, show that the converse of Theorem 1.1.17 is false.

Exercise 1.1.9. Let $n = 1$ and find a continuous scalar function $a(t)$ such that the zero solution of $x' = a(t)x$ is Liapunov stable, but not uniformly stable.

Theorem 1.1.18. *The zero solution of* (1.1.10) *is Liapunov stable if and only if the real part of every characteristic root* α *has* Re $\alpha \leq 0$ *and, if* α *is a root with* Re $\alpha = 0$ *and* α *has multiplicity* $m > 1$, *then there are* m *linearly independent characteristic vectors belonging to* α.

Proof. Let $x = 0$ be Liapunov stable for (1.1.10). Then all solutions of (1.1.10) are bounded for $t \geq 0$. Let α be a characteristic root of A and $x(t)$ a characteristic solution obtained from α. If Re $\alpha > 0$, then from the form of $x(t)$, it must be unbounded; thus, Re $\alpha \leq 0$. Suppose α is a root with Re $\alpha = 0$, α has multiplicity $m > 1$, and suppose that there are fewer than m linearly independent characteristic vectors belonging to α. Then there is a solution $x(t) = (P_1(t)e^{\alpha t}, \ldots, P_n(t)e^{\alpha t})^T$ with at least one $P_i(t)$ a polynomial of degree greater than zero. Thus, $x(t)$ is unbounded.

   To prove the converse, notice that Re $\alpha \leq 0$ and characteristic vectors of the specified type imply that all solutions are bounded.

Theorem 1.1.19. *The zero solution of* (1.1.10) *is asymptotically stable if and only if each characteristic root* α *of* A *satisfies* Re $\alpha < 0$.

Proof. The zero solution of (1.1.10) is asymptotically stable if and only if each solution tends to zero as $t \to \infty$. This will happen if and only if Re $\alpha < 0$ for each root of A, owing to the form of the linear basis in Theorem 1.1.14.

Exercise 1.1.10. Let A be constant and suppose each root α_i of A satisfies Re $\alpha_i < -\alpha$ for some $\alpha > 0$. Show that $|Z(t, 0)| \leq K \exp(-\alpha t)$ for some $K > 0$ and all $t \geq 0$.

Exercise 1.1.11. Let

$$A = \begin{pmatrix} -1 & 0 \\ +1 & -1 \end{pmatrix}$$

and find α and K for Exercise 1.1.10. Can you find α exactly? Also, find K if we ask $|Z(t, 0)| \leq K \exp(-\alpha t)$.

1.1.5   The Second-Order Case, Transformations

In this section we attempt to illustrate the theory by considering a general second-order system with constant coefficients. At the same time we prepare the reader for the material on Jordan forms by introducing some transformations. We consider the system $x' = Ax$ written as

(1.1.13)
$$x' = \begin{pmatrix} a & b \\ c & d \end{pmatrix} x$$

with a, b, c, d real constants. It will be assumed that det $A \neq 0$ so that there is a unique constant solution, $x = 0$.

It is convenient to make a transformation $x = Py$ in (1.1.13) where P is a constant nonsingular matrix to be specified later. When $x = Py$ is applied to (1.1.13) we obtain $x' = Py' = Ax = APy$ so that

(1.1.14)
$$y' = P^{-1}APy.$$

Several cases will be distinguished.

Let α_1 and α_2 be the characteristic roots of A.

Case I   The characteristic roots are real.

Case IA $\alpha_1 \neq \alpha_2$.

Case IB $\alpha_1 = \alpha_2$. (Note that as $\det(A - \alpha_1 I) = 0$ the rank of $(A - \alpha_1 I)$ is zero or one.)

Case IB(i)   The rank of $(A - \alpha_1 I)$ is zero.

Case IB(ii)   The rank of $(A - \alpha_1 I)$ is one.

Case II   The characteristic roots are complex.

Suppose Case IA holds. Let $u = (u_1, u_2)^T$ and $v = (v_1, v_2)^T$ be characteristic vectors belonging to α_1 and α_2, respectively, with the property that $u_1 v_2 - v_1 u_2 = 1$. In (1.1.14) define $P = (u, v)$ so that

$$P = \begin{pmatrix} u_1 & v_1 \\ u_2 & v_2 \end{pmatrix} \quad \text{and} \quad P^{-1} = \begin{pmatrix} v_2 & -v_1 \\ -u_2 & u_1 \end{pmatrix}.$$

A calculation shows that

$$P^{-1}AP = \begin{pmatrix} \alpha_1 & 0 \\ 0 & \alpha_2 \end{pmatrix}.$$

Hence, in Case IA, then (1.1.14) becomes

(1.1.14)$_{\text{IA}}$
$$y_1' = \alpha_1 y_1, \qquad y_2' = \alpha_2 y_2$$

and there are linearly independent solutions

$$\begin{pmatrix} e^{\alpha_1 t} \\ 0 \end{pmatrix} \quad \text{and} \quad \begin{pmatrix} 0 \\ e^{\alpha_2 t} \end{pmatrix}.$$

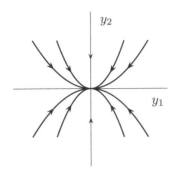

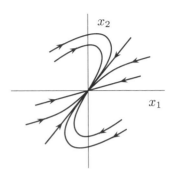

Fig. 1.4   Equation (1.1.14),   **Fig. 1.5**   Equation (1.1.13),
   $\alpha_2 < \alpha_1 < 0$. $\alpha_1 < \alpha_2 < 0$.

Linear combinations are solutions and, if $Z(t, 0)$ is the PMS of $(1.1.14)_{\text{IA}}$, then $W(t, 0) = PZ(t, 0)$ is the PMS of $(1.1.13)$. We then have the portraits of Figs. 1.4 and 1.5 in which the arrows indicate the direction of motion for increasing t.

The reader may construct portraits for the case in which α_1 and α_2 are positive. Note that there are two rectilinear characteristic solutions. All other solutions cluster to the solution with the root having least magnitude. Clearly, the zero solution is asymptotically stable. In this case, $x = 0$ is called a *node*. We have Figs. 1.6 and 1.7 when $\alpha_1 < 0 < \alpha_2$. The zero solution is called a *saddle point* in this case, and it is unstable.

In Case IB(i) we have rank zero for $(A - \alpha_1 I)$ and so $(1.1.13)$ is

$$x_1' = \alpha_1 x_1, \qquad x_2' = \alpha_1 x_2$$

which has solutions

$$\begin{pmatrix} e^{\alpha_1 t} \\ 0 \end{pmatrix} \qquad \text{and} \qquad \begin{pmatrix} 0 \\ e^{\alpha_1 t} \end{pmatrix}.$$

The portrait is then very simple and is shown in Fig. 1.8. This is called a *star node*, and it is asymptotically stable.

In case IB(ii) the rank of $(A - \alpha_1 I)$ is one. Thus, we may suppose that $a = d = \alpha_1$ and $c = 0$ for, if not, then choose $P = (u, v)$ where u is a characteristic vector belonging to α_1 and v is any other vector chosen so that $\det P \neq 0$. Then $P^{-1}AP$ will be of the stipulated form. Now follow with a new transformation with

$$P = \begin{pmatrix} 1 & 1 \\ 0 & 1/b \end{pmatrix}.$$

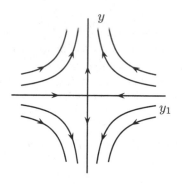

Fig. 1.6 Equation (1.1.14),
$\alpha_1 < 0 < \alpha_2$.

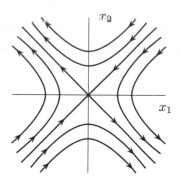

Fig. 1.7 Equation (1.1.13),
$\alpha_1 < 0 < \alpha_2$.

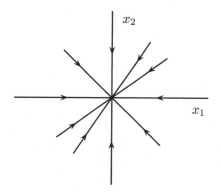

Fig. 1.8 Equation (1.1.13), $\alpha_1 = \alpha_2 < 0$.

Then (1.1.14) is

$$y_1' = \alpha_1 y_1 + y_2, \qquad y_2' = \alpha_1 y_2$$

and this has $(e^{\alpha_1 t}, 0)^T$ as a characteristic solution. As the coefficient matrix is triangular, one can solve it directly by integrating the second equation for $y_2 = e^{\alpha_1 t}$, so that $y_1' = \alpha_1 y_1 + e^{\alpha_1 t}$ can be solved by the variation of parameters formula. However, to illustrate our theory we proceed to find a primitive solution. Thus, we substitute $[(c_1 + d_1 t)e^{\alpha_1 t}, (c_2 + d_2 t)e^{\alpha_1 t}]^T$ into (1.1.14) to obtain

$$[d_1 + \alpha_1 c_1 + \alpha_1 d_1 t]e^{\alpha_1 t} = [(\alpha_1 c_1 + \alpha_1 d_1 t) + (c_2 + d_2 t)]e^{\alpha_1 t}$$

and
$$[d_2 + \alpha_1 c_2 + \alpha_1 d_2 t]e^{\alpha_1 t} = [\alpha_1 c_2 + \alpha_1 d_2 t]e^{\alpha_1 t}.$$

Upon canceling $e^{\alpha_1 t}$ and equating coefficients in t we obtain: $d_1 + \alpha_1 c_1 = \alpha_1 c_1 + c_2$, $\alpha_1 d_1 = \alpha_1 d_1 + d_2$, $d_2 + \alpha_1 c_2 = \alpha_1 c_2$, and $\alpha_1 d_2 = \alpha_1 d_2$. These yield $d_1 = c_2$ and $d_2 = 0$. Now $d_1 = c_2$ is arbitrary and c_1 is arbitrary, subject to the requirement that the characteristic vector $(1,0)^T$ and the vector $(c_1, c_2)^T$ be linearly independent. Thus, we choose $c_1 = 0$ and $c_2 = 1 = d_1$. We then have $(e^{\alpha_1 t}, 0)^T$ and $(te^{\alpha_1 t}, e^{\alpha_1 t})^T$ as the two linearly independent solutions. The portrait is that of a *degenerate node* and is shown in Fig. 1.9.

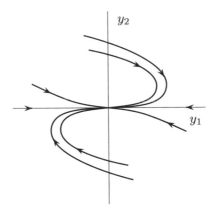

Fig. 1.9   Equation (1.1.14), $\alpha_1 < 0$.

In Case II, if $\alpha_1 = \alpha + i\beta$, then $\alpha_2 = \alpha - i\beta$ for α and β real. There is then a matrix P for which $P^{-1}AP$ is

$$\begin{pmatrix} \alpha & \beta \\ -\beta & \alpha \end{pmatrix}.$$

This is the real Jordan form for complex roots. Note that this case could have been considered under Case IA, but the resulting system $y' = P^{-1}APy$ would have been complex.

If $u+iv$ is a characteristic vector belonging to $\alpha+i\beta$, then one may verify that neither u nor v is zero. Also, it may be assumed that $u_1 v_2 - u_2 v_1 = 1$. We then take

$$P = (u, v) = \begin{pmatrix} u_1 & v_1 \\ u_2 & v_2 \end{pmatrix}.$$

Now $A(u+iv) = (\alpha+i\beta)(u+iv) = (\alpha u - \beta v) + i(\alpha v + \beta u)$. By the definition of equality of complex numbers we have $Au = (\alpha u - \beta v)$ and $Av = \alpha v + \beta u$.

Then $AP = (Au, Av)$ and

$$P^{-1}AP = \begin{pmatrix} \alpha & \beta \\ -\beta & \alpha \end{pmatrix}.$$

Certainly, $\alpha \pm i\beta$ are still the characteristic roots of $P^{-1}AP$ and $(1, i)^T$ is a characteristic vector belonging to $\alpha + i\beta$. Taking real and imaginary parts of the characteristic solution we obtain

$$(e^{\alpha t}\cos\beta t, -e^{\alpha t}\sin\beta t)^T$$

and

$$(e^{\alpha t}\sin\beta t, e^{\alpha t}\cos\beta t)^T$$

as two linearly independent real solutions. Representative portraits are given in Figs. 1.10 and 1.11 for $\alpha = 0$. These are called *vortex* points or *centers*. The corresponding portraits for (1.1.13) are ellipses. The zero solution is stable, but not asymptotically stable.

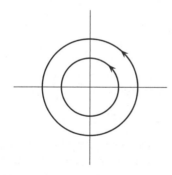 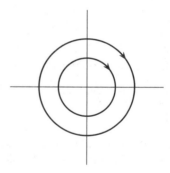

Fig. 1.10 Equation (1.1.14),   **Fig. 1.11** Equation (1.1.14),
$\alpha = 0$, $\beta < 0$. $\alpha = 0$, $\beta > 0$.

When $\alpha < 0$, the portraits are given in Figs. 1.12 and 1.13. In these cases, the origin is called a *focus*. It is asymptotically stable.

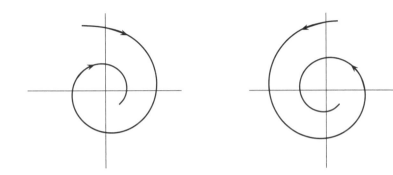

Fig. 1.12   Equation (1.1.14),   **Fig. 1.13**   Equation (1.1.14),
        $\alpha < 0$, $\beta > 0$. $\alpha < 0$, $\beta < 0$.

1.1.6   Transformations

We have seen several similarities between the system

$$(1.1.1) \qquad\qquad x' = A(t)x$$

and a scalar equation, say $x' = a(t)x$. Perhaps the most notable similarities are seen in the Jacobi identity and in the variation of parameters formula. When $A(t)$ and $a(t)$ are independent of t and when e^{At} is properly defined, then the principal solution matrices of the system (1.1.10) and the scalar equation are formally the same.

Definition 1.1.8. *If C is any square matrix, then*

$$(1.1.15) \qquad\qquad e^C = I + \sum_{n=1}^{\infty} \frac{C^n}{n!}.$$

This series converges if the individual elements converge (in the sequence of partial sums) or if it converges in the sense of the matrix norm.

Theorem 1.1.20. *If A is constant, then $Z(t,0) = e^{At}$ is the PMS of*

$$(1.1.10) \qquad\qquad x' = Ax.$$

Proof. Let $e_1, \ldots, e_n$ be the standard unit basis vectors for R^n and let $x_j(t)$ be the solution of (1.1.10) with $x_j(0) = e_j$. Consider the sequence of functions $\{x_{jk}(t)\}$ defined by the successive approximations:

$$x_{j0}(t) = e_j,$$

$$x_{j1}(t) = e_j + \int_0^t Ae_j ds = (I + At)e_j,$$

(1.1.16)

$$\vdots$$

$$x_{jk}(t) = (I + At + \cdots + (A^k t^k / k!))e_j.$$

Now for any $n \times n$ matrix C, Ce_j is the jth column of C. Hence, the right side of (1.1.16) is the jth column of the kth partial sum of the series for e^{At}. But we know from the proof of Theorem 1.1.2 that the left-hand side of (1.1.16) converges to the jth column of the PMS of (1.1.10) and, hence, the right side does also. Thus, in one step, we have shown that e^{At} converges and that it is the PMS of (1.1.10).

Corollary. $(e^{At})' = Ae^{At}$.

Remark 1.1.5. The reader may wish to extend Theorem 1.1.20 and prove that if the matrix $A(t)$ in (1.1.1) commutes with its integral for all t on an interval $[t_0, b)$, then the PMS of (1.1.1) is $Z(t, t_0) = \exp \int_{t_0}^t A(s)ds$.

The material on characteristic solutions and primitive solutions given in Section 1.1.3 may be obtained in a different manner by considering $e^{At}x_i$ where x_i is suitably chosen. One obtains $e^{At}x_i = e^{\alpha_i t}e^{Bt}x_i$ where B is nilpotent on x_i, yielding a closed form solution. A nice discussion of this type is given by Brauer and Nohel (1969, Chapter 2).

The object now is to make a transformation $x = Py$ (with P nonsingular) mapping (1.1.10) into

(1.1.17) $$y' = P^{-1}APy$$

where P is chosen so that $P^{-1}AP$ is in Jordan canonical form. Then the PMS of (1.1.17) will be $Y(t, 0) = e^{P^{-1}APt}$. As $x = Py$, a linear basis for (1.1.10) will be the columns of $PY(t, 0)$ and, hence, the PMS of (1.1.10) will be $Z(t, 0) = PY(t, 0)P^{-1}$ which yields $Z(t, 0) = Pe^{P^{-1}APt}P^{-1}$. But $Z(t, 0) = e^{At}$, and so we have obtained

(1.1.18) $$Pe^{P^{-1}APt}P^{-1} = e^{At}.$$

We refer the reader to Halmos (1958) for a proof of the following result.

Theorem 1.1.21. Jordan canonical form *If A is an $n \times n$ constant matrix (real or complex), then there exists a constant nonsingular matrix P such that $P^{-1}AP =$*

$$(1.1.19) \qquad J = \begin{bmatrix} J_0 & & & 0 \\ & J_1 & & \\ & & \ddots & \\ 0 & & & J_q \end{bmatrix}$$

where J_0 is a diagonal matrix with elements $\alpha_1, \ldots, \alpha_k$ and each J_i is an $m_i \times m_i$ matrix of the form

$$J_i = \begin{bmatrix} \alpha_{k+i} & 1 & 0 & \cdots & & 0 \\ 0 & \alpha_{k+i} & 1 & 0 & \cdots & 0 \\ & & \ddots & \ddots & & \vdots \\ & & & \ddots & \ddots & 1 \\ 0 & & & 0 & \cdots & \alpha_{k+i} \end{bmatrix}$$

for $i = 1, \ldots, q$, $m_i \geq 2$, $k + m_1 + \cdots + m_q = n$. Also, the α_r are not necessarily distinct.

In Section 1.1.5 we saw how to compute P (and, hence, J) for second-order systems. It should be stressed that some of that work went beyond the above result. In particular, the real Jordan matrix for complex roots would have been simply a pair of complex entries in the diagonal matrix J_0 above. Although A may be real, in Theorem 1.1.21 the matrix $P^{-1}AP$ will often be complex. Thus, the conclusions of Propositions 1.1.2–1.1.4 of Section 1.1.3 will not hold for $y' = Jy$.

The transformation $x = Py$ maps (1.1.10) into a system $y' = Jy$ which is a set of *entirely independent* systems of linear differential equations. The PMS reflects this as we see that

$$e^{Jt} = \begin{bmatrix} e^{J_0 t} & & & 0 \\ & e^{J_1 t} & & \\ & & \ddots & \\ 0 & & & e^{J_q t} \end{bmatrix}.$$

In order to more fully utilize the representation of $Z(t,0)$ as e^{At} and to display the solutions of $y' = Jy$ we present the following result.

Theorem 1.1.22. *If A and B are constant square matrices which commute, then*

$$e^{A+B} = e^A e^B \quad and \quad Be^{At} = e^{At}B.$$

Proof. Let $X(t) = Be^{At} - e^{At}B$. Then $X'(t) = BAe^{At} - Ae^{At}B = A[Be^{At} - e^{At}B] = AX(t)$; thus, $X(t)$ is a matrix of solutions of (1.1.10) and, as $X(0) = 0$, we have $X(t) \equiv 0$. This shows that $Be^{At} = e^{At}B$. Next, let $y' = [A + B]y$ with PMS $Y(t, 0) = e^{(A+B)t}$ and let $W(t) = e^{At}e^{Bt}$. Then $W'(t) = Ae^{At}e^{Bt} + e^{At}Be^{Bt} = [A + B]W(t)$; hence, $W(t)$ is a solution matrix of $y' = [A + B]y$ and, as $W(0) = I$, we have $W(t) \equiv e^{(A+B)t}$. That is, $e^{At}e^{Bt} = e^{(A+B)t}$.

Remark 1.1.6. The form of the PMS of (1.1.10) as $Z(t, 0) = e^{At}$, and the results of the last theorem yield three important properties:

(i) $Z(t, t_0) = e^{A(t-t_0)}$,

(ii) $Z^{-1}(t, 0) = e^{-At} = Z(-t, 0)$, and

(iii) $Z(t, t_0)Z^{-1}(s, t_0) = Z(t - s, 0)$.

To verify these, (i) holds by the autonomous principle; (ii) holds as $e^{At}e^{-At} = I$ yields e^{-At} as the (unique) inverse of e^{At}; and, finally, $Z(t, t_0)Z^{-1}(s, t_0) = e^{A(t-t_0)}e^{-A(s-t_0)} = e^{A(t-s)} = Z(t - s, 0)$.

Equation (iii) is important in certain applications using the variation of parameters formula.

We are now in a position to display e^{Jt}. As e^{Jt} is, in fact, decomposed into the $e^{J_i t}$ for $i = 0, \ldots, q$, we need only display $e^{J_i t}$. It should be clear that

$$e^{J_0 t} = \begin{bmatrix} e^{\alpha_1 t} & 0 & \cdots & & 0 \\ 0 & e^{\alpha_2 t} & 0 & & \vdots \\ & & \ddots & & \\ \vdots & & & \ddots & \\ 0 & \cdots & & 0 & e^{kt} \end{bmatrix}.$$

Now for $i > 0$, $J_i = \alpha_{k+i}I + B_i$ where B_i is an $m_i \times m_i$ matrix having 1's immediately above the main diagonal and zeros elsewhere. Hence,

$$e^{J_i t} = e^{\alpha_{k+i}It}e^{B_i t} = e^{\alpha_{k+i}t}e^{B_i t}.$$

From the form of B_i we see that it is nilpotent. In fact, B_i^2 is obtained from B_i by raising each 1 to the position immediately above it. Also, B_i^3 is obtained from B_i^2 by again raising each 1. From the series definition of $e^{B_i t}$ and the fact that B_i is $m_i \times m_i$, it follows that

$$e^{B_i t} = \begin{bmatrix} 1 & t & t^2/2! & \cdots & t^h/h! \\ 0 & 1 & t & \cdots & t^{h-1}/(h-1)! \\ \vdots & & \ddots & & \\ 0 & & \cdots & & 1 \end{bmatrix}$$

in which we have let $m_i - 1 = h$ for notational purposes. This gives a full and complete representation of the solutions of $y' = Jy$.

The weakness of the method is the difficulty in finding P. However, selection of P can be accomplished if the characteristic roots of A are known. The simple case occurs when the (not necessarily distinct) characteristic roots $\alpha_1, \ldots, \alpha_n$ of A have linearly independent characteristic vectors $u_1, \ldots, u_n$, respectively. Then P is formed by selecting $P = (u_1, \ldots, u_n)$, as we now show.

$$P^{-1}AP = P^{-1}(Au_1, \ldots, Au_n)$$
$$= P^{-1}(\alpha_1 u_1, \ldots, \alpha_n u_n)$$
$$= (\alpha_1 P^{-1} u_1, \ldots, \alpha_n P^{-1} u_n).$$

Now u_i is the ith column of P and so $P^{-1} u_i$ is the ith column of $P^{-1}P$ which is e_i. This verifies the statement.

Calculation of P when there are fewer than n linearly independent characteristic vectors is difficult, cumbersome, and beyond the scope of this discussion. It involves a careful selection of a suitable basis for the subspace generated by the vectors satisfying $(A - \alpha_i I)^{m_i} x = 0$.

The transformation to a Jordan form is presented here in order to show the minimal ingredients involved in the PMS so that the exact structure will be seen. However, virtually everything, both practical and theoretical, that can be accomplished here with the Jordan form can also be done with a triangular form. Furthermore, constructing a transformation $x = Py$ which will map (1.1.10) into $y' = Ry$ with R an upper triangular matrix is not at all difficult.

Theorem 1.1.23. *Let A be an $n \times n$ matrix. There exists a matrix P for which $P^{-1}AP$ is an upper triangular matrix.*

Proof. We proceed by induction on n. Let $n = 2$, let α_1 be a root of A, let x_1 be a characteristic vector of A belonging to α_1, and let x_2 be any other vector chosen so that $P_2 = (x_1, x_2)$ is nonsingular. Then $P_2^{-1}AP_2 = P_2^{-1}(Ax_1, Ax_2) = P_2^{-1}(\alpha_1 x_1, Ax_2) = (\alpha_1 P_2^{-1} x_1, P_2^{-1} Ax_2)$. As x_1 is the first column of P_2, $P_2^{-1} x_1 = e_1$ and this establishes the result for $n = 2$.

Assume now that if A is $n \times n$, then there is a matrix P_n with the property that $P_n^{-1}AP_n$ is upper triangular. Then let A be an $(n+1) \times (n+1)$ matrix and we will use the inductive hypothesis to obtain P_{n+1} for which $P_{n+1}^{-1}AP_{n+1}$ is upper triangular. Again, let α_1 be a characteristic root of A, x_1 a characteristic vector belonging to α_1, and choose n vectors $x_2, \ldots, x_{n+1}$

such that the matrix $\bar{P}_{n+1} = (x_1, \ldots, x_{n+1})$ is nonsingular. Then

$$\bar{P}_{n+1}^{-1} A \bar{P}_{n+1} = \begin{bmatrix} \alpha_1 & C_{12} & \cdots & C_{1,n+1} \\ 0 & C_{22} & \cdots & C_{2,n+1} \\ & & \vdots & \\ 0 & C_{n+1,2} & \cdots & C_{n+1,n+1} \end{bmatrix}$$

$$= \begin{bmatrix} \alpha_1 & C_{12} & \cdots & C_{1,n+1} \\ 0 & & & \\ \vdots & & C_n & \\ 0 & & & \end{bmatrix}$$

where C_n is an $n \times n$ matrix. Now by the inductive assumption there is a matrix P_n for which $P_n^{-1} C_n P_n$ is upper triangular. Construct a matrix

$$D_{n+1} = \begin{bmatrix} 1 & 0 & \cdots & 0 \\ 0 & & & \\ \vdots & & P_n & \\ 0 & & & \end{bmatrix}.$$

Then $D_{n+1}^{-1}(\bar{P}_{n+1}^{-1} A \bar{P}_{n+1})D_{n+1}$ is upper triangular and $P_{n+1} = \bar{P}_{n+1}D_{n+1}$. This completes the proof.

Once a system (1.1.10) is transformed into a system $y' = Ry$ with R triangular, then one may solve the latter system using the variation of parameters formula. The diagonal elements are the characteristic roots $\alpha_1, \ldots, \alpha_n$. First, $(e^{\alpha_1 t}, 0, \ldots, 0)^T$ is a solution. Next, $(y_1, e^{\alpha_2 t}, 0, \ldots, 0)^T$ is a solution where y_1 is the solution of $y_1' = \alpha_1 y_1 + r_{12}e^{\alpha_2 t}$ which is the first row of $y' = Ry$ when $y_2 = e^{\alpha_2 t}$ and $y_i = 0$ for $i > 2$. Continue with $y_i = 0$ if $i > 3$, $y_3 = e^{\alpha_3 t}$, and y_1 and y_2 are found after this transformation is substituted into the first two rows of $y' = Ry$. Continue and obtain a basis. This gives us a method of finding a linear basis without appealing to any (unproved) theorems of linear algebra.

The technique of solving equations via characteristic and primitive solutions appears to be the easiest and most direct method, whereas the Jordan form is preferable for the theoretical structure.

1.1.7　Transformations, Floquet Theory, the Adjoint

Mapping (1.1.10) into a simpler form $y' = Jy$ is the first stage of a problem still generating much interest.

Definition 1.1.9. *Let $A(t)$ be continuous on $[t_0, \infty)$ and let P be an $n \times n$ matrix of differentiable functions on $[t_0, \infty)$ with the property that $|P(t)| \leq M$ and $|P^{-1}(t)| \leq M$ for $t_0 \leq t \leq \infty$ and some $M > 0$. Then $x = P(t)y$ mapping (1.1.1) into*

$$(1.1.20) \qquad y' = [P^{-1}(t)A(t)P(t) - P^{-1}(t)P'(t)]y$$

is called a Liapunov transformation.

The computation verifying (1.1.20) will be given in the proof of the next theorem. It is easily verified that, owing to the bound on $|P|$ and $|P^{-1}|$, all stability properties of (1.1.1) are preserved by Liapunov transformations.

The object is to find conditions on $A(t)$ to ensure the existence of $P(t)$ so that the equation in y is tractable. A particularly nice result of this type forms the basis of Floquet theory.

Theorem 1.1.24. Floquet *Suppose that A is continuous on $(-\infty, \infty)$ and $A(t+T) = A(t)$ for all t and some $T > 0$. Then there exists a differentiable matrix P on $(-\infty, \infty)$ and a constant matrix R such that $P(t+T) = P(t)$, $P(0) = I$, and the transformation $x = P(t)y$ maps (1.1.1) into*

$$(1.1.21) \qquad\qquad\qquad y' = Ry.$$

Proof. Let $Z(t, 0) = Z(t)$ and let $W(t) = Z(t + T)$. Then $W'(t) = Z'(t + T) = A(t + T)Z(t + T) = A(t)W(t)$ and, hence, the columns of W are solutions of (1.1.1). Therefore, there is a nonsingular matrix C with $Z(t + T) = Z(t)C$. Now there is a matrix R such that $e^{RT} = C$. (cf. Bellman, 1953, p. 29). Next, define $P(t)$ by $Z(t) = P(t)e^{Rt}$. We then have $Z(t+T) = P(t + T)e^{R(t+T)}$ or $Z(t)C = P(t + T)e^{Rt}e^{RT}$ so that $Z(t) = P(t + T)e^{Rt}$, and this yields $P(t) = P(t + T)$. As $P(t)$ is periodic and nonsingular, the existence of the required bound M is clear. As $Z(t, 0) = P(t)e^{Rt}$ and, as e^{Rt} is the PMS of $y' = Ry$, it follows that $x = P(t)y$ maps (1.1.1) into $y' = Ry$. However, we now compute directly the formula in the Liapunov transformation definition as it was not verified there. We let $x = P(t)y$ and verify that $y' = Ry$ with $R = P^{-1}AP - P^{-1}P'$. We have $x' = P'y + Py' = P'y + PRy = [P'P^{-1} + PRP^{-1}]Py = Ax$. Thus, $A = P'P^{-1} + PRP^{-1}$ so that $P^{-1}[A - P'P^{-1}]P = R$, as specified.

Although $P(t)$ can seldom be found in practice, this result has many important consequences. In the variation of parameters formula, $Z(t, 0)Z^{-1}(s, 0) = P(t)e^{Rt}e^{-Rs}P^{-1}(s) = P(t)e^{R(t-s)}P^{-1}(s)$, and this can be used to great advantage as we will subsequently see. To mention but one other use, the form of $Z(t, 0)$ can be used to show that, for $A(t)$ periodic, Liapunov stability implies uniform stability.

Example 1.1.1. *revisited* Epstein (1962) gave a particularly nice proof of the result in Example 1.1.1 using the Floquet theorem. We have $A(t + T) = A(t)$ and $A(-t) = -A(t)$. Now notice that $Z(t, 0)$ is an even matrix function. That is, $Z'(-t, 0) = -A(-t)Z(-t, 0)$ or $Z'(-t, 0) = AZ(-t, 0)$ and so $Z(-t, 0)$ is a solution matrix; however, as $Z(-t, 0)|_{t=0} = I$, we have $Z(-t, 0) = Z(t, 0)$. Now by Floquet's theorem, $Z(-\frac{1}{2}T + T, 0) = Z(-\frac{1}{2}T, 0)C$ or $Z(\frac{1}{2}T, 0) = Z(-\frac{1}{2}T, 0)C$ so that $Z(\frac{1}{2}T, 0) = Z(\frac{1}{2}T, 0)C$ yielding $C = I$. Thus, $Z(t + T) = Z(t)$ for all t.

We now develop the adjoint system of (1.1.1). First, note that if $W(t)$ is any $n \times n$ differentiable nonsingular matrix, then $W(t)W^{-1}(t) = I$ so $W'(t)W^{-1}(t) + W(t)[W^{-1}(t)]' = 0$ or $[W^{-1}(t)]' = -W^{-1}(t)W'(t)W^{-1}(t)$. Now, if the columns of W are linearly independent solutions of (1.1.1), then $[W^{-1}(t)]' = -W^{-1}(t)A(t)$ and so $[W^{-1}(t)^T]' = -A(t)^T W^{-1}(t)^T$.

Definition 1.1.10. *The equation*

$$(1.1.22) \qquad\qquad z' = -A^T(t)z$$

is the adjoint *of* (1.1.1).

Theorem 1.1.25. *If the columns of $W(t)$ are a linear basis for (1.1.1) and if the columns of $M(t)$ are a linear basis for the adjoint, then $M^T(t)W(t) = C$ where C is constant and nonsingular.*

Proof. By the preceding discussion, $W^{-1}(t)^T$ is a solution of the adjoint and so $W^{-1}(t)^T = M(t)K$ for some constant matrix K. Thus, $I = W(t)^T M(t)K$ or $W(t)^T M(t) = K^{-1}$ and $M^T(t)W(t) = C$. This completes the proof.

Corollary. *If $A(t)$ is skew symmetric, then solutions of (1.1.1) have constant Euclidean length.*

Proof. The adjoint of (1.1.1) is $z' = -A^T(t)z$ and $-A^T(t) = A(t)$ as A is skew symmetric. Thus, for any matrix W whose columns are a linear basis of (1.1.1), we have $W^T(t)W(t) = C$. This completes the proof.

Corollary. *If $A(t)$ is skew symmetric, then $Z(t, 0)$ is orthogonal; that is,*

$$Z^{-1}(t, 0) = Z^T(t, 0).$$

Corollary. *If $A(t)$ is skew symmetric, then the zero solution of (1.1.1) is uniformly stable.*

Recall that any $n \times n$ matrix may be expressed as a sum of symmetric and skew symmetric matrices:

$$A(t) = \frac{1}{2}(A + A^T) + \frac{1}{2}(A - A^T).$$

Theorem 1.1.26. *If $P(t)$ is the principal matrix solutions of*

$$p' = \frac{1}{2}(A(t) - A^T(t))p$$

then the transformation $x = P(t)y$ is a Liapunov transformation mapping (1.1.1) *into*

$$y' = \frac{1}{2}P^T(t)(A(t) + A^T(t))P(t)y.$$

Proof. As $P^{-1}(t) = P^T(t)$ and the columns of P have constant length, $x = P(t)y$ is a Liapunov transformation. Now $P'P^{-1} = \frac{1}{2}(A - A^T)$ and, if $y' = Ry$, then $A = P'P^{-1} + PRP^{-1}$ and so $\frac{1}{2}(A + A^T) + \frac{1}{2}(A - A^T) = \frac{1}{2}(A - A^T) + PRP^{-1}$ or $R = \frac{1}{2}P^{-1}(A + A^T)P = \frac{1}{2}P^T(A + A^T)P$, as was to be shown.

This result shows that we may always assume that the coefficient matrix is symmetric. The reader may wish to verify that for $n = 2$ and $A^T(t) = -A(t)$, then $P(t)$ can be found by integration.

As a final transformation theorem, we show that (1.1.1) can always be mapped into a system $y' = B(t)y$ in such a fashion that the solutions of $w' = 2B(t)w$ are known. Tantalizing as it appears, the connecting links between solutions of $y' = By$ and $w' = 2Bw$ have not been clearly established.

Theorem 1.1.27. *Let $A_1(t)$ and $A_2(t)$ be continuous matrices on an interval (a, b) with $A(t) = A_1(t) + A_2(t)$ and let*

$$h' = 2A_1(t)h \quad and \quad z' = 2A_2(t)z$$

have PMS's $H(t)$ and $Z(t)$, respectively, at $t = t_0 \in (a, b)$. The transformation $x = H(t)y$ maps (1.1.1) *into*

$$y' = B(t)y$$

and $z = H(t)w$ maps $z' = 2A_2(t)z$ into

$$w' = 2B(t)w$$

where $B = H^{-1}[A - H'H^{-1}]H$.

Proof. Now $x = Hy$ implies that $x' = H'y + Hy' = [H'H^{-1} + HBH^{-1}]Hy$. Since $x' = Ax$ we have $A = H'H^{-1} + HBH^{-1}$ or $B = H^{-1}[A - H'H^{-1}]H$. But $A = A_1 + A_2$ while $A_1 = \frac{1}{2}H'H^{-1}$ and so $2B = H^{-1}[2A_2 - H'H^{-1}]H$. This completes the proof.

It is, of course, simple to choose A_1 and A_2 so that solutions for $h' = 2A_1(t)h$ and $z' = 2A_2(t)z$ are readily found.

1.1.8 Liapunov's Direct Method I, the Routh-Hurwitz Criterion

In 1892 a Russian mathematician, A.M. Liapunov, published a major work which has had enormous impact on the study of differential equations. The study was translated into French and was reproduced as "Annals of Mathematics Study No. 17," Princeton University Press, in 1907 under the title "Problème général de la stabilité du mouvement" by A.M. Liapounoff. However, it attracted little interest until much later.

By 1940 the classical theory of ordinary differential equations was essentially dead in the Western world. In large measure, the ideas of Liapunov, coupled with the impetus of sophisticated needs arising from technical problems of World War II, generated a rebirth of the subject which then grew into one of the most active and important areas of modern mathematics.

The idea of Liapunov was simple in the extreme. Suppose that $f : [t_0, \infty) \times D \to R^n$ where D is an open set containing $x = 0$ and $f(t, 0) \equiv 0$. Let solutions of the initial value problem for

$$(1.0.1) \qquad\qquad x' = f(t, x)$$

exist for each $(t_1, x_0) \in [t_0, \infty) \times D$.

If $V : [t_0, \infty) \times D \to [0, \infty)$ is any differentiable scalar function and if $\phi : [t_0, \infty) \to D$ is any differentiable function, then $V(t) \stackrel{\text{def}}{=} V(t, \phi(t))$ is a scalar function of t, and we can compute V' by the chain rule:

$$V'(t) = \frac{\partial V}{\partial x_1}\phi_1'(t) + \cdots + \frac{\partial V}{\partial x_n}\phi_n'(t) + \frac{\partial V}{\partial t}.$$

Thus, it appears that in order to find V' we need not know $\phi(t)$, but rather $\phi'(t)$. Hence, if ϕ is an unknown solution of (1.0.1) we know that $\phi_i'(t) = f_i(t, \phi(t))$ and so for $V(t, \phi(t))$ we have

$$V'(t, \phi(t)) = \frac{\partial V}{\partial x_1}(t, \phi(t))f_1(t, \phi(t)) + \cdots$$

$$+ \frac{\partial V}{\partial x_n}(t, \phi(t))f_n(t, \phi(t)) + \frac{\partial V}{\partial t}.$$

Written in this form, it is now evident that we still need to know $\phi(t)$ to compute V' *exactly*. However, suppose V were chosen so shrewdly that for any $(t, x) \in [t_0, \infty) \times D$ we have

$$\frac{\partial V}{\partial x_1}(t, x)f_1(t, x) + \cdots + \frac{\partial V}{\partial x_n}(t, x)f_n(t, x) + \frac{\partial V}{\partial t} \leq 0.$$

Then we could certainly conclude that

$$V'(t, \phi(t)) \leq 0.$$

We know that if $V(t)$ is any scalar differentiable function with $V'(t) \leq 0$, then $V(t)$ is nonincreasing. Hence, if $V(t, x)$ is a function which grows with the growth of $|x|$, and if one can show that for any solutions ϕ of (1.0.1) we have $V'(t, \phi(t)) \leq 0$, then we can conclude from the nongrowth of $V(t, \phi(t))$ that $|\phi(t)|$ does not grow very much.

The situation is considerably simplified when V is independent of t and this is the assumption which we initially make.

Definition 1.1.11. *Let D be an open set in R^n with $0 \in D$, let $f(t, 0) = 0$, and let $V : D \rightarrow [0, \infty)$ have continuous first partial derivatives. If for $(t, x) \in [t_0, \infty) \times D$ we have*

$$V'(x) = \frac{\partial V}{\partial x_1}(x)f_1(t, x) + \cdots + \frac{\partial V}{\partial x_n}(x)f_n(t, x) \leq 0,$$

then V is an autonomous Liapunov function for (1.0.1).

Definition 1.1.12. *An autonomous function $V : D \rightarrow [0, \infty)$ is positive definite if $V(0) = 0$ and $V(x) > 0$ if $x \neq 0$.*

It will later be shown that if $f : [t_0, \infty) \times D \rightarrow R^n$ with f continuous and D open, then for each $(t_1, x_0) \in [t_0, \infty) \times D$, each solution $x(t, t_1, x_0)$ exists on an interval (or can be continued as a solution to a maximal interval) $[t_1, \alpha)$ and, unless $\alpha = \infty$, then $x(t, t_1, x_0)$ approaches the boundary of D as $t \rightarrow \alpha^-$. However, as we do not yet have this result and, as this is a particularly appropriate point to introduce certain results, we add an extraneous hypothesis concerning the existence of solutions on $[t_1, \infty)$. The hypothesis is not needed for linear systems, but the proofs are identical for linear and nonlinear systems, making it pointless to restrict the results to linear systems.

Theorem 1.1.28. *Suppose there is a positive definite autonomous Liapunov function for (1.0.1) and suppose that each solution $x(t, t_1, x_0)$ which remains in D exists on $[t_1, \infty)$. Then $x = 0$ is uniformly stable.*

Proof. Let $\varepsilon > 0$ be given. We must find $\delta > 0$ such that $|x_0| < \delta$ and $t \geq t_1 \geq t_0$ implies $|x(t, t_1, x_0)| < \varepsilon$. We suppose $\varepsilon < d(0, D^c)$. The boundary of a set U will be denoted by ∂U and $S(x_1, m)$ will denote a spherical neighborhood centered at x_1 with radius m. Now $\partial S(0, \varepsilon)$ is compact and V is continuous so V has a positive minimum, say m, there. As $V(0) = 0$ and V is continuous, there exists $\delta > 0$ such that $|x_0| < \delta$ implies $V(x_0) < m$. (Clearly, then $|x_0| < \varepsilon$.) Let $x(t, t_1, x_0)$ be any solution of (1.0.1) defined on any interval $[t_1, \alpha)$ with $|x_0| < \delta$. Then $V'(x(t, t_1, x_0)) \leq 0$ so long as $x(t, t_1, x_0)$ in R^n is defined and in D. Hence, $V(x(t, t_1, x_0)) \leq V(x_0) < m$ so long as the solution is defined and in D. The path of $x(t, t_1, x_0)$ is connected and starts inside $S(0, \varepsilon)$ so it can never reach $\partial S(0, \varepsilon)$ as $V(x) \geq m$ on $\partial S(0, \varepsilon)$. Thus $|x(t, t_1, x_0)| < \varepsilon$ on $[t_1, \alpha)$ and so $\alpha = \infty$.

Example 1.1.2. Let

$$x_1' = x_2, \qquad x_2' = -x_1 - x_2$$

and let $V(x) = x_1^2 + x_2^2$. Then

$$\begin{aligned} V'(x) &= 2x_1 x_1' + 2x_2 x_2' \\ &= 2[x_1 x_2 - x_1 x_2 - x_2^2] \leq 0. \end{aligned}$$

Hence, $x = 0$ is uniformly stable.

There are several geometrical interpretations which can be very helpful in constructing Liapunov functions and proving theorems.

First, if V is positive definite with $V' \leq 0$ and if $n = 2$, then consider curves defined by $V(x) = C$. These are often simple closed curves (a different one for each sufficiently small C) around $x = 0$. Also, the vector $N = (\partial V/\partial x_1, \partial V/\partial x_2)$ should be the outward drawn normal to the curve. Then $V'(x) = N \cdot f(t, x) = |N| |f(t, x)| \cos \theta \leq 0$ and so the tangent vector f to the solution forms an angle satisfying $\pi \geq |\theta| \geq \pi/2$ with N and, hence, the solution does not go outside the curve $V(x) = C$. For a given $\varepsilon > 0$, find C so that the curve $V(x) = C$ is just inside $S(0, \varepsilon)$. Then find $\delta > 0$ so that $S(0, \delta)$ is just inside $V(x) = C$. If $|x_0| < \delta$, then $x(t, t_1, x_0)$ cannot cross $V(x) = C$ as $N \cdot F(t, x) \leq 0$ and so $|x(t, t_1, x_0)| < \varepsilon$ for $t \geq t_1$. See Fig. 1.14.

The geometry of the curves in the next interpretation is the same. As $V(x)$ is positive definite, in a neighborhood of $x = 0$ it is a cup which has no holes in it; we imagine it to be a paraboloid of revolution. As $V'(x(t)) \leq 0$, $x(t)$ cannot wander far from zero because then $V(x(t))$ would be climbing the cup. See Fig. 1.15.

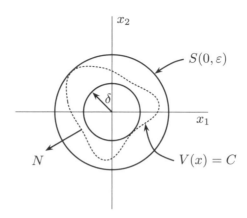

Fig. 1.14

Definition 1.1.13. *A continuous positive definite function* $W : R^n \to [0, \infty)$ *is called a* wedge.

Theorem 1.1.29. *Suppose there is a positive definite autonomous Lia-punov function for* (1.0.1) *and suppose that each solution* $x(t, t_1, x_0)$ *which remains in* D *exists on* $[t_1, \infty)$. *If there is a wedge* W *such that for* $(t, x) \in [t_0, \infty) \times D$ *we have* $V'(x) \leq -W(x)$, *then* $x = 0$ *is asymptotically stable.*

Proof. We have already shown that $x = 0$ is uniformly stable. It remains to be shown that for each $t_1 \geq t_0$, there exists $\eta > 0$ such that $|x_0| < \eta$ implies that $x(t, t_1, x_0) \to 0$ as $t \to \infty$. In this case, η is independent of t_1. Let $\varepsilon \leq d(0, \partial D)/2$ and let a δ of uniform stability be selected. We claim that η may be chosen as δ. To see this, let $|x_0| < \delta$ so that $|x(t, t_1, x_0)| < \varepsilon$ if $t \geq t_1$ and so $V'(x(t, t_1, x_0)) \leq -W(x(t, t_1, x_0))$. Thus, if we let $x(t) = x(t, t_1, x_0)$ then $V(x(t))$ is monotone decreasing with limit $C \geq 0$. If $C > 0$ then as $V(0) = 0$, there exists $\gamma > 0$ with $V(x) < C/2$ if $|x| \leq \gamma$. Hence, for $t \geq t_1$ we have $|x(t)| > \gamma$ and, as W is positive definite and continuous with $|x(t)| \leq \varepsilon$, $W(x(t)) \geq \mu$ for some $\mu > 0$. Therefore, $V'(x(t)) \leq -\mu$ for $t \geq t_1$ and so $V(x(t)) \leq V(x(t_1)) - \mu(t - t_1) \to -\infty$ as $t \to \infty$, a contradiction to $V \geq 0$. We therefore conclude that $C = 0$. As V is positive definite, it is clear that $x(t) \to 0$. Indeed, if not, then there is a $p > 0$ and a sequence $\{t_n\} \uparrow +\infty$ with $|x(t_n)| \geq p$. If $m = \min_{p \leq \|x\| \leq \varepsilon} V(x)$, then m is positive as V is positive definite, and $V(x(t_n)) \geq m$ contradicts $V(x(t)) \to 0$. This completes the proof.

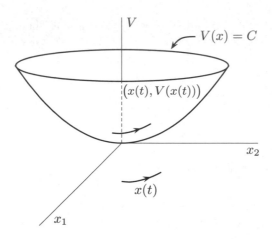

Fig. 1.15

The following theorem gives the basic result on the construction of Liapunov functions. For reference, a symmetric matrix C is positive definite if $x^T C x > 0$ for $x \neq 0$.

Theorem 1.1.30. *For a given positive definite symmetric matrix C, the equation*

$$(1.1.23) \qquad\qquad A^T B + BA = -C$$

can be solved for a positive definite symmetric matrix B if and only if every characteristic root of A has a negative real part.

Proof. Suppose that $A^T B + BA = -C$ can be solved for a positive definite symmetric matrix B. Then the function $V(x) = x^T B x$ is a positive definite autonomous Liapunov function for (1.1.10) as $V'(x) = (x^T)' B x + x^T B x' = x^T A^T B x + x^T B A x = x^T [A^T B + BA] x = -x^T C x \leq 0$. As $x^T C x$ is a wedge, by Theorem 1.1.29, $x = 0$ is asymptotically stable. Hence all characteristic roots of A must have negative real parts.

Suppose all characteristic roots of A have negative real parts and define

$$B = \int_0^\infty e^{A^T t} C e^{At}\, dt.$$

As $|e^{At}| \leq K e^{-\alpha t}$ for $t \geq 0$ and some positive constants K and α, the integral certainly converges. For each value of t, $x^T e^{A^T t} C e^{At} x =$

$(e^{At}x)^T C(e^{At}x) = y^T Cy$ and, if $x \neq 0$, then $y \neq 0$; hence, as C is positive definite, $y^T Cy > 0$ for $x \neq 0$. Thus, B is positive definite. Now

$$A^T B + BA = \int_0^\infty A^T e^{A^T t} C e^{At}\, dt + \int_0^\infty e^{A^T t} C e^{At} A\, dt$$

$$= \int_0^\infty \frac{d}{dt}(e^{A^T t} C e^{At})\, dt = (e^{A^T t} C e^{At})\Big|_0^\infty = -C.$$

This completes the proof.

Exercise 1.1.13. Let $C = I$ and $A = \begin{pmatrix} -1 & 0 \\ 2 & -3 \end{pmatrix}$. Find a symmetric B with $A^T B + BA = -I$.

Exercise 1.1.14. With $V(x) = x^T Bx$ and B chosen in Exercise 1.1.13, show that V is an autonomous Liapunov function in some open set D containing $x = 0$ for the system

$$x_1' = -x_1 + f_1(t, x),$$
$$x_2' = 2x_1 - 3x_2 + f_2(t, x)$$

where $\lim_{|x| \to 0} |f_i(t, x)|/|x| = 0$ uniformly for $0 \leq t < \infty$ and the $f_i(t, x)$ are continuous on $[0, \infty) \times S(0, 1)$. Then find a wedge W with $V'(x) \leq -W(x)$ on D.

From Exercise 1.1.14 we see that knowledge about solutions of $x' = Ax$ may yield information about solutions of $x' = Ax + f(t, x)$ provided that we can find a suitable Liapunov function for (1.1.10). For a given pair A and C, the equation $A^T B + BA = -C$ is a set of algebraic equations in B which can be solved regardless of the magnitude of n. The matrix C can always be conveniently chosen as I to ensure that it is positive definite. We need a verifiable criteria to establish when the characteristic roots of A have negative real parts so that we can then be sure *in advance* that B will be positive definite. The Routh-Hurwitz theorem provides this criterion. We give it here without proof. A proof may be found in Gantmacher (1960, p. 193). We expand $\det(A - \alpha I) = 0$ into its polynomial form:

(1.1.24) $\alpha^n + a_1 \alpha^{n-1} + \cdots + a_n = 0.$

Theorem 1.1.31.   Routh-Hurwitz   *All of the characteristic roots of* (1.1.24) *have negative real parts if and only if $D_k > 0$ for $k = 1, 2, \ldots, n$*

where

$$D_1 = a_1, \quad D_2 = \begin{vmatrix} a_1 & a_3 \\ 1 & a_2 \end{vmatrix}, \dots, D_k = \begin{bmatrix} a_1 & a_3 & a_5 & \cdots & a_{2k-1} \\ 1 & a_2 & a_4 & \cdots & a_{2k-2} \\ 0 & a_1 & a_3 & \cdots & a_{2k-3} \\ 0 & 1 & a_2 & \cdots & a_{2k-4} \\ 0 & 0 & a_1 & \cdots & a_{2k-5} \\ 0 & 0 & 1 & \cdots & a_{2k-6} \\ \vdots & \vdots & \vdots & & \vdots \\ 0 & & & \cdots & a_k \end{bmatrix}$$

with $a_j = 0$ *for* $j > n$.

Exercise 1.1.15. Deduce the conditions on the coefficients of (1.1.24) for the characteristic roots to have negative real parts when $n = 3$ and $n = 4$.

It frequently happens that a function $V(t, x)$ which we wish to use for a Liapunov function is continuous in (t, x) and locally Lipschitz in x, but is not differentiable. A typical example is

$$V(t, x) = |x|.$$

This offers no real difficulty and we explain in Section 4.2 of Chapter 4 how to deal with it even when V is a functional and the equation is a functional differential equation. But we will need some information before then so we offer the following sketch.

Consider (1.0.1) and suppose there is a continuous function $V : [t_0, \infty) \times D \to R$ which is locally Lipschitz in x. Then define

$$V'_{(1.0.1)}(t, x) = \limsup_{h \to 0^+} \{V(t + h, x + hf(t, x)) - V(t, x)\}/h.$$

Next, suppose $x(t)$ is a solution of (1.0.1) which remains in D on some interval $[t, t + h_0]$ for $h_0 > 0$. Then the upper right-hand derivative of $V(t, x(t))$ is

$$V'(t, x(t)) = \limsup_{h \to 0^+} \{V(t + h, x(t + h)) - V(t, x(t))\}/h.$$

It is an exercise in inequalities and the Lipschitz condition (cf. Yoshizawa, 1966, p. 3) to show that for $x = x(t)$ then

$$V'_{(1.0.1)}(t, x) = V'(t, x(t)).$$

In particular, if

$$V'_{(1.0.1)}(t, x) \leq W(t, x)$$

for some continuous function W, then for a solution $x(t)$ in D on an interval $[t_1, t]$ we may integrate that last inequality and have

$$V(t, x(t)) - V(t_1, x(t_1)) \leq \int_{t_1}^{t} W(s, x(s))\, ds.$$

1.1.9   Matrix Comparison Theorems, Liouville Transformations

We consider a system

(1.1.25) $$x' = (A(t) + B(t))x$$

with A and B $n \times n$ matrices of continuous functions on $[t_0, \infty)$. It is assumed that much is known about solutions of

(1.1.26) $$y' = A(t)y$$

and that $B(t)$ is small in some sense. The object then is to find a relationship between solutions of (1.1.25) and (1.1.26).

Theorem 1.1.32. *Let A be constant with all solutions of (1.1.26) bounded on $[t_0, \infty)$ and let $\int_{t_0}^{\infty} |B(s)|\, ds < \infty$. Then all solutions of (1.1.25) are bounded on $[t_0, \infty)$.*

Proof. We give the proof for $t_0 = 0$ and show that each solution $x(t, 0, x_0)$ of (1.1.25) is bounded. Denote by $Z(t)$ the PMS of (1.1.26) at $t = 0$. If $x(t)$ is any solution of (1.1.25) then

$$x'(t) = Ax(t) + B(t)x(t) \overset{\text{def}}{=} Ax(t) + d(t)$$

and so by the variation of parameters formula,

$$x(t) = Z(t)\left[x(0) + \int_0^t Z^{-1}(s)B(s)x(s)\, ds\right]$$
$$= Z(t)x(0) + \int_0^t Z(t - s)B(s)x(s)\, ds.$$

Then we obtain

$$|x(t)| \leq |Z(t)|\,|x(0)| + \int_0^t |Z(t - s)||B(s)|\,|x(s)|\, ds$$
$$\leq K|x(0)| + \int_0^t K|B(s)|\,|x(s)|\, ds$$

where $|Z(u)| \leq K$ if $u \geq 0$. We apply Gronwall's inequality to obtain

$$|x(t)| \leq K|x(0)| \exp\left[K \int_0^t |B(s)|\, ds\right] \leq K|x(0)|P$$

for some $P > 0$. This completes the proof.

Note that the proof hinged on being able to assign a bound to $Z(t)Z^{-1}(s)$ from the bound on $Z(t)$. For a general $A(t)$, this is impossible. If $A(t + T) = A(t)$ for all t and some $T > 0$, then $Z(t) = P(t)e^{Rt}$ so $Z(t)Z^{-1}(s) = P(t)e^{R(t-s)}P^{-1}(s)$ with P periodic. Thus, in this case, $Z(t)$ bounded yields $Z(t)Z^{-1}(s)$ bounded. Hence, in exactly the same manner we have the following result.

Theorem 1.1.33. *Let $A(t)$ be periodic of period $T > 0$ and let the PMS of (1.1.26) be bounded on $[t_0, \infty)$. If $\int_{t_0}^\infty |B(s)|\, ds < \infty$, then all solutions of (1.1.25) are bounded on $[t_0, \infty)$.*

Theorem 1.1.34. *Let $A(t)$ be constant and let all the characteristic roots of A have negative real parts. Then there exists $\alpha > 0$ such that $|B(t)| \leq \alpha$ implies all solutions of (1.1.25) tend to zero as $t \to \infty$.*

Proof. Again, we prove this for $t_0 = 0$. There exists $K > 0$ and $\beta > 0$ with $|e^{At}| \leq Ke^{-\beta t}$ for $t \geq 0$. Thus, as before, we have any solution $x(t)$ of (1.1.25) written as

$$\begin{aligned}
x(t) &= e^{At}\left[x(0) + \int_0^t e^{-As}B(s)x(s)\, ds\right] \\
&= e^{At}x(0) + \int_0^t e^{A(t-s)}B(s)x(s)\, ds
\end{aligned}$$

so that

$$|x(t)| \leq Ke^{-\beta t}|x(0)| + \int_0^t Ke^{-\beta(t-s)}|B(s)|\,|x(s)|\, ds$$

or

$$|x(t)|e^{\beta t} \leq K|x(0)| + \int_0^t K|B(s)|e^{\beta s}|x(s)|\, ds.$$

Application of Gronwall's inequality yields

$$|x(t)|e^{\beta t} \leq K|x(0)| \exp \int_0^t K|B(s)|\, ds,$$

or

$$|x(t)| \leq K|x(0)| \exp \int_0^t [K|B(s)| - \beta]\, ds.$$

Choose $\alpha > 0$ with $\alpha < \beta/K$. Then $|B(s)| \le \alpha$ implies

$$|x(t)| \le K|x(0)| \exp \int_0^t (\alpha K - \beta)\, ds \to 0$$

as $t \to \infty$. This completes the proof.

Exercise 1.1.16. Let $A = \begin{pmatrix} -2 & 0 \\ 3 & -4 \end{pmatrix}$ and find α and K for the previous theorem. Can you find α exactly?

Exercise 1.1.17. In the previous theorem ask that $A(t + T) = A(t)$ and that all solutions tend to zero. Obtain the same conclusion.

Theorem 1.1.35. *Let all solutions of* (1.1.26) *be bounded for $t \ge t_0$ and suppose $\int_{t_0}^t \operatorname{tr} A(s)\, ds \ge -M > -\infty$ for all $t \ge t_0$. If $\int_{t_0}^t |B(s)|\, ds < \infty$, then all solutions of* (1.1.25) *are bounded.*

Proof. With the same conventions as before,

$$x(t) = Z(t)x(0) + \int_0^t Z(t)Z^{-1}(s)B(s)x(s)\, ds.$$

Thus, as the elements of $Z^{-1}(s)$ are finite sums of finite products of elements of $Z(s)$ divided by $\det Z(s) = \exp \int_0^s \operatorname{tr} A(u)\, du \ge e^{-M}$, we see that $Z^{-1}(s)$ is bounded. Therefore, the proof may be completed exactly as in Theorem 1.1.32.

Exercise 1.1.18. Let $A(t + T) = A(t)$ for all t and some $T > 0$ and suppose that all solutions of (1.1.26) tend to zero as $t \to \infty$. If $B(t) = B_1(t) + B_2(t)$ in which $\int_{t_0}^\infty |B_1(t)|\, dt < \infty$, then show that there exists $\alpha > 0$ such that $|B_2(t)| \le \alpha$ implies that all solutions of (1.1.25) tend to zero as $t \to \infty$.

Improvements are found in Bellman (1953) and Bihari (1957).

Second-order equations have always played a central role in differential equations owing to the fact that they are often tractable, but nontrivial, and also the fact that Newton's second law of motion for systems having one degree of freedom gives rise to second-order differential equations.

We consider the scalar equation

(1.1.27) $y'' + a(t)y' + b(t)y = 0$

with $a(t)$ and $b(t)$ continuous on some interval, say for convenience $[t_0, \infty)$, and show that it can be mapped into two different standard forms which are often convenient. Exercises are then given which are intended to indicate how the preceding theorems apply.

Theorem 1.1.36. Liouville's first transformation *Let $b(t) > 0$. Then the transformation $y(t) = z(s)$ with $s = \int_{t_0}^{t} \sqrt{b(u)}\, du$ maps (1.1.27) into*

$$(1.1.28) \qquad \ddot{z} + \{[b'(t)/2b^{3/2}(t)] + [a(t)/\sqrt{b(t)}]\}\dot{z} + z = 0$$

in which $\dot{} = d/ds$ and $z = z(s)$.

Proof. We have

$$y'(t) = \dot{z}(s)s' = \dot{z}\sqrt{b(t)}$$

and

$$y''(t) = \ddot{z}(s)b(t) + \dot{z}(s)b'(t)/2\sqrt{b(t)}$$

so that (1.1.27) becomes (1.1.28) as asserted.

Caution! If $\int_{t_0}^{\infty} \sqrt{b(t)}\, dt < \infty$, then in (1.1.28) we cannot let $s \to \infty$. A similar remark holds if the basic interval is $[t_0, \alpha)$. Also, as $b(t) > 0$, s is a strictly monotone increasing function of t and so the inverse function exists enabling us to write (1.1.28) as an equation in s throughout.

Theorem 1.1.37. Liouville's second transformation *The transformation $y = z\exp(\frac{1}{2})\int_{t_0}^{t} -a(s)\, ds$ maps (1.1.27) into*

$$(1.1.29) \qquad z'' + \left[-\frac{a'(t)}{2} - \frac{a^2(t)}{4} + b(t)\right]z = 0.$$

Proof. We let $y = zv(t)$ where v is to be determined. Then $y' = z'v + zv'$ and $y'' = z''v + 2z'v' + zv''$, so that (1.1.27) becomes $z''v + 2z'v' + zv'' + a(t)[z'v + zv'] + b(t)zv = 0$. Set the coefficient of z' equal to zero, obtaining $[2v' + a(t)v] = 0$ which has one solution $v(t) = \exp\frac{1}{2}\int_{t_0}^{t} -a(s)\, ds$. This yields (1.1.29).

Caution! If $\int_{t_0}^{\infty} a(t)\, dt = +\infty$, then one need only show that solutions of (1.1.29) are bounded to show that solutions of (1.1.27) tend to zero.

Exercise 1.1.19. In (1.1.27), let $b(t) \equiv 1$ and $a(t) > 0$. Change to a system: $y = x_1$; $(x_1' = x_2, x_2' = -a(t)x_2 - x_1)^T$. Define a Liapunov function $V(x_1, x_2) = x_1^2 + x_2^2$ and deduce that all solutions are bounded. Also, if $a(t)$ is periodic, show that all solutions tend to zero. Use the first result and Theorem 1.1.35 to give conditions on $a(t)$, $a_1(t)$, and $b_1(t)$ to ensure that all solutions of

$$y'' + [a_1(t) + a(t)]y' + [1 + b_1(t)]y = 0$$

are bounded. Use the second result and Exercise 1.1.18 to improve your conclusion.

Exercise 1.1.20. Let $a(t) \geq 0$, $b(t) > 0$, and $b(t)$ nondecreasing. Use the first Liouville transformation, followed by the first part of Exercise 1.1.19, to show that each solution $y(t)$ of (1.1.27) (although possibly not $y'(t)$) is bounded.

1.2   Periodic Solutions of Linear Differential Equations

We now approach the central question of the book. Under what conditions does a differential equation have a periodic solution?

Definition 1.2.1. *A system*

$$(1.2.1) \qquad\qquad\qquad x' = A(t)x$$

with $A(t + T) = A(t)$ for all t and some $T > 0$ is called noncritical relative to T *if* (1.2.1) *has no periodic solution of period T other than the zero solution.*

Theorem 1.2.1. *Let*

$$(1.2.2) \qquad\qquad\qquad x' = A(t)x + p(t)$$

where $p : (-\infty, \infty) \to R^n$, A and p are continuous on $(-\infty, \infty)$, both are periodic of period T, and let $x' = A(t)x$ be noncritical relative to T. Then (1.2.2) *has a unique periodic solution of period T.*

Proof. Again, as A and p are periodic and solutions are unique, we need only show that there exists x_0 with $x(T, 0, x_0) = x_0$. Also, as the difference of two solutions of (1.2.2) is a solution of (1.2.1) and as (1.2.1) has no nontrivial solution of period T, there is at most one periodic solution of (1.2.2).

If $Z(t)$ is the PMS of (1.2.1) at $t = 0$, then a solution of (1.2.2) is given by $x(t) = x(t, 0, x_0) = Z(t)[x_0 + \int_0^t Z^{-1}(s)p(s)\, ds]$. If $x(t)$ is to be periodic, then $x(T) = x_0$ or $x_0 = Z(T)[x_0 + \int_0^T Z^{-1}(s)p(s)\, ds]$ from which we obtain

$$(1.2.3) \qquad\qquad (Z^{-1}(T) - I)x_0 = \int_0^T Z^{-1}(s)p(s)\, ds.$$

If $(Z^{-1}(T) - I)$ has an inverse, then we can solve the last equation for x_0, thereby completing the proof.

To prove that $(Z^{-1}(T) - I)$ is nonsingular we consider the general solution of (1.2.1), say $x(t) = Z(t)x_1$. We claim that if $Z^{-1}(T) - I$ is singular (and hence has zero as a characteristic root) with characteristic vector x_1 belonging to zero, then $Z(t)x_1$ is periodic. That is, $Z(t)x_1$ is periodic of period T if and only if $Z(T)x_1 = x_1$ or $x_1 = Z^{-1}(T)x_1$ or $(Z^{-1}(T) - I)x_1 = 0$. But the last equation is true as $(Z^{-1}(T) - I)x_1 = 0x_1$. This contradicts the assumption that there is no such periodic solution, as $x_1 \neq 0$ is a characteristic vector. Hence, (1.2.3) may be solved for x_0 yielding a periodic solution

$$x(t) = Z(t)\left[(Z^{-1}(T) - I)^{-1}\int_0^T Z^{-1}(s)p(s)\,ds + \int_0^t Z^{-1}(s)p(s)\,ds\right].$$

Under stronger conditions on (1.2.1) there is a very attractive way to consider the question of the existence of periodic solutions. It gives a compact expression for the solution and it extends in a beautiful way to Volterra integrodifferential equations. Thus, although it produces less than the last result, it is well worth studying here. To conjecture the result we proceed as follows.

First, if all characteristic roots of A have negative real parts and if $p(t)$ is any continuous function, then the difference of two solutions of

(*) $$x' = Ax + p(t)$$

satisfies

(**) $$y' = Ay$$

and, hence, that difference tends to zero exponentially, even though the solutions of (*) themselves may be unbounded. Next, if we add that $p(t)$ is bounded on $[0, \infty)$, then solutions of (*) are all bounded and so those solutions are all converging exponentially to some bounded function, say ϕ. Can we find ϕ? When p is periodic or almost periodic, then ϕ is easy to find. We simply use the variation of parameters formula and our knowledge that solutions of (*) converge exponentially to ϕ. In fact, a solution of (*) is expressed as

$$x(t, 0, x_0) = e^{At}x_0 + \int_0^t e^{A(t-s)}p(s)\,ds$$

in which we substitute $t + nT$ for t, to obtain

$$x(t + nT, 0, x_0) = e^{A(t+nT)}x_0 + \int_0^{t+nT} e^{A(t+nT-s)}p(s)\,ds.$$

The first term on the right converges to zero (for t on compact sets) as $n \to \infty$, while the change of variable $s = u + nT$ converts the second term on the right to

$$\int_{-nT}^t e^{A(t-u)}p(u)\,du,$$

taking into account the periodicity of p. We then let $n \to \infty$ and see that

$$\phi(t) = \int_{-\infty}^{t} e^{A(t-u)} p(u) \, du.$$

Theorem 1.2.2. *In* $(1.2.2)$ *let* A *be constant and let all characteristic roots of* A *have negative real parts. Then*

$$(1.2.4) \qquad\qquad x(t) = \int_{\infty}^{t} e^{A(t-s)} p(s) \, ds$$

is a T*-periodic solution of* $(1.2.2)$ *to which all other solutions converge.*

Proof. Because the characteristic roots of A have negative real parts there are positive constants K and α with

$$|e^{A(t-s)} p(s)| \leq |p(s)| K e^{-\alpha(t-s)}$$

for $t \geq s$. Since p is bounded, the integral in $(1.2.4)$ converges. Moreover,

$$x'(t) = p(t) + \int_{-\infty}^{t} A e^{A(t-s)} p(s) \, ds = p(t) + Ax(t)$$

so that $(1.2.4)$ is a solution. If we let $t - s = u$ then

$$x(t) = \int_{0}^{\infty} e^{Au} p(t - u) \, du$$

from which the periodicity of x is an obvious consequence of that of p. To see that all other solutions converge to this periodic solution, note that the difference of solutions of $(1.2.2)$ is a solution of $(1.2.1)$ with A constant so that difference tends to zero exponentially. This completes the proof.

Exercise 1.2.1. Show that if A is constant and if all characteristic roots of A have positive real parts, then

$$x(t) = -\int_{t}^{\infty} e^{A(t-s)} p(s) \, ds$$

is a T-periodic solution of $(1.2.2)$.

For a more general result, proceed as follows. Suppose that A is constant and no characteristic root of A has zero real part. Then there is a matrix Q with

$$Q^{-1} A Q = \begin{pmatrix} B_1 & 0 \\ 0 & B_2 \end{pmatrix}$$

where all characteristic roots of B_1 have positive real parts and all characteristic roots of B_2 have negative real parts. The transformation $x = Qy$ maps (1.2.2) into

$$y' = Q^{-1}AQy + Q^{-1}p(t).$$

A T-periodic solution of this equation is

$$y(t) = \begin{bmatrix} -\int_t^\infty e^{B_1(t-u)} p_1(u)\, du \\ \int_{-\infty}^t e^{B_2(t-u)} p_2(u)\, du \end{bmatrix}$$

and $x = Qy(t)$ is a T-periodic solution of (1.2.2). Here, $Q^{-1}p = (p_1, p_2)^T$.

Exercise 1.2.2. If $A(t+T) = A(t)$ and $p(t+T) = p(t)$, then $h' = A(t)h$ has the PMS $P(t)e^{Jt}$ where P is T-periodic and J is constant. If J has no characteristic root with zero real parts, then display a T-periodic solution for $x' = A(t)x + p(t)$.

1.3 Linear Volterra Equations

Let A be an $n \times n$ constant matrix, let B be an $n \times n$ matrix of functions, and let p be a vector function with B and p continuous on some interval, say $[0, \beta)$, with $\beta \le \infty$. Then the equation

(1.3.1) $$x' = Ax + \int_0^t B(t-s)x(s)\, ds + p(t)$$

is a Volterra integrodifferential equation of convolution type.

In the standard literature (1.3.1) and the ordinary differential equation

(1.2.2) $$x' = Ax + p(t)$$

are treated as entirely separate problems. But the thesis of Burton (1983b) is that their solution spaces are essentially indistinguishable and that the knowledge of one system can be transferred to the other. In fact, there is every reason to consider them together.

It is our view here that the periodic theory for the two systems is indistinguishable. In Theorem 1.2.2 we saw that (1.2.2) had the periodic solution

$$x(t) = \int_{-\infty}^t e^{A(t-s)} p(s)\, ds.$$

And we will see here that, under certain conditions, solutions of (1.3.1) converge to the periodic function

$$x(t) = \int_{-\infty}^t Z(t-s)p(s)\, ds$$

for a certain matrix Z. Moreover, the periodic solution

$$x(t) = \int_{-\infty}^{t} P(t)e^{J(t-s)}P^{-1}(s)p(s)\,ds$$

found in Exercise 1.2.2 for the system

$$x' = A(t)x + p(t)$$

will extend to the periodic solution

$$x(t) = \int_{-\infty}^{t} R(t,s)p(s)\,ds$$

for the nonconvolution system

$$x' = A(t)x + \int_{-\infty}^{t} C(t,s)x(s)\,ds + p(t)$$

for a certain matrix $R(t,s)$.

This section is devoted to showing the relationships between linear ordinary differential equations and Volterra equations. The work concerns a more general form of (1.3.1) which includes the general linear integral equation

$$(1.3.2) \qquad\qquad x(t) = F(t) + \int_{0}^{t} D(t,s)x(s)\,ds$$

in which $F : [0, \beta) \to R^n$ is continuous and D is an $n \times n$ matrix of functions continuous for $0 \leq s \leq t < \beta$.

Equation (1.3.1) is said to be *hereditary* or an equation with memory. Seldom does the future behavior of a process depend only on its initial position, say $x(t_0) = x_0$. Instead, it usually depends on the past history. This was noted by Picard (1907, p. 15) (cf. Davis, 1962, p. 112) concerning elasticity in the early part of this century and applies to a great variety of problems including some in mathematical biology and economics.

Generally, one is given an *initial interval* $[0, t_0]$ and a continuous *initial function* $\phi : [0, t_0] \to R^n$. The problem is to find a solution $x(t)$ of (1.3.1) for $t \geq t_0$ with $x(t) = \phi(t)$ for $0 \leq t \leq t_0$. However, we will later show that when ϕ is given it is possible to translate (1.3.1) into

$$(1.3.3) \qquad y'(t) = Ay(t) + \int_{0}^{t} B(t-s)y(s)\,ds + g(t),$$

where ϕ is incorporated into g, and the problem is to find a solution of (1.3.3) for $t \geq 0$ subject to $y(0) = \phi(t_0)$.

This means that it is *completely general* to ask only for solutions of (1.3.1) on $[0, \infty)$ subject to the initial condition $x(0) = x_0$, keeping in mind that $p(t)$ contains the initial function ϕ.

We now give the details of obtaining (1.3.3) from (1.3.1). For the given ϕ on $[0, t_0]$, Eq. (1.3.1) is expressed as

$$(1.3.4) \quad x'(t) = Ax(t) + \int_0^{t_0} B(t-s)\phi(s)\,ds + \int_{t_0}^t B(t-s)x(s)\,ds + p(t)$$

subject to $x(t_0) = \phi(t_0)$. We seek a solution $x(t)$ only for $t \geq t_0$. Let $y(t) = x(t + t_0)$ in (1.3.4) and obtain

$$y'(t) = x'(t + t_0)$$
$$= Ay(t) + \int_0^{t_0} B(t + t_0 - s)\phi(s)\,ds$$
$$+ \int_{t_0}^{t+t_0} B(t + t_0 - s)x(s)\,ds + p(t + t_0).$$

In the last integral let $s = u + t_0$ to obtain

$$y'(t) = Ay(t) + \int_0^t B(t - u)y(u)\,du$$
$$+ \left\{ p(t + t_0) + \int_0^{t_0} B(t + t_0 - s)\phi(s)\,ds \right\}$$

which is (1.3.3) when we define

$$(1.3.5) \qquad g(t) = p(t + t_0) + \int_0^{t_0} B(t + t_0 - s)\phi(s)\,ds.$$

A satisfactory point of view, then, is to consider (1.3.1) as an initial value problem in which we want a solution $x(t)$ on $[0, \beta)$ subject to $x(0) = x_0$ where

$$(1.3.6) \qquad\qquad x_0 \text{ is an arbitrary constant vector}$$

and

$$(1.3.7) \qquad p \text{ is an arbitrary continuous function on } [0, \beta).$$

The nature of $B(t)$ may greatly restrict the functions g which may be generated by (1.3.5) for arbitrary continuous functions ϕ when p is fixed.

The surprising fact is that the dimension of the solution space of (1.3.3) under the generous conditions of (1.3.6) and (1.3.7) is precisely the same as that of the system of ordinary differential equations

$$(1.3.8) \qquad\qquad w' = Pw + q(t)$$

in which P is an $n \times n$ constant matrix and $q : [0, \beta) \to R^n$ is continuous.

To completely solve (1.3.8) we require an $n \times n$ matrix of functions $W(t)$ whose columns are solutions of

$$(1.3.9) \qquad\qquad w' = Pw$$

satisfying $W(0) = I$. Any solution of (1.3.8) on $[0, \infty)$ is then expressed by the variation of parameters formula

$$(1.3.10) \qquad w(t) = W(t)w(0) + \int_0^t W(t - s)q(s)\,ds.$$

Thus, the solution of (1.3.8) is determined by n functions (the columns of W) plus a quadrature. Although the space from which q is drawn is infinite dimensional, it would seem to strain a point to think of (1.3.10) as infinite dimensional.

Equation (1.3.1), subject to (1.3.6) and (1.3.7), is solved in exactly the same way. We find n vector functions $z_1(t), \ldots, z_n(t)$ which are columns of an $n \times n$ matrix $Z(t)$ with $Z(0) = I$ and each column is a solution of (1.3.1) when $p(t) \equiv 0$; that is,

$$(1.3.11) \qquad Z'(t) = AZ(t) + \int_0^t B(t - s)Z(s)\,ds.$$

The solution of (1.3.1) on $[0, \beta)$ is then given by

$$(1.3.12) \qquad x(t) = Z(t)x(0) + \int_0^t Z(t - s)p(s)\,ds.$$

It turns out that proofs of several of the properties (but not the variation of parameters formula) of a nonconvolution form of (1.3.1) may be obtained easily. Thus, we extend (1.3.1) to

$$(1.3.13) \qquad x'(t) = E(t)x + \int_0^t C(t, s)x(s)\,ds + p(t)$$

where E is an $n \times n$ matrix of continuous functions on $[0, \beta)$, C is an $n \times n$ matrix of functions continuous for $0 \leq s \leq t < \beta$, and p is as in (1.3.1).

We convert (1.3.13) to an integral equation by integrating from 0 to t:

$$x(t) = x(0) + \int_0^t E(s)x(s)\,ds$$
$$+ \int_0^t \int_0^u C(u, s)x(s)\,ds\,du + \int_0^t p(s)\,ds.$$

Next, interchange the order of integration (cf. Fulks, 1969, pp. 311–313) in the iterated integral.

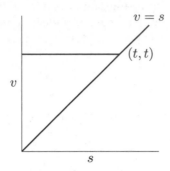

Fig. 1.16

The region of integration is the triangle in Fig. 1.16. The new limits are given by

$$\int_{s=0}^{s=t} \quad \int_{u=s}^{u=t}$$

so that the equation becomes

$$x(t) = x(0) + \int_0^t p(s)\, ds$$
$$+ \int_0^t \left[E(s) + \int_s^t C(u,s)\, du \right] x(s)\, ds$$

which we express as our old equation

$$(1.3.2) \qquad\qquad x(t) = F(t) + \int_0^t D(t,s) x(s)\, ds.$$

Theorem 1.3.1. *If $F : [0, \beta) \to R^n$ is continuous and D is an $n \times n$ matrix of functions continuous for $0 \leq s \leq t < \beta$, then there is a function $x(t)$ on $[0, \beta)$ satisfying (1.3.2).*

Proof. Let $0 < L < \beta$ be given. We will show that the sequence of functions defined inductively on $[0, L]$ by

$$x_1(t) = F(t)$$
$$(1.3.14)$$
$$x_{n+1}(t) = F(t) + \int_0^t D(t,s) x_n(s)\, ds, \qquad n \geq 1,$$

converges uniformly to a limit function, say $x(t)$. Thus, if we take the limit in (1.3.14) as $n \to \infty$, it may be passed under the integral yielding (1.3.2).

As in the case of ordinary differential equations we form a series

$$(1.3.15) \qquad x_1(t) + \sum_{n=1}^{\infty} (x_{n+1}(t) - x_n(t)), \qquad 0 \leq t \leq L,$$

and notice that the sequence of partial sums is just $\{x_n(t)\}$. We have

$$(1.3.16) \qquad |F(t)| \leq K \qquad \text{and} \qquad |D(t,s)| \leq M$$

for $0 \leq s \leq t \leq L$ for appropriate K and M by continuity.

We will show by induction that

$$(1.3.17) \qquad |x_{n+1}(t) - x_n(t)| \leq KM^n t^n / n!.$$

For $n = 1$, we consider (1.3.14) and have

$$|x_2(t) - F(t)| = \left| \int_0^t D(t,s)F(s)\,ds \right| \leq KMt$$

and so (1.3.17) is true for $n = 1$. Suppose then that (1.3.17) is true for $n = k$:

$$|x_{k+1}(t) - x_k(t)| \leq KM^k t^k / k!.$$

We wish to show that (1.3.17) holds for $n = k + 1$. By (1.3.14) we have

$$
\begin{aligned}
|x_{k+2}(t) - x_{k+1}(t)| &= \left| \int_0^t D(t,s)x_{k+1}(s)\,ds - \int_0^t D(t,s)x_k(s)\,ds \right| \\
&\leq \int_0^t |D(t,s)|\,|x_{k+1}(s) - x_k(s)|\,ds \\
&\leq M \int_0^t KM^k s^k / k!\,ds \\
&= KM^{k+1} s^{k+1} / (k+1)! \Big|_{s=0}^{s=t} \\
&= KM^{k+1} t^{k+1} / (k+1)!,
\end{aligned}
$$

as required.

Now $\sum_{k=0}^{\infty} K(Mt)^k / k! = Ke^{Mt}$ and that series converges uniformly on every interval $[0, t]$. Thus, by a comparison test (essentially the M-test), our series (1.3.15) converges uniformly on $[0, L]$. As L is arbitrary in $[0, \beta)$ this completes the proof.

For use in uniqueness we remind the reader of Gronwall's inequality which was proved in Section 1.1.

Gronwall's Inequality. *Let $0 < \alpha \leq \infty$ and $h : [0, \alpha) \rightarrow [0, \infty)$ be continuous and let c be a nonnegative constant. If $r : [0, \alpha) \rightarrow [0, \infty)$ is a continuous function satisfying*

$$(1.3.18) \qquad r(t) \leq c + \int_0^t h(s)r(s)\, ds, \quad 0 \leq t < \alpha,$$

then

$$(1.3.19) \qquad r(t) \leq c \exp \int_0^t h(s)\, ds.$$

Theorem 1.3.2. *Under the conditions of Theorem 1.3.1, there is only one solution of (1.3.2).*

Proof. By way of contradiction, assume that $x(t)$ and $y(t)$ both satisfy (1.3.2) on some interval $[0, L]$. Then

$$x(t) - y(t) = \int_0^t D(t, s)(x(s) - y(s))\, ds$$

so that if $|D(t, s)| \leq M$ for $0 \leq s \leq t \leq L$, then

$$|x(t) - y(t)| \leq \int_0^t M |x(s) - y(s)|\, ds.$$

This has the form (1.3.18) with $h(t) = M$ and $c = 0$. Thus, $x(t) = y(t)$ and the proof is complete.

Note 1. Now review the construction of (1.3.2) from (1.3.13) to see that Theorems 1.3.1 and 1.3.2 claim that for each x_0, there is one and only one solution $x(t)$ of (1.3.13) on $[0, \beta)$ satisfying $x(0) = x_0$.

Note 2. We have at the same time, obtained existence and uniqueness theory for $x' = E(t)x$.

Theorem 1.3.3. *Consider (1.3.13) with $p(t) \equiv 0$:*

$$(1.3.20) \qquad x'(t) = E(t)x + \int_0^t C(t, s)x(s)\, ds.$$

   (i) *There are n linearly independent solutions $z_1(t), \ldots, z_n(t)$ of (1.3.20) on $[0, \beta)$. If $Z(t)$ is the matrix with columns $z_1, \ldots, z_n$, respectively, then $Z(0) = I$.*

  (ii) *Every solution of (1.3.20) on $[0, \beta)$ may be expressed as a linear combination of those n solutions. In fact, if $x(t)$ is the solution, then*

$$x(t) = Z(t)x(0).$$

(iii) *The difference of two solutions of* (1.3.13) *is a solution of* (1.3.20).

(iv) *If x_p is any solution of* (1.3.13) *on* $[0, \beta)$, *then every solution of* (1.3.13) *on* $[0, \beta)$ *may be expressed as*

$$(1.3.21) \qquad x(t) = x_p(t) + \sum_{k=1}^{n} c_k z_k(t)$$

for appropriate constants c_k. In fact,

$$(1.3.22) \qquad x(t) = Z(t)[x(0) - x_p(0)] + x_p(t).$$

Proof. Let $e_1, \ldots, e_n$ be the standard unit orthogonal constant vectors. By Theorem 1.3.1 and Note 1 there is a solution $z_i(t)$ on $[0, \beta)$ for each i with $z_i(0) = e_i$. Thus, the matrix $Z(t)$ will satisfy $Z(0) = I$. Now the $z_i(t)$ are linearly independent on $[0, \beta)$; for if there is a nontrivial linear relation

$$c_1 z_1(t) + \cdots + c_n z_n(t) \equiv 0 \quad \text{on} \quad [0, \beta),$$

then at $t = 0$ we have

$$c_1 e_1 + \cdots + c_n e_n = 0,$$

a contradiction. This proves (i).

To prove (ii), notice that if $x(t)$ is a solution of (1.3.20) on $[0, \beta)$, it is uniquely determined by $x(0)$ according to Theorem 1.3.2 and Note 1. But such a solution is $x(t) = Z(t)x(0)$, for this relation is merely a linear combination of solutions of (1.3.20), and that is certainly a solution of (1.3.20).

A brief calculation proves (iii).

To prove (iv), notice that if $x(t)$ and $x_p(t)$ are solutions of (1.3.13), then $x(t) - x_p(t)$ is a solution of (1.3.20) by (iii) and, hence,

$$x(t) - x_p(t) = Z(t)[x(0) - x_p(0)]$$

by (ii). This completes the proof.

Definition 1.3.1. *The matrix $Z(t)$ defined in the proof of Theorem 1.3.3(i) is called the* principal matrix solution (PMS) *of* (1.3.20).

Although (1.3.22) is the general solution of (1.3.13), it is of much interest to compute $x_p(t)$ from $Z(t)$. And this can be done in the convolution case with $E(t)$ constant.

1.3.1   The Linear Convolution Equations

There are several ways of arriving at the variation of parameters formula. We first give a formal presentation using Laplace transforms. That presentation carries with it many nontrivial problems to be resolved, but its merit is that it is so straightforward and natural that the broad outlines may be followed by the very unsophisticated reader.

Most elementary treatments of ordinary differential equations cover the basic aspects of Laplace transforms. A good (and elementary) discussion is given by Churchill (1958). Some very fundamental questions are treated by Bellman and Cooke (1963) concerning Volterra equations. But for our purposes the essential properties may be summarized as follows.

Definition 1.3.2. *A function* $r : [0, \infty) \to R$ *is said to be of* exponential order *if there are positive constants* K *and* α *with* $|r(t)| \leq Ke^{\alpha t}$ *on* $[0, \infty)$.

In (i)–(vi) below it is assumed that all functions of t under discussion are continuous and of exponential order.

(i) If $h : [0, \infty) \to R$, then the Laplace transform of h is

$$L(h) = H(s) = \int_0^\infty e^{-st} h(t) \, dt.$$

(ii) If $D = (d_{ij}(t))$ is a matrix (or vector), then $L(D(t)) = (L(d_{ij}(t)))$.

(iii) If c is a constant and $h_1, h_2 : [0, \infty) \to R$, then

$$L(ch_1 + h_2) = cL(h_1) + L(h_2).$$

(iv) If $D(t)$ is an $n \times n$ matrix of functions on $[0, \infty)$ and if $h : [0, \infty) \to R^n$, then $L(\int_0^t D(t - s)h(s) \, ds) = L(D)L(h)$.

(v) If $h' : [0, \infty) \to R$, then

$$L(h'(t)) = sL(h(t)) - h(0).$$

(vi) If $h_1, h_2 : [0, \infty) \to R$ are of exponential order and continuous, then $L(h_1) \equiv L(h_2)$ implies $h_1(t) \equiv h_2(t)$.

For this result we suppose $\beta = \infty$. As we see in the proof, this is not a crucial assumption.

Theorem 1.3.4. *Consider* (1.3.1) *and suppose that* B *and* p *are continuous on* $[0, \infty)$. *If* $Z(t)$ *is the principal matrix solution of*

$$(1.3.23) \qquad x'(t) = Ax(t) + \int_0^t B(t - s)x(s) \, ds$$

and if $x(t)$ is a solution of (1.3.1) *on* $[0, \infty)$, *then*

$$(1.3.24) \qquad x(t) = Z(t)x(0) + \int_0^t Z(t-s)p(s)\,ds.$$

Proof. We first suppose that B and p are in $L^1[0, \infty)$ and of exponential order. If we convert (1.3.1) to an integral equation as we did in (1.3.13) we have

$$x(t) = x(0) + \int_0^t p(s)\,ds + \int_0^t \left[A + \int_s^t B(u-s)\,du\right]x(s)\,ds$$

and, as B and p are L^1, it follows that $|x(t)| \leq |x(0)| + K + \int_0^t K|x(s)|\,ds$. Thus,

$$|x(t)| \leq [|x(0)| + K]e^{Kt}$$

by Gronwall's inequality. That is, both $x(t)$ and $Z(t)$ are of exponential order so we may take their Laplace transforms. The equation is

$$Z'(t) = AZ(t) + \int_0^t B(t-s)Z(s)\,ds$$

and, upon transforming both sides, we obtain

$$sL(Z) - Z(0) = AL(Z) + L(B)L(Z)$$

using (iii)–(v). Thus,

$$[sI - A - L(B)]L(Z) = Z(0) = I$$

and, as the right side is nonsingular, so is $[sI - A - L(B)]$ for appropriate s. [Actually, $L(Z)$ is an analytic function of s in the half-plane Re $s \geq \alpha$ where $|Z(t) \leq Ke^{\alpha t}$. See Churchill (1958, p. 171).] We then have

$$L(Z) = [sI - A - L(B)]^{-1}.$$

Now, transform both sides of (1.3.1):

$$sL(x) - x(0) = AL(x) + L(B)L(x) + L(p)$$

or

$$[sI - A - L(B)]L(x) = x(0) + L(p)$$

so that

$$\begin{aligned}
L(x) &= [sI - A - L(B)]^{-1}[x(0) + L(p)] \\
&= L(Z)x(0) + L(Z)L(p) \\
&= L(Zx(0)) + L\left(\int_0^t Z(t-s)p(s)\,ds\right) \\
&= L\left(Z(t)x(0) + \int_0^t Z(t-s)p(s)\,ds\right).
\end{aligned}$$

Because x, Z, and p are of exponential order and continuous, by (vi) we have (1.3.24). Thus, the proof is complete for B and p being in $L^1[0, \infty)$ and of exponential order.

In the general case (i.e., B and p not in L^1 or not of exponential order), for each $T > 0$ define continuous $L^1[0, \infty)$ functions, p_T and B_T by

$$p_T(t) = \begin{cases} p(t) & \text{if } 0 \le t \le T \\ p(T)[1/\{(t-T)^2 + 1\}] & \text{if } t > T \end{cases}$$

and

$$B_T(t) = \begin{cases} B(t) & \text{if } 0 \le t \le T \\ B(T)[1/\{(t-T)^2 + 1\}] & \text{if } t > T. \end{cases}$$

Consider (1.3.1) and

$$(1.3.1)_T \qquad x'(t) = Ax(t) + p_T(t) + \int_0^t B_T(t-s)x(s)\,ds$$

with $x(0) = x_0$ for both. As the equations are identical on $[0, T]$, so are their solutions; and this is true for each $T > 0$. As (1.3.24) holds for $(1.3.1)_T$ for each $T > 0$, it holds for (1.3.1) on each interval $[0, T]$. This completes the proof.

Note. From this point forward we take $\beta = \infty$ and consider solutions on $[0, \infty)$.

Theorems 1.3.1–1.3.4 show a complete correspondence between the solution spaces of

$$(1.3.1) \qquad x'(t) = Ax(t) + f(t) + \int_0^t B(t-s)x(s)\,ds$$

and

$$(1.3.8) \qquad w'(t) = Pw(t) + q(t).$$

The variation of parameters formula for (1.3.8) is

$$(1.3.10) \qquad w(t) = W(t)w(0) + \int_0^t W(t-s)q(s)\,ds$$

and, when all the characteristic roots of P have negative real parts, then a fundamental property of W for use in (1.3.10) is that $\int_0^\infty |W(t)|\,dt < \infty$ and $W(t) \to 0$ as $t \to \infty$.

In this same way, a fundamental property of Z for use in (1.3.24) is that $\int_0^\infty |Z(t)|\,dt < \infty$. Unfortunately, a necessary and sufficient condition for that to hold is not at all simple. Grossman and Miller (1973) showed that for A constant and B in $L^1[0,\infty)$, then Z is in $L^1[0,\infty)$ if and only if

$$\det[sI - A - L(B)]$$

does not vanish for Re $s \geq 0$, a transcendental relation.

We now give a very simple result which shows $Z \in L^1[0,\infty)$ and, at the same time, introduces the extension of Liapunov's direct method to Volterra equations.

Theorem 1.3.5. *Let*

$$(1.3.25) \qquad x'(t) = Ax(t) + \int_0^t B(t-s)x(s)\,ds$$

be a scalar equation with $A < 0$, $B(t) > 0$, B continuous, $\int_0^\infty B(t)\,dt < \infty$, and $A + \int_0^\infty B(s)\,ds \neq 0$. All solutions of (1.3.25) are in $L^1[0,\infty)$ if and only if $A + \int_0^\infty B(t)\,dt < 0$.

Proof. Notice first that if $x(0) > 0$, then $x(t) > 0$ for all $t \geq 0$. For if that is false, then there exists $t_1 > 0$ with $x(t) > 0$ on $[0, t_1)$ and $x(t_1) = 0$. Thus, $x'(t_1) \leq 0$. But from (1.3.25) we have

$$x'(t_1) = Ax(t_1) + \int_0^{t_1} B(t_1 - s)x(s)\,ds$$

$$= \int_0^{t_1} B(t_1 - s)x(s)\,ds > 0,$$

a contradiction. As $-x(t)$ is also a solution, if $x(0) < 0$, then $x(t) < 0$ for all $t \geq 0$.

From this it follows that $|x(t)|$ is continuously differentiable whenever $x(t)$ is a solution of (1.3.25) on $[0,\infty)$.

Consider the scalar functional (called a Liapunov functional)

$$V(t, x(\cdot)) = |x(t)| + \int_0^t \int_t^\infty B(u - s)\,du\,|x(s)|\,ds$$

where $x(t)$ is any solution of (1.3.25) on $[0,\infty)$. Although we do not know $x(t)$, we do know $x'(t)$ (from (1.3.25)) and so we may compute $V'(t, x(\cdot))$.

Suppose first that $A + \int_0^\infty B(t)\,dt = -\alpha < 0$. Then

$$V'(t, x(\cdot)) \leq A|x(t)| + \int_0^t B(t-s)|x(s)|\,ds$$

$$+ \int_t^\infty B(u - t)\,du\,|x(t)| - \int_0^t B(t-s)|x(s)|\,ds.$$

In the second integral let $u - t = v$ and obtain

$$V'(t, x(\cdot)) \leq \left[A + \int_0^\infty B(v)\, dv \right] |x(t)| = -\alpha |x(t)|.$$

If we integrate both sides from 0 to t we have

$$0 \leq V(t, x(\cdot)) \leq V(0, x(\cdot)) - \alpha \int_0^t |x(s)|\, ds$$

implying $\int_0^\infty |x(s)|\, ds < \infty$.

Next, suppose that $A + \int_0^\infty B(t)\, dt = \alpha > 0$. We will display a solution not in $L^1[0, \infty)$. Given any $t_0 > 0$ we can find x_0 so large that the solution $x(t, x_0) \geq 1$ on $[0, t_0]$. One proceeds as follows. As $B(t) > 0$, if $x_0 > 1$, then so long as $x(t, x_0) > 1$, we have that

$$x'(t) \geq Ax(t)$$

so that

$$x'(t)/x(t) \geq A$$

implying

$$\ln[x(t)/x_0] \geq At$$

or

$$x(t)/x_0 \geq e^{At}.$$

Pick t_0 with $A + \int_0^{t_0} B(s)\, ds \geq \alpha/2$ and $x_0 > e^{-At_0}$. Then $x(t_0) > 1$.

For this choice of x_0, if there is a first $t_1 > t_0$ with $x(t_1) = 1$, then $x'(t_1) \leq 0$ and

$$x'(t_1) = Ax(t_1) + \int_0^{t_1} B(t_1 - s)x(s)\, ds$$

$$\geq A + \int_0^{t_1} B(t_1 - s)\, ds$$

$$= A + \int_0^{t_1} B(s)\, ds$$

$$> A + \int_0^{t_0} B(s)\, ds \geq \alpha/2 > 0,$$

a contradiction. This completes the proof.

The extension of the theory of Liapunov functions to Liapunov functionals may be found quite fully developed in Burton (1983a). We will not introduce it so extensively here, but we will see several aspects of it in Section 1.6. In that section we extract certain constant matrices from

$$x' = Ax + \int_0^t B(t-s)x(s)\,ds$$

whose algebraic properties govern the behavior of solutions, much the same as the characteristic roots of A govern the behavior of solutions of $x' = Ax$. Theorem 1.3.5 is, of course, a case in point: under certain conditions the algebraic properties of

$$(1.3.26) \qquad\qquad A + \int_0^\infty B(s)\,ds$$

govern the behavior of solutions of the integrodifferential equation.

We now consider three questions. When can an ordinary differential equation be expressed as a Volterra equation? When can a Volterra equation be expressed as an ordinary differential equation? Are there substantial differences in properties of the PMS for ordinary differential equations and Volterra equations?

We think of a scalar equation

$$(1.3.27) \qquad\qquad x' = p(t)x + q(t)$$

as being utterly trivial; its general solution is given by quadratures. And it is not surprising that a second-order scalar differential equation

$$(1.3.28) \qquad\qquad x'' + p(t)x' + q(t)x = 0$$

can be expressed as a first-order scalar integrodifferential equation of the form of (1.3.13). Indeed, if $p(t)$ is constant, then an integration of (1.3.28) immediately yields

$$(1.3.29) \qquad x'(t) = x'(0) - p(t)[x(t) - x(0)] - \int_0^t q(s)x(s)\,ds.$$

However, it should come as a great surprise that a linear nth-order scalar ordinary differential equation can be expressed as a scalar equation of the form

$$(1.3.30) \qquad\qquad x(t) = F(t) + \int_0^t C(t,s)x(s)\,ds.$$

Because of the detail in the general case, it is worth seeing in two steps.

A Liouville transformation will eliminate the middle term in (1.3.28). Thus, we consider the scalar initial value problem

$$(1.3.31) \qquad x'' = p(t)x + q(t), \quad x(0) = a, \quad x'(0) = b$$

with $p(t)$ and $q(t)$ continuous scalar functions on $[0, T]$. Integrate (1.3.31) twice from 0 to $t < T$, obtaining successively

$$x'(t) = b + \int_0^t q(s)\,ds + \int_0^t p(s)x(s)\,ds$$

and

$$x(t) = a + bt + \int_0^t \int_0^u q(s)\,ds\,du$$
$$+ \int_0^t \int_0^u p(s)x(s)\,ds\,du.$$

If we interchange the order of integration in the last integral we have

$$x(t) = a + bt + \int_0^t \int_0^u q(s)\,ds\,du + \int_0^t \int_s^t p(s)x(s)\,du\,ds$$

or

$$x(t) = a + bt + \int_0^t \int_0^u q(s)\,ds\,du + \int_0^t (t-s)p(s)x(s)\,ds.$$

Let us now consider the nth-order scalar equation

$$(1.3.32) \qquad x^{(n)} + a_1(t)x^{(n-1)} + \cdots + a_n(t)x = f(t)$$

with the $a_i(t)$ and $f(t)$ continuous on $[0, T]$. Define $z(t) = x^{(n)}(t)$, integrate from 0 to $t < T$, and obtain

$$x^{(n-1)}(t) = x^{(n-1)}(0) + \int_0^t z(s)\,ds.$$

Successive integrations from 0 to t followed by interchanges in the order of integration will yield

$$x^{(n-2)}(t) = x^{(n-2)}(0) + tx^{(n-1)}(0) + \int_0^t (t-s)z(s)\,ds,$$
$$x^{(n-3)}(t) = x^{(n-3)}(0) + tx^{(n-2)}(0) + [t^2/2!]x^{(n-1)}(0)$$
$$+ \int_0^t [(t-s)^2 z(s)/2!]\,ds,$$

$$\vdots$$

$$x(t) = x(0) + tx'(0) + \cdots + [t^{n-1}/(n-1)!]x^{(n-1)}(0)$$
$$+ \int_0^t [(t-s)^{n-1}/(n-1)!]z(s)\,ds.$$

If we replace these values of x and its derivative in (1.3.32) we have a scalar integral equation for z of the form

$$(1.3.33) \qquad z(t) = G(t) + \int_0^t C(t,s)z(s)\,ds.$$

Of course, $G(t)$ contains the n constants $x(0), \ldots, x^{(n-1)}(0)$.

Another significant difference between ordinary differential equations and (1.3.1) is the fact that if

$$(1.3.34) \qquad x' = P(t)x$$

with $P(t)$ an $n \times n$ matrix of continuous functions on $[0, \infty)$, and if $Z(t)$ is the principal matrix solution of (1.3.34) then det $Z(t)$ is never zero. That is simply a form of the statement that the Wronskian is either identically zero or never zero.

However, if $Z(t)$ is the principal matrix solution of

$$(1.3.35) \qquad x'(t) = Ax(t) + \int_0^t B(t-s)x(s)\,ds$$

with A constant and B continuous, then det $Z(t)$ may vanish for many values of t.

Theorem 1.3.6. *Suppose that (1.3.35) is a scalar equation with $A \leq 0$ and $B(t) \leq 0$ on $[0, \infty)$. If there exists $t_1 > 0$ such that*

$$\int_{t_1}^t \int_0^{t_1} B(u-s)\,ds\,du \to -\infty \quad as \quad t \to \infty,$$

then there exists $t_2 > 0$ such that if $x(0) = 1$, then $x(t_2) = 0$.

Proof. If the theorem is false, then $x(t)$ has a positive minimum, say x_1, on $[0, t_1]$. Then for $t \geq t_1$ we have

$$x'(t) \leq \int_0^{t_1} B(t-s)x(s)\,ds + \int_{t_1}^t B(t-s)x(s)\,ds$$

$$\leq \int_0^{t_1} B(t-s)x_1\,ds$$

implying, upon integration, that

$$x(t) \leq x_1 + \int_{t_1}^t \int_0^{t_1} B(u-s)x_1\,ds\,du \to -\infty$$

as $t \to \infty$, a contradiction. This completes the proof.

One of the consequences of this result is that a Jacobi's identity will probably not be true for a Volterra equation. Recall that Jacobi's identity says that if $Z(t)$ is the principal matrix solution of (1.3.9), then

$$\det Z(t) = \exp \int_0^t \operatorname{tr} P(s)\, ds,$$

from which the nonvanishing of $\det Z(t)$ follows.

We turn now to the question of just when can an equation of the type of (1.3.35) be written as an ordinary differential equation. The question is answered in part by the following simple principal. Using the principle and the Routh-Hurwitz criterion it is then possible to obtain a nice set of integrodifferential equations that can be completely solved for stability properties. And these can be a very convenient guide for our intuition.

Let (1.3.1) be a scalar equation

$$(1.3.36) \qquad x' = Ax + \int_0^t B(t-s)x(s)\, ds + p(t)$$

with A constant, while B and p have n continuous derivatives.

Principle. *If $B(t)$ is a solution of a linear nth-order homogeneous ordinary differential equation with constant coefficients, then (1.3.36) may be reduced to a linear $(n+1)$st order ordinary differential equation with constant coefficients.*

We illustrate the principle with an example. Suppose that

$$(1.3.37) \qquad B''(t) + B(t) = 0$$

and differentiate (1.3.36) twice obtaining:

$$x'' = Ax' + B(0)x + \int_0^t B'(t-s)x(s)\, ds + p'(t)$$

and

$$(1.3.38) \qquad \begin{aligned} x''' =& Ax'' + B(0)x' + B'(0)x \\ & + \int_0^t B''(t-s)x(s)\, ds + p''(t). \end{aligned}$$

If we add (1.3.36) and (1.3.38) we obtain

$$x''' + x' = A(x + x'') + B(0)x' + B'(0)x + p(t) + p''(t).$$

The homogeneous equation is

$$(1.3.39) \qquad x''' - Ax'' + (1 - B(0))x' - (A + B'(0))x = 0$$

with characteristic equation

$$\alpha^3 - A\alpha^2 + (1 - B(0))\alpha - (A + B'(0)) = 0$$

so that solutions of (1.3.39) will tend to zero exponentially by the Routh-Hurwitz criterion in case

$$A < 0, \quad B(0) < 1, \quad A + B'(0) < 0$$

and

$$A(B(0) - 1) > -(A + B'(0)).$$

This may be summarized as

(1.3.40)
$$A < 0, \quad B(0) < 1, \quad A + B'(0) < 0$$
$$AB(0) + B'(0) > 0.$$

As a particular example, one may note that all solutions of

(1.3.41)
$$x' = -2x + \int_0^t \sin(t - s)x(s)\, ds$$

tend to zero exponentially.

We caution the reader that not every solution of (1.3.39) on $[0, \infty)$ is a solution of

$$x' = Ax + \int_0^t B(t - s)x(s)\, ds$$

on $[0, \infty)$. The former may have three arbitrary constants, but the latter has at most one.

The following lemma is used frequently.

Lemma 1.3.1. *Consider Eq. (1.3.23) with A constant and both B and Z being $L^1[0, \infty)$. Then $Z(t) \to 0$ as $t \to \infty$.*

Proof. We have

$$Z' = AZ + \int_0^t B(t - s)Z(s)\, ds$$

and we notice that this integral is an L^1-function, say $g(t)$, because B and Z are. If we integrate from 0 to t we obtain

$$Z(t) = Z(0) + A \int_0^t Z(s)\, ds + \int_0^t g(s)\, ds.$$

This says that $Z(t)$ has a limit as $t \to \infty$. But that limit must be zero, and the proof is complete.

When we reduce a Volterra equation to a linear homogeneous ordinary differential equation with constant coefficients whose solutions tend to zero, then the decay is always exponential. It is natural to inquire if decay of solutions of

$$x' = Ax + \int_0^t B(t-s)x(s)\,ds$$

is always exponential, whenever the solutions tend to zero. Recall that in our study of periodic solutions of $x' = Ax + p(t)$ it was crucial that $\int_0^\infty |e^{At}|\,dt < \infty$. In our study of periodic solutions of Volterra equations it will be crucial to ask that $B \in L^1[0,\infty)$ and that solutions satisfy $\int_0^\infty |x(t)|\,dt < \infty$. Our initial inquiry now becomes: if $B \in L^1[0,\infty)$, if $\int_0^\infty |x(t)|\,dt < \infty$, and if $x(t) \to 0$ as $t \to \infty$, then does $x(t)$ decay exponentially? It does not. In fact, the decay of x is linked closely to the decay of B. Moreover, a good measure of the decay rate of x is given by $\int_t^\infty |x(u)|\,du$ and this is extensively discussed by Burton *et al.* (1985b). The following is a typical result.

We consider the scalar equation

(1.3.42) $$z' = Az + \int_0^t C(t-s)z(s)\,ds$$

with A constant and C continuous.

Theorem 1.3.7. *Let* (1.3.42) *be scalar,* $C(t) \geq 0$, *and* $C \in L^1[0,\infty)$. *Suppose that* $z(t)$ *is a nontrivial* L^1-*solution of* (1.3.42). *Then there exists* $\beta > 0$ *with* $\int_t^\infty |z(u)|\,du \geq \beta \int_t^\infty C(u)\,du$ *for* $t \geq 1$.

Proof. Let $z(t)$ be a nonzero solution of (1.3.42). Since $-z(t)$ is also a solution, assume $z(0) > 0$. Now either $z(t) > 0$ for all $t \geq 0$ or there is a $t_1 > 0$ such that $z(t_1) = 0$ and $z(t) > 0$ on $[0, t_1)$ with $z'(t_1) \leq 0$. In the second case we have

$$z'(t_1) = Az(t_1) + \int_0^{t_1} C(t_1 - s)z(s)\,ds$$
$$= \int_0^{t_1} C(t_1 - s)z(s)\,ds.$$

If $C(t) \not\equiv 0$ on $[0, t_1]$, then the last integral is positive, contradicting $z'(t_1) \leq 0$. If $C(t) \equiv 0$ on $[0, s)$, $s \geq t_1$, then $z(t) = z(0)e^{At}$ for all $t \in [0, s]$. In any case, then, $z(t) > 0$ for all $t \geq 0$. Integration of (1.3.42) from t to ∞ yields [because $z(\infty) = 0$]

$$-z(t) = A\int_t^\infty z(s)\,ds + \int_t^\infty \int_0^u C(u-s)z(s)\,ds\,du$$

or

$$-A \int_t^\infty z(s)\, ds \geq \int_t^\infty \int_0^u C(u-s)z(s)\, ds\, du$$

$$= \int_t^\infty \int_0^t C(u-s)z(s)\, ds\, du$$

$$+ \int_t^\infty \int_t^u C(u-s)z(s)\, ds\, du$$

$$\geq \int_0^t \int_t^\infty C(u-s)z(s)\, du\, ds$$

$$= \int_0^t \int_{t-s}^\infty C(v)\, dv\, z(s)\, ds$$

$$\geq \int_0^t z(s)\, ds \int_t^\infty C(v)\, dv \geq \int_0^1 z(s)\, ds \int_t^\infty C(v)\, dv$$

for $t \geq 1$. Since A must be negative for $z \in L^1$ and $C(t) \geq 0$, the proof is complete.

1.4   Periodic Solutions of Convolution Equations

We have studied periodic solutions of

(a) $$x' = Ax + p(t)$$

with A constant and p being T-periodic. At first it seems natural to extend the study to

(b) $$x' = Ax + \int_0^t B(t-s)x(s)\, ds + p(t),$$

but there are immediate difficulties. If $x(t)$ is a T-periodic solution of either (a) or (b), then certainly $x(t+T)$ is also a T-periodic solution since $x(t) = x(t+T)$. In the search for periodic solutions it becomes crucial to know that if $x(t)$ is *any solution*, so is $x(t+T)$. Such a requirement is obviously satisfied for (a), but it does not hold for (b). Yet, we are puzzled that there are examples of equations of the form of (b) with periodic solutions. The puzzlement is finally relieved when we learn that there are only a "few" such examples. We present these results later in this section. Ultimately we learn that (b) has solutions which approach periodic solutions of

(c)
$$x' = Ax + \int_{-\infty}^{t} B(t - s)x(s)\, ds + p(t)$$

in a certain "limiting" fashion. Consequently, (c) is called a *limiting equation* and we begin to see the need to ask that $B \in L^1[0, \infty)$. The study concerns four related equations.

Let A be an $n \times n$ real constant matrix, C an $n \times n$ real matrix of functions continuous for $0 \leq t < \infty$, and let $p : (-\infty, \infty) \to R^n$ be continuous. Consider the systems

(1.4.1)
$$z' = Az + \int_{0}^{t} C(t - s)z(s)\, ds, \qquad\qquad z \in R^n,$$

(1.4.2)
$$Z' = AZ + \int_{0}^{t} C(t - s)Z(s)\, ds, \qquad Z(0) = I,$$

(1.4.3)
$$y' = Ay + \int_{0}^{t} C(t - s)y(s)\, ds + p(t), \qquad y \in R^n,$$

and

(1.4.4)
$$x' = Ax + \int_{-\infty}^{t} C(t - s)x(s)\, ds + p(t), \qquad x \in R^n.$$

These are all linear integrodifferential equations of *convolution* type. Equation (1.4.1) is *homogeneous*, (1.4.2) is called the *resolvent equation*, $Z(t)$ is the resolvent or the *principal matrix solution*, (1.4.3) and (1.4.4) are *non-homogeneous* or *forced equations*, and $p(t)$ is the *forcing function*. We have already proved the following facts.

There exists a unique solution matrix $Z(t)$ of (1.4.2) on $[0, \infty)$.

(i) For each $z_0 \in R^n$ there is a unique solution $z(t, 0, z_0)$ satisfying (1.4.1) on $[0, \infty)$ with $z(t, 0, z_0) = Z(t)z_0$.

(ii) For each $y_0 \in R^n$ there is a unique solution $y(t, 0, y_0)$ satisfying (1.4.3) on $[0, \infty)$ with $y(t, 0, y_0) = Z(t)y_0 + \int_0^t Z(t - s)p(s)\, ds$.

The following theorem will complete the summary of fundamental properties of this set of equations.

Theorem 1.4.1. *Assume $p(t + T) = p(t)$ for all t and some $T > 0$.*

(iii) Let $C \in L^1[0, \infty)$ and let $y(t)$ be a bounded solution of (1.4.3) on $[0, \infty)$. Then there is an increasing sequence of integers $\{n_j\}$ such that

$\{y(t + n_j T)\}$ converges uniformly on compact subsets of $(-\infty, \infty)$ to a function $x(t)$ which is a solution of (1.4.4) on $(-\infty, \infty)$:

$$y(t + n_j T) = Z(t + n_j T)y_0 + \int_0^{t+n_j T} Z(t + n_j T - s)p(s)\, ds \to x(t).$$

(iv) If $Z(t) \to 0$ as $t \to \infty$, if $y(t)$ is a bounded solution of (1.4.3), and if $C \in L^1[0, \infty)$ then

$$x(t) = \int_{-\infty}^t Z(t - s)p(s)\, ds$$

is a T-periodic solution of (1.4.4).

(v) If C and $Z \in L^1[0, \infty)$, then

$$x(t) = \int_{-\infty}^t Z(t - s)p(s)\, ds$$

is a T-periodic solution of (1.4.4).

Proof. Part (iii) is a consequence of a result of Miller (1971a) and will also follow from Theorem 1.5.1. Thus at this point we assume the validity of (iii) and prove (iv). We have $C \in L^1$, $Z(t) \to 0$ as $t \to \infty$, and $y(t)$ a bounded solution of (1.4.3) so

$$
\begin{aligned}
y(t + n_j T) &= Z(t + n_j T)y_0 + \int_0^{t+n_j T} Z(t + n_j T - s)p(s)\, ds \\
&= Z(t + n_j T)y_0 + \int_0^{t+n_j T} Z(u)p(t - u)\, du \\
&\to \int_0^{\infty} Z(u)p(t - u)\, du,
\end{aligned}
$$

a solution of (1.4.4) according to (iii) and which is certainly T-periodic.

For (v) we note that $Z \in L^1$ and $C \in L^1$ implies Z is bounded and tends to zero as $t \to \infty$; thus, the formula in (ii) shows that all solutions of (1.4.3) are bounded. The result now follows from (iv).

Note. It seems likely that (iv) and (v) are the same; that is, one may be able to show that (iv) implies $Z \in L^1[0, \infty)$.

Recall that we began the discussion of $Z \in L^1[0, \infty)$ in Section 1.3. It was shown in Theorem 1.3.5 that for (1.4.1) scalar, $C(t) \geq 0$, $C \in L^1[0, \infty)$, and $A + \int_0^{\infty} C(u)\, du \neq 0$, then $Z \in L^1[0, \infty)$ if and only if $A + \int_0^{\infty} C(u)\, du < 0$.

It was also remarked that Grossman and Miller (1973) have shown that for the general case of (1.4.2) when $C \in L^1[0, \infty)$, and A is constant, then $Z \in L^1[0, \infty)$ if and only if

$$\det[Is - A - C^*(s)] \neq 0 \quad \text{for} \quad \text{Re } s \geq 0$$

where C^* is the Laplace transform of C.

Much effort has gone into verifying the condition in particular cases, but it is a transcendental equation and the roots are very difficult to locate. The following result, offered here without proof, is obtained by Brauer (1978).

Example 1.4.1. Let (1.4.1) be a scalar equation, A constant, C of one sign on $[0, \infty)$, $C \in L^1[0, \infty)$, and let $L = \int_0^\infty t|C(t)| \, dt < \infty$.

(a) If $A + \int_0^\infty C(t) \, dt \geq 0$, then $Z \notin L^1[0, \infty)$.

(b) If $C(t) > 0$ and $A + \int_0^\infty C(t) \, dt < 0$, then $Z \in L^1[0, \infty)$.

(c) If $C(t) < 0$ and $A + \int_0^\infty C(t) \, dt < 0$, then for L sufficiently small $Z \in L^1[0, \infty)$.

Jordan (1979) has extended Brauer's theorem to systems which are diagonal dominant.

We remark that the results of Brauer and Jordan refer to uniform asymptotic stability, a concept equivalent to $Z \in L^1[0, \infty)$ under the present conditions (cf. Burton, 1983a, p. 47). Additional conditions for $Z \in L^1[0, \infty)$ using Liapunov's direct method will be given in Section 1.6.

Example 1.4.2. Levin (1963) has a beautiful result for a nonlinear form of the scalar equation

$$z' = - \int_0^t C(t - s)z(s) \, ds$$

where $C(t) > 0$, $C'(t) \leq 0$, $C''(t) \geq 0$, $C'''(t) \leq 0$, and $C(t) \not\equiv C(0)$. He shows that $z^{(i)}(t) \to 0$ as $t \to \infty$ for $i = 0, 1, 2$.

Levin uses the functional

$$V(t, z(\cdot)) = z^2 + C(t)\left(\int_0^t z(s) \, ds\right)^2 - \int_0^t C'(t - s)\left(\int_s^t z(u) \, du\right)^2 ds$$

where $z(t)$ is a solution and shows that

$$V'(t, z(\cdot)) \leq C'(t)\left(\int_0^t z(s) \, ds\right)^2 - \int_0^t C''(t - s)\left(\int_s^t z(u) \, du\right)^2 ds \leq 0,$$

implying immediately that $z(t)$ is bounded. It requires some hard analysis to show that $z^{(i)}(t) \to 0$ as $t \to \infty$.

Problem 1.4.1. Under the conditions of Example 1.4.2 show that

$$y' = -\int_0^t C(t-s)y(s)\,ds + p(t)$$

has a bounded solution when p is T-periodic. Use Theorem 1.4.1(iv) to conclude that when $C \in L^1[0,\infty)$ then

$$x' = -\int_{-\infty}^t C(t-s)x(s)\,ds + p(t)$$

has a periodic solution. Alternatively, examine Levin's proof to determine if the conclusion $Z \in L^1[0,\infty)$ can be extracted. Then apply Theorem 1.4.1(v) for the T-periodic conclusion.

The proof of Theorem 1.4.1(iii) indicates that $C \in L^1[0,\infty)$ is needed. Yet, there are times when it might be avoided, as indicated in the next example. A general theory dropping that condition would be most welcome. One might only be able to conclude that solutions of (1.4.4) approach a periodic solution.

Example 1.4.3. Consider the scalar equation

$$y' = -\int_0^t [a^2 + C(t-s)]y(s)\,ds + p(t)$$

where $a \neq 0$, $C \in L^1[0,\infty)$, C satisfies the conditions of Example 1.4.2, and p is T-periodic with a continuous derivative. Then

$$y'' = -[a^2 + C(0)]y - \int_0^t C'(t-s)y(s)\,ds + p'(t)$$

so that if $y = y_1$ and $y_2 = y_1'$, then we obtain the system

$$y_1' = y_2,$$

$$y_2' = -[a^2 + C(0)]y_1 - \int_0^t C'(t-s)y_1(s)\,ds + p'(t)$$

or

$$\begin{pmatrix} y_1 \\ y_2 \end{pmatrix}' = \begin{pmatrix} 0 & 1 \\ -a^2 - C(0) & 0 \end{pmatrix} \begin{pmatrix} y_1 \\ y_2 \end{pmatrix} - \int_0^t \begin{pmatrix} 0 & 0 \\ C'(t-s) & 0 \end{pmatrix} \begin{pmatrix} y_1(s) \\ y_2(s) \end{pmatrix} ds + \begin{pmatrix} 0 \\ p'(t) \end{pmatrix}$$

and the kernel of the new system is $L^1[0,\infty)$. Moreover, if for $p(t) \equiv 0$ we have from Example 1.4.2 that $z(t)$ and $z'(t) \to 0$ as $t \to \infty$, then this information can be used in the new system.

Problem 1.4.2. Prove that the resulting system has a bounded solution, or prove that z and z' are $L^1[0, \infty)$. Then apply Theorem 1.4.1(iv) or (v) to conclude the existence of a T-periodic solution in Example 1.4.3.

Problem 1.4.3. Generalize Example 1.4.3 and Problem 1.4.2 obtaining results parallel to Theorem 1.4.1(iii)–(v) without $C \in L^1[0, \infty)$. You may be able to conclude only that solutions of (1.4.4) approach a periodic solution.

In order for one to use the machinery of fixed-point theory in the search for periodic solutions, one needs to be able to verify that if $\phi(t)$ is a solution of the equation of interest, then $\phi(t+T)$ is also a solution. That is a property that is easy to verify in (1.4.4), but it cannot be verified in (1.4.1)–(1.4.3). Yet, it is known that some equations of the form of (1.4.3) do have periodic solutions. For example, $y = \cos t + \sin t$ is a solution of the scalar equation

$$y' = ay + b \int_0^t e^{-(t-s)} y(s)\, ds - (1 + a + b)\sin t + (1 - a)\cos t$$

with a and b being nonzero constants. Investigators have sought to prove a general result on periodic solutions of (1.4.3), without success. The fact is that no such result exists, as we now show (cf. Burton, 1984).

From Theorem 1.4.1, if C and $Z \in L^1[0, \infty)$ then a solution of (1.4.3) may be expressed as

$$y(t) = Z(t)y(0) + \int_0^t Z(t - s)p(s)\, ds$$

and

$$y(t) \to \int_0^t Z(t - s)p(s)\, ds \overset{\text{def}}{=} \psi(t).$$

But

(1.4.5)
$$\int_{-\infty}^t Z(t - s)p(s)\, ds \overset{\text{def}}{=} \phi(t)$$

is T-periodic and satisfies (1.4.4) with

$$\phi(t) = \int_{-\infty}^0 Z(t - s)p(s)\, ds + \int_0^t Z(t - s)p(s)\, ds$$
$$= \int_{-\infty}^0 Z(t - s)p(s)\, ds + \psi(t),$$

while $\int_{-\infty}^0 Z(t - s)p(s)\, ds \to 0$ as $t \to \infty$. We therefore conclude

Theorem 1.4.2. *If C and $Z \in L^1[0, \infty)$ and $p(t + T) \equiv p(t)$, then each solution of (1.4.3) converges to the T-periodic solution of (1.4.4)*

$$\phi(t) = \int_{-\infty}^{t} Z(t - s)p(s)\,ds.$$

Theorem 1.4.3. *If C and $Z \in L^1[0, \infty)$ and if (1.4.3) has a T-periodic solution $y(t)$, then $y(t) = \int_{-\infty}^{t} Z(t-s)p(s)\,ds$ and $\int_{-\infty}^{0} C(t-s)y(s)\,ds \equiv 0$.*

Proof. If $y(t)$ is a T-periodic solution of (1.4.3), since $y(t)$ converges to the T-periodic solution (1.4.5) of Eq. (1.4.4), it must be that $y(t)$ is given by (1.4.5). Comparing (1.4.3) and (1.4.4) we conclude that $\int_{-\infty}^{0} C(t - s) y(s)\,ds \equiv 0$ when $y(t)$ is given by (1.4.5). This completes the proof.

Theorem 1.4.4. *If C and $Z \in L^1[0, \infty)$ and if (1.4.3) has a T-periodic solution, then*

$$h(t) = \int_{t}^{\infty} Z(u)p(t - u)\,du$$

is a solution of (1.4.1) (which is independent of p).

Proof. If (1.4.3) has a T-periodic solution it is given by $\phi(t)$ in (1.4.5). But $\int_{0}^{t} Z(t - s)p(s)\,ds$ is also a solution of (1.4.3) and so the difference

$$\int_{-\infty}^{0} Z(t - s)p(s)\,ds$$

is a solution of (1.4.1). Let $s = t - u$ to complete the proof.

Exercise 1.4.1. Let (1.4.3) be scalar and let C and $Z \in L^1[0, \infty)$. Show that if p is a nonzero constant and if (1.4.3) has a T-periodic solution, then $Z(t) = e^{-\alpha t}$ for some constant $\alpha > 0$.

Note. Under the condition of Theorem 1.4.4 we have $h(t)$ a solution of (1.4.1), while every solution of (1.4.1) can be expressed as $Z(t)b$ for some constant vector b; hence,

(1.4.6) $$Z(t)b = \int_{t}^{\infty} Z(u)p(t - u)\,du.$$

Principle 1.4.1. *Let*

$$h(t) = \int_{t}^{\infty} Z(u)p(t - u)\,du$$

be a solution of (1.4.1). If p satisfies a linear homogeneous ordinary differential equation with constant coefficients, then Z also satisfies such an equation.

Corduneau (1981, 1982) and Seifert (1981) use different techniques to obtain almost periodic solutions. Grimmer (1979) uses a summation technique to reduce the problem to a finite delay equation for which a periodic solution exists. Furumochi (1981, 1982) also looks at the problem from that point of view.

1.4.1 Noncritical Convolution Systems

This section contains recent work of Langenhop (1984) concerning the extension of the theory of noncritical systems to Volterra equations. That work actually concerns complex systems with almost periodic coefficients; however, we restrict the discussion to periodicity and real systems.

Let

$$(1.4.7) \qquad x'(t) = \int_0^\infty [dE(s)]x(t-s) + f(t)$$

where f is T-periodic with T sufficiently small. Here, $x \in R^n$, $f : (-\infty, \infty) \to R^n$ is continuous, and E is an $n \times n$ matrix of functions continuous from the left and of bounded total variation on $[0, \infty)$; that is,

$$(1.4.8) \qquad 0 < \gamma = \int_0^\infty |dE(s)| \, ds < \infty.$$

The assumptions on E allow for (1.4.7) to include such equations as

$$(1.4.9) \qquad x'(t) = B_0 x(t) + B_1 x(t-r) + \int_{-\infty}^t C(t-s)x(s) \, ds + f(t)$$

in which B_0 and B_1 are constant matrices, $r > 0$, and the continuous matrix function C satisfies

$$(1.4.10) \qquad \int_0^\infty |C(\sigma)| d\sigma < \infty.$$

Obviously, additional terms with finite delays can also be included.

Results regarding the existence and uniqueness of periodic solutions of (1.4.9) with f periodic or a.p. are well known for the case $B_1 = 0$, $C(\sigma) \equiv 0$. These are expounded in Hale (1969, Chapter IV). More general equations which include nonlinear terms have been treated by Seifert (1981) in the a.p. context.

For $T > 0$ we let $(\mathcal{P}_T, \| \cdot \|)$ denote the Banach space of continuous functions $f : (-\infty, \infty) \to R^n$ with period T,

$$(1.4.11) \qquad\qquad f(t + T) \equiv f(t),$$

with the norm

$$(1.4.12) \qquad\qquad \|f\| = \sup\{|f(t)| : -\infty < t < \infty\}.$$

By virtue of (1.4.8) the relation

$$(1.4.13) \qquad\qquad (Ax)(t) = \int_0^\infty [dE(s)]x(t - s)$$

defines a bounded linear operator $A : \mathcal{P}_T \to \mathcal{P}_T$ with

$$(1.4.14) \qquad\qquad \|A\| \leq \gamma.$$

Using D to denote differentiation, $x'(t) = (Dx)(t)$, we may consider (1.4.7) in the form

$$(1.4.15) \qquad\qquad (D - A)x = f.$$

The existence and uniqueness of a solution $x \in \mathcal{P}_T$ of (1.4.7) for each $f \in \mathcal{P}_T$ is thus equivalent to the operator $D - A$ with domain $\mathcal{P}'_T = \{x \in \mathcal{P}_T : x' \in \mathcal{P}_T\}$ being one-one and having range $\mathcal{P}_T$.

The mean value of a function $f \in \mathcal{P}_T$ will be denoted by $m(f)$:

$$(1.4.16) \qquad\qquad m(f) = \frac{1}{T} \int_0^T f(t)\, dt.$$

We may interpret $m(f)$ as a constant function in $\mathcal{P}_T$ or an element of R^n. Our assumptions on E assure that

$$(1.4.17) \qquad m(Af) = Am(f) = Mm(f), \quad f \in \mathcal{P}_T,$$

where M is the $n \times n$ constant matrix

$$(1.4.18) \qquad\qquad M = \int_0^\infty dE(s).$$

If $x \in \mathcal{P}_T$ is a solution of (1.4.7) with $f \in \mathcal{P}_T$, then $x' \in \mathcal{P}_T$ and $m(x') = 0$ so from (1.4.7) and (1.4.17) we get

$$(1.4.19) \qquad\qquad Mm(x) + m(f) = 0.$$

If M is singular, there are functions $f \in \mathcal{P}_T$ (appropriate constants will do) for which (1.4.19) cannot hold. Hence we must require M to be nonsingular in order for $D - A$ to have $\mathcal{P}_T$ as range. Then (1.4.19) implies

$$(1.4.20) \qquad\qquad m(x) = -M^{-1}m(f).$$

We shall say that the operator $D - A$ is $\mathcal{P}_T$-noncritical if it is one-one on $\mathcal{P}'_T$. This is a slight change in terminology and emphasis for an analogous concept applied to ordinary differential equations by Hale (1969, Chapter IV). Observe that if $D - A$ is $\mathcal{P}_T$-noncritical, then M must be nonsingular. In the following material we give conditions which imply that $D - A$ is $\mathcal{P}_T$-noncritical.

Theorem 1.4.5. *Under the conditions described for the matrix function E, including (1.4.8), if M is nonsingular and $0 < T < 2/\gamma$, then for each $f \in \mathcal{P}_T$ Eq. (1.4.7) has one and only one solution $x \in \mathcal{P}_T$, i.e., then $D - A$ is $\mathcal{P}_T$-noncritical.*

Before giving the essentials of the proof it is convenient to introduce some further notation. The subspace $\mathcal{P}_T^0$ of $\mathcal{P}_T$ defined by

$$(1.4.21) \qquad \mathcal{P}_T^0 = \{g \in \mathcal{P}_T : m(g) = 0\}$$

is of importance since by (1.4.20) we have $m(x) = 0$ if and only if $m(f) = 0$. Moreover, by (1.4.17), we see that A maps $\mathcal{P}_T^0$ into $\mathcal{P}_T^0$. We also introduce the operator $J : \mathcal{P}_T^0 \to \mathcal{P}_T$ by

$$(1.4.22) \qquad (Jg)(t) = \int_0^t g(s)\, ds$$

and the operator $L : \mathcal{P}_T^0 \to \mathcal{P}_T^0$ by

$$(1.4.23) \qquad Lg = Jg - m(Jg).$$

Clearly, both J and L are linear. Finally, for any function $f \in \mathcal{P}_T$ we define its correspondent $\tilde{f} \to \mathcal{P}_T^0$ by

$$(1.4.24) \qquad \tilde{f} = f - m(f).$$

Proof of Theorem 1.4.5. Suppose first that $f \in \mathcal{P}_T$ and let $x \in \mathcal{P}_T$ be a solution of (1.4.7). Since M is nonsingular, we have

$$(1.4.25) \qquad \tilde{x} = x + M^{-1}m(f) = x + m(M^{-1}f)$$

by (1.4.20). Since $Am(M^{-1}f) = Mm(M^{-1}f) = m(f)$ by (1.4.17) it follows that $A\tilde{x} + \tilde{f} = Ax + f$. But $\tilde{x}' = x'$ so we have

$$(1.4.26) \qquad \tilde{x}' = A\tilde{x} + \tilde{f}.$$

Using (1.4.22), we may write

$$(1.4.27) \qquad \tilde{x} = \tilde{x}(0) + J\tilde{x}'$$

and, taking the mean value, we get

(1.4.28) $0 = \tilde{x}(0) + m(J\tilde{x}')$

since $\tilde{x} \in \mathcal{P}_T^0$ and $\tilde{x}(0)$ is constant. Combining (1.4.26)–(1.4.28), we get

(1.4.29) $\tilde{x} = LA\tilde{x} + L\tilde{f}$

by the definition of L in (1.4.23). Thus x is a fixed point in $\mathcal{P}_T^0$ of the map $H : \mathcal{P}_T^0 \to \mathcal{P}_T^0$ defined by

(1.4.30) $Hg = LAg + L\tilde{f}.$

We note that if $DX = Ax + f$ and $Dy = Ay + f$ for $x, y \in \mathcal{P}_T$ and if $\tilde{y} = \tilde{x}$, then (1.4.20) implies $y = x$. Thus the correspondence $x \to \tilde{x}$ is one-one on the solutions of (1.4.7) in $\mathcal{P}_T$ to the fixed points of H in $\mathcal{P}_T^0$.

   Conversely, suppose $\phi \in \mathcal{P}_T^0$ is a fixed point of H. Then $\phi = J(A\phi + \tilde{f}) - m(J(A\phi + \tilde{f}))$. Differentiating this, we get $\phi' = A\phi + \tilde{f}$. If we now define $x = \phi - M^{-1}m(f)$, then $x \in \mathcal{P}_T$. It suffices to show that H has one and only one fixed point in $\mathcal{P}_T^0$. This is guaranteed by $\|LA\| < 1$ since then H would be a contraction on $\mathcal{P}_T^0$.

   Now we have

(1.4.31) $\|LA\| \leq \|L\|\,\|A\| \leq \gamma\|L\|$

by (1.4.14). Hence the theorem follows once we show that $\|L\| \leq T/2$. To this end we note by (1.4.23) and (1.4.16) that if $g \in \mathcal{P}_T^0$, then

(1.4.32)
$$(Lg)(t) = \int_0^t g(s)\,ds - \frac{1}{T}\int_0^T \int_0^u g(s)\,ds\,du$$
$$= \int_0^T k(t,s)g(s)\,ds$$

where

(1.4.33) $k(t,s) = \begin{cases} s/T, & 0 \leq s \leq t, \\ (s-T)/T, & t < s \leq T. \end{cases}$

Hence for $0 \leq t \leq T$ we have

(1.4.34) $\int_0^T |k(t,s)|\,ds = [t^2 + (T-t)^2]/2T.$

The right side of (1.4.34) is easily seen to be at most $T/2$ for $t \in [0, T]$ so it follows that

$$|Lg(t)| \leq \int_0^T |k(t,s)|\,|g(s)|\,ds \leq (T/2)\|g\|.$$

Hence $\|L\| \leq T/2$ and the proof is complete.

Remark. The condition of the theorem can be replaced by $0 < T < 4/\gamma$.

When $D - A$ is $\mathcal{P}_T$-noncritical, then $D - A$ is invertible on $\mathcal{P}_T$. If we define

$$(1.4.35) \qquad K = (D - A)^{-1},$$

then for each $f \in \mathcal{P}_T$ the function $x = Kf$ is the unique solution of (1.4.7) in $\mathcal{P}_T$. Clearly K is linear since $D - A$ is linear. We now prove that K is bounded with the restriction on T being replaced by the hypothesis that $D - A$ is $\mathcal{P}_T$-noncritical.

Theorem 1.4.6. *Let E be as described above so that (1.4.8) holds and let A be defined by (1.4.13). If $D - A$ is $\mathcal{P}_T$-noncritical, then $K = (D - A)^{-1}$ is a bounded linear operator on $\mathcal{P}_T$ with range $\mathcal{P}_T'$.*

Proof. If $f \in \mathcal{P}_T$, then $x = Kf \in \mathcal{P}_T$ satisfies $x' = Ax + f \in \mathcal{P}_T$ so $x \in \mathcal{P}_T'$. On the other hand, for any $x \in \mathcal{P}_T'$ we may define $f = x' - Ax \in \mathcal{P}_T$ and then $x = Kf$. Thus the range of K is $\mathcal{P}_T'$.

Now let $\{f_k\}$ be a sequence in $\mathcal{P}_T$ and let $x_k = Kf_k$. Integrating $x_k' = Ax_k + f_k$ from 0 to t, we get

$$(1.4.36) \qquad x_k(t) = x_k(0) + \int_0^t (Ax_k)(\sigma)d\sigma + \int_0^t f_k(\sigma)d\sigma.$$

Suppose $f \in \mathcal{P}_T$ and $x \in \mathcal{P}_T$ are such that $\|f_k - f\| \to 0$ and $\|x_k - x\| \to 0$ as $k \to \infty$. The resulting uniform convergence of $\{f_k(s)\}$ and $\{x_k(s)\}$ for $s \in R$ and the conditions on E imply for any fixed t that

$$\int_0^t f_k(\sigma)d\sigma \to \int_0^t f(\sigma)d\sigma$$

and

$$\int_0^t (Ax_k)(\sigma)d\sigma \to \int_0^t (Ax)(\sigma)d\sigma$$

as $k \to \infty$. Hence, letting $k \to \infty$ in (1.4.36), we get

$$(1.4.37) \qquad x(t) = x(0) + \int_0^t (Ax)(\sigma)d\sigma + \int_0^t f(\sigma)d\sigma.$$

Continuity of Ax and f then gives $Dx = Ax + f$. But x and f are in $\mathcal{P}_T$ and $D - A$ is $\mathcal{P}_T$-noncritical so $x = Kf$. Thus K is a closed operator. Since it is linear and defined on the Banach space $\mathcal{P}_T$, it is bounded (by the closed graph theorem).

The preceding result can now be used as in Hale (1969), Chapter IV) to obtain existence of periodic solutions for suitable perturbations of (1.4.7) when $D - A$ is $\mathcal{P}_T$-noncritical. Consider, for example, the equation

$$(1.4.38) \qquad x'(t) = \int_0^t [dE(s)]x(t - s) + G(x)(t) + f(t)$$

where $G : \mathcal{P}_T \to \mathcal{P}_T$ satisfies the regularity conditions given below. These conditions allow for $G(x)(t)$ to be a sum of terms of the type $F(x(t))$, $F(x(t-r))$ and $\int_0^\infty C(s)F(x(t-s))\,ds$ in which $F : R^n \to R^n$ is continuously differentiable and $F(0) = 0$. [If $F(0) \neq 0$ one can replace F by $F^* = F - F(0)$ and add a corresponding constant term to $f(t)$ in (1.4.38).]

About $G : \mathcal{P}_T \to \mathcal{P}_T$ we assume

$$(1.4.39) \qquad\qquad\qquad G(0) = 0$$

and that there exist a $\Gamma \geq 0$ and a continuous nondecreasing function $w : [0, \infty) \to [0, \infty)$ with $w(0) = 0$ such that

$$(1.4.40) \qquad\qquad \|G(\sigma) - G(\psi)\| \leq [\Gamma + w(u)]\|\phi - \psi\|$$

when $\phi, \psi \in \mathcal{P}_T$ satisfy $\|\phi\| \leq u$ and $\|\psi\| \leq u$. Observe that (1.4.39) and (1.4.40) imply

$$(1.4.41) \qquad\qquad \|G(\phi)\| \leq [\Gamma + w(\|\phi\|)]\|\phi\|$$

for $\phi \in \mathcal{P}_T$.

Theorem 1.4.7. *Let E be as described above so that (1.4.8) holds and let A be defined by (1.4.13). Suppose $D - A$ is $\mathcal{P}_T$-noncritical and $G : \mathcal{P}_T \to \mathcal{P}_T$ satisfies the conditions in the preceding paragraph. If $K = (D - A)^{-1}$ on $\mathcal{P}_T$ and $\Gamma < 1/\|K\|$, then there are positive numbers α, β such that if $f \in \mathcal{P}_T$ and $\|f\| \leq \alpha$, then (1.4.38) has a unique solution $x_f \in \mathcal{P}_T$ satisfying $\|x_f\| \leq \alpha\beta$; moreover, $\|x_f\| \leq \beta\|f\|$.*

Proof. For $f \in \mathcal{P}_T$ out hypotheses imply, by virtue of Theorem 1.4.6, that

$$(1.4.42) \qquad\qquad Q(\phi) = KG(\phi) + Kf, \qquad \phi \in \mathcal{P}_T,$$

defines a mapping $Q : \mathcal{P}_T \to \mathcal{P}_T$. We see then that $x \in \mathcal{P}_T$ is a solution of (1.4.38) if and only if x is a fixed point of Q. The boundedness of K from Theorem 1.4.6 and condition (1.4.40) then imply

$$(1.4.43) \qquad\qquad \|Q(\phi) - Q(\psi)\| \leq \|K\|[\Gamma + w(u)]\|\phi - \psi\|$$

for $\phi, \psi \in \mathcal{P}_T$ with $\|\phi\| \leq u$ and $\|\psi\| \leq u$. Similarly, from (1.4.41) we conclude that

$$(1.4.44) \qquad\qquad \|Q(\phi)\| \leq \|K\|[\Gamma + w(\|\phi\|)]\|\phi\| + \|K\|\,\|f\|.$$

But, by hypothesis, $w(u)$ is continuous, nondecreasing, $w(0) = 0$, and $\Gamma \|K\| < 1$ so there is a $u_0 > 0$ such that

$$(1.4.45) \qquad \mu_0 = \|K\|[\Gamma + w(u_0)] < 1.$$

Moreover,

$$(1.4.46) \qquad \|K\|[\Gamma + w(u)] \leq \mu_0, \qquad 0 \leq u \leq u_0.$$

Now let $\rho = (1 - \mu_0)/\|K\|$ and note that $\rho > 0$. If $\|\phi\| \leq u_0$, then from (1.4.44) and (1.4.46) we get

$$(1.4.47) \qquad \|Q(\phi)\| \leq \mu_0\|\phi\| + \|K\| \|f\|.$$

In particular, if also $\|f\| \leq \rho u_0$, then the right side of (1.4.47) is at most u_0 and Q maps the closed ball $\mathcal{B}_T(u_0) = \{\phi \in \mathcal{P}_T : \|\phi\| \leq u_0\}$ into itself. Relations (1.4.43) and (1.4.45) imply that Q is a contraction on $\mathcal{B}_T(u_0)$. If we take $\beta = 1/\rho$, $\alpha = u_0/\beta = \rho u_0$, then we may conclude that for each $f \in \mathcal{P}_T$ satisfying $\|f\| \leq \alpha$ there is a unique fixed point $x_f \in \mathcal{P}_T$ of Q satisfying $\|x_f\| \leq u_0 = \alpha\beta$. From (1.4.47) and the relation $x_f = Q(x_f)$ we get

$$\|x_f\| \leq \mu_0\|x_f\| + \|K\| \|f\|$$

which implies $\|x_f\| \leq (1/\rho)\|f\| = \beta\|f\|$.

1.5 Periodic Solutions of Nonconvolution Equations

Linear systems of ordinary differential equations with periodic coefficients are, of course, deceptive. According to Floquet theory anything that can be proved for

$$x' = Ax + p(t)$$

with A constant can be proved for

$$x' = A(t)x + p(t)$$

with $A(t)$ periodic because the PMS of

$$Q' = A(t)Q$$

is $Q(t) = P(t)e^{Jt}$ with P being periodic and J constant. But for A constant, then e^{At} is known; whereas for $A(t)$ periodic we must know $Q(t)$ before we can find J. The real beauty of Floquet theory is that a very usable variation of parameters formula still holds:

$$x(t) = Q(t)x_0 + \int_0^t P(t)e^{J(t-s)}P^{-1}(s)p(s)\,ds.$$

And an equally nice formula holds in the Volterra convolution case. But for Volterra systems without constant coefficients or of nonconvolution type, then the theory is in very poor condition. We still obtain a variation of parameters formula, but the kernel comes from a different equation and its utility is drastically reduced. In the convolution case

$$x' = Ax + \int_0^t C(t-s)x(s)\,ds + p(t)$$

with A constant we have x bounded if $Z \in L^1[0,\infty)$ and p is bounded. But, so far, that property does not seem to extend to nonconvolution systems and the boundedness problems become enormous. The main results here are found in Burton (1985a).

Let A be an $n \times n$ matrix of functions continuous on $(-\infty, \infty)$, let $p : (-\infty, \infty) \to R^n$ be continuous, and let $C(t,s)$ be an $n \times n$ matrix of functions continuous for $-\infty < s \le t < \infty$. We will be interested in

$$(1.5.1) \qquad z' = A(t)z + \int_0^t C(t,s)z(s)\,ds, \qquad z \in R^n,$$

$$(1.5.2) \qquad \begin{aligned} \partial R(t,s)/\partial s &= -R(t,s)A(s) - \int_s^t R(t,u)C(u,s)\,du, \\ R(t,t) &= I, \end{aligned}$$

$$(1.5.3) \qquad y' = A(t)y + \int_0^t C(t,s)y(s)\,ds + p(t), \qquad y \in R^n,$$

and

$$(1.5.4) \qquad x' = A(t)x + \int_{-\infty}^t C(t,s)x(s)\,ds + p(t), \qquad x \in R^n.$$

The goal is to give conditions to ensure that (1.5.4) has a T-periodic solution. Equation (1.5.2) is called the *resolvent equation* and $R(t,s)$ is called the *resolvent*.

Theorem 1.5.1. *There exists a unique $n \times n$ matrix $Z(t)$ satisfying (1.5.1) on $[0,\infty)$ with $Z(0) = I$.*

(i) *For each $z_0 \in R^n$ there is a unique solution $z(t, 0, z_0)$ of (1.5.1) on $[0, \infty)$ and*

$$z(t, 0, z_0) = Z(t)z_0.$$

(ii) *There is a unique $n \times n$ matrix $R(t, s)$ satisfying (1.5.2) for $0 \leq s \leq t < \infty$. For each $y_0 \in R^n$ there is a unique solution $y(t, 0, y_0)$ of (1.5.3) on $[0, \infty)$ with*

(1.5.5) $$y(t, 0, y_0) = Z(t)y_0 + \int_0^t R(t, s)p(s)\, ds,$$

(1.5.6) $$Z(t) = R(t, 0),$$

and if $C(t + T, s + T) = C(t, s)$ and $A(t + T) = A(t)$ then

(1.5.7) $$R(t + T, s + T) = R(t, s).$$

(iii) *Let*

(1.5.8)
$$A(t + T) = A(t), \qquad C(t + T, s + T) = C(t, s),$$
$$p(t + T) = p(t).$$

Suppose also that there is an $M > 0$ such that for each $[a, b]$ if $t \in [a, b]$ then

(1.5.9) $$\lim_{n \to \infty} \int_{-nT}^t |C(t, s)|\, ds = \int_{-\infty}^t |C(t, s)|\, ds \leq M$$

is continuous. Under these conditions if $y(t)$ is a bounded solution of (1.5.3) on $[0, \infty)$, then there is a sequence of integers $\{n_j\}$ such that $\{y(t + n_j T)\}$ converges to a solution $x(t)$ of (1.5.4) on $(-\infty, \infty)$ and the convergence is uniform on compact subsets of $(-\infty, \infty)$.

(iv) *Let the conditions of (iii) hold and let $Z(t) \to 0$ as $t \to \infty$. Then (1.5.4) has a T-periodic solution*

(1.5.10) $$x(t) = \int_{-\infty}^t R(t, s)p(s)\, ds.$$

Proof. Part (i) was proved in Section 1.3.

Proof of (ii). This is a fundamental result of Grossman and Miller (1970). We outline the main parts, leaving out some of the detail. Define

$$(1.5.11) \qquad \psi(t,s) = A(t) + \int_s^t C(t,u)\,du, \qquad 0 \le s \le t < \infty,$$

and

$$(1.5.12) \qquad R(t,s) = I + \int_s^t R(t,u)\psi(u,s)\,du, \qquad 0 \le s \le t < \infty,$$

with $C(t,s) = \psi(t,s) = R(t,s) = 0$ if $s \ge t \ge 0$. If one assumes for the moment that R is differentiable, then it follows by direct substitution that $R(t,s)$ satisfies (1.5.2). Note in particular from (1.5.12) that $R(t,t) = I$ is satisfied. The following sequence of steps (without details) are the crucial ones for the existence and smoothness of $R(t,s)$. We have

$$|\psi(t,s)| \le |A(t)| + \int_0^t |C(t,s)|\,ds \stackrel{\text{def}}{=} \alpha(t),$$

$$(1.5.13) \qquad |R(t,s)| \le \alpha_0(t) \stackrel{\text{def}}{=} 1 + \left(\int_0^t \alpha(s)\,ds\right) \exp\left(\int_0^t \alpha(s)\,ds\right),$$

and this in (1.5.2) yields the existence of $\partial R(t,s)/\partial s$ and that

$$(1.5.14) \qquad |\partial R(t,s)/\partial s| \le \alpha_0(t)|A(s)| + \int_s^t \alpha_0(t)|C(u,s)|\,du$$

$$\le \alpha_0(S)\left(|A(s)| + \int_s^S |C(u,s)|\,du\right)$$

if $0 \le s \le t \le S$. Continuity of $R(t,s)$ in t for fixed s follows from (1.5.13) by dominated convergence. Inequality (1.5.14) shows $R(t,s)$ continuous in s uniformly for $0 \le s \le t \le S$.

Thus, $R(t,s)$ is continuous in (t,s). That sketch establishes the existence of the solution $R(t,s)$ of (1.5.2). We now show that the variation of parameters (1.5.5) is valid. Let $y(0) = y_0$ and let $y(t) = y(t,0,y_0)$ be the unique solution of (1.5.3) on $[0,S]$. Fix $t \in (0,S]$ and integrate by parts to get

$$\int_0^t \{R(t,s)y'(s) + (\partial R(t,s)/\partial s)y(s)\}\,ds$$

$$= R(t,t)y(t) - R(t,0)y(0)$$

or

$$\int_0^t [R(t,s)y'(s) + (\partial R(t,s)/\partial s)y(s)]\,ds$$

$$= R(t,t)y(t) - R(t,0)y_0 = y(t) - R(t,0)y_0$$

since $R(t, t) = I$. Because $y(t)$ satisfies (1.5.3) we write this as

$$
y(t) = R(t, 0)y_0 + \int_0^t \left\{ R(t, s)\left[p(s) + A(s)y(s) \right. \right.
$$
$$
\left. \left. + \int_0^s C(s, u)y(u)\, du \right] + (\partial R(t, s)/\partial s)y(s) \right\} ds.
$$

Changing the order of integration yields

$$
\int_0^t \int_0^s R(t, s)C(s, u)y(u)\, du\, ds = \int_0^t \int_u^t R(t, s)C(s, u)y(u)\, ds\, du
$$
$$
= \int_0^t \int_s^t R(t, u)C(u, s)y(s)\, du\, ds
$$

so that we have

$$
y(t) - R(t, 0)y_0 - \int_0^t R(t, s)p(s)\, ds
$$
$$
= \int_0^t \left[R(t, s)A(s) + (\partial R(t, s)/\partial s) + \int_s^t R(t, u)C(u, s)\, du \right] y(s)\, ds.
$$

The integral on the right is zero because of (1.5.2).

We now have $y(t) = R(t, 0)y_0 + \int_0^t R(t, s)p(s)\, ds$ and need to show that $R(t, 0) = Z(t)$. But this formula for y is valid for $p = 0$ so that $y(t) = R(t, 0)y_0 = Z(t)y_0$ for every y_0. Thus, $R(t, 0) = Z(t)$ and (1.5.6) is true. We have yet to show that $R(t, s)$ is unique. Suppose there are two such matrices, say $R(t, s)$ and $W(t, s)$, satisfying (1.5.2) so that for every y_0 and every continuous p we have

$$
y(t, 0, y_0) = R(t, 0)y_0 + \int_0^t R(t, s)p(s)\, ds
$$

and

$$
y(t, 0, y_0) = W(t, 0)y_0 + \int_0^t W(t, s)p(s)\, ds.
$$

Now $y(t, 0, y_0)$ is unique, while $R(t, 0) = Z(t) = W(t, 0)$. Thus, for each continuous p we have

$$
\int_0^t R(t, s)p(s)\, ds = \int_0^t W(t, s)p(s)\, ds
$$

for $0 \le t < \infty$; but there is a (t_1, s_1) with $R(t_1, s_1) \ne W(t_1, s_1)$. There is then a pair $r_{ij}(t_1, s_1) \ne w_{ij}(t_1, s_1)$ where $R(t, s) = (r_{ij}(t, s))$ and $W(t, s) = (w_{ij}(t, s))$. We may save notation by considering W and R

as scalar. Then $\int_0^{t_1} R(t_1, s)p(s)\,ds = \int_0^{t_1} W(t_1, s)p(s)\,ds$ for every continuous p and we note that $s_1 \leq t_1$ because $R = W = 0$ for $t < s$. Thus, $\int_0^{t_1}[R(t_1, s) - W(t_1, s)]p(s)\,ds = 0$ for every continuous p. To be definite, let $R(t_1, s_1) - W(t_1, s_1) = \alpha > 0$. Choose a continuous p with $p(s_1) = 1$, $p(s) \geq 0$, and $p(s) = 0$ for $0 \leq s \leq s_1 - \varepsilon$ and $s \geq s_1 + \varepsilon$ for an $\varepsilon > 0$ chosen so that $\varepsilon < s_1$ and so that $R(t_1, s) - W(t_1, s)$ is positive near s_1. We will then have the integral positive, yielding a contradiction.

We now show that $R(t + T, s + T) = R(t, s)$. From (1.5.2) we have

$$\partial R(t + T, s + T)/\partial s = -R(t + T, s + T)A(s + T)$$

$$-\int_{s+T}^{t+T} R(t + T, u)C(u, s + T)du$$

$$= -R(t + T, s + T)A(s)$$

$$-\int_s^t R(t + T, v + T)C(v + T, s + T)\,dv$$

$$= -R(t + T, s + T)A(s)$$

$$-\int_s^t R(t + T, v + T)C(v, s)\,dv.$$

This shows that $R(t + T, s + T)$ is also a solution of (1.5.2) and, since $R(t, t) \equiv I$, $R(t + T, t + T) = I$. By uniqueness $R(t + T, s + T) = R(t, s)$. This completes the proofs of (ii).

Proof of (iii). Now $y(t)$ is bounded so $\{y(t + nT)\}$ is bounded and by the integrability condition on $C(t, s)$ we see that the sequence has a bounded derivative for $t > -nT$. Thus, it is uniformly bounded and equicontinuous on $[-1, 1]$, so it has a subsequence $\{y(t + n_jT)\}$ converging uniformly on $[-1, 1]$. Then $\{y(t + n_jT)\}$ is uniformly bounded and equicontinuous on $[-2, 2]$. In this way we obtain a subsequence, say $\{y(t + n_jT)\}$ again, converging uniformly on any fixed interval $[-k, k]$ to a continuous function $x(t)$.

Now for $t \geq -n_jT$ we have

$$y'(t + n_jT) = A(t + n_jT)y(t + n_jT)$$

$$+ \int_0^{t+n_jT} C(t + n_jT, s)y(s)\,ds + p(t + n_jT)$$

$$= A(t)y(t + n_jT)$$

$$+ \int_{-n_jT}^t C(t, s)y(s + n_jT)\,ds + p(t).$$

Let $[a, b]$ be given and let $t \in [a, b]$. Taking the limit as $j \to \infty$, the right side yields

$$A(t)x(t) + \int_{-\infty}^{t} C(t, s)x(s)\, ds + p(t)$$

by the Lebesgue-dominated convergence theorem. Thus, the limit as $j \to \infty$ on the right exists; and we need to show $\lim_{j \to \infty} y'(t + n_j T) = x'(t)$. Now $\{y'(t + n_j T)\}$ is continuous and it will suffice to show that it converges uniformly. Let $n_k < n_j$, let $-n_k T < -P < b$, and let $|y(t)| \leq L$. Then

$$|y'(t + n_j T) - y'(t + n_k T)|$$

$$\leq |A(t)|\, |y(t + n_j T) - y(t + n_k T)|$$

$$+ \int_{-P}^{t} |C(t, s)|\, |y(s + n_j T) - y(s + n_k T)|\, ds$$

$$+ \int_{-n_k T}^{-P} |C(t, s)|\, |y(s + n_j T) - y(s + n_k T)|\, ds$$

$$+ \int_{-n_j T}^{-n_k T} |C(t, s)|\, |y(s + n_j T)|\, ds$$

$$\leq (|A(t)| + M) \sup_{s \in [-P, b]} |y(s + n_j T) - y(s + n_k T)|$$

$$+ 2L \int_{-n_k T}^{-P} |C(t, s)|\, ds + L \int_{-n_j T}^{-n_k T} |C(t, s)|\, ds.$$

The uniform convergence now follows from that of $\{y(t + n_j T)\}$ and the convergence of the integral of $|C(t, s)|$. Thus, $\{y'(t + n_j T)\}$ converges to $x'(t)$ and we have $y(t + n_j T) \to x(t)$, where $x(t)$ satisfies (1.5.4) on $(-\infty, \infty)$. This proves (iii).

Proof of (iv). We have $y(t, 0, y_0) = Z(t)y_0 + \int_0^t R(t, s)p(s)\, ds$. Then

$$y(t + n_j T, 0, y_0) = Z(t + n_j T)y_0 + \int_0^{t + n_j T} R(t + n_j T, s)p(s)\, ds$$

$$= Z(t + n_j T)y_0 + \int_{-n_j T}^{t} R(t + n_j T, s + n_j T)p(s)\, ds$$

$$= Z(t + n_j T)y_0 + \int_{-n_j T}^{t} R(t, s)p(s)\, ds$$

$$\to \int_{-\infty}^{t} R(t, s)p(s)\, ds = x(t).$$

A change of variable shows that $x(t + T) = x(t)$, as required.

Remark 1.5.1. The conditions

$$C(t+T, s+T) = C(t,s) \quad \text{and} \quad R(t+T, s+T) = R(t,s)$$

do not, of course, imply any type of periodicity. In Exercise 1.2.2 we have

$$R(t,s) = P(t)e^{Jt}(P(s)e^{Js})^{-1}$$
$$= P(t)e^{J(t-s)}P^{-1}(s)$$

and the conditions on R hold.

Remark 1.5.2. It turns out that in parts (iii) and (iv) it is relatively easy to prove that $Z(t) \to 0$ as $t \to \infty$; but it is a major area of research to give good conditions ensuring that (1.5.3) has a bounded solution.

Example 1.5.0. Consider (1.5.3) with y a scalar and with A, C, and p satisfying Theorem 1.5.1(iii), including (1.5.9). Suppose also that $\int_t^\infty |C(u,s)|\, du$ is continuous for $0 \le s \le t < \infty$. If

$$A(t) + \int_t^\infty |C(u,t)|\, du \le -\alpha < 0$$

then $Z(t) \to 0$ as $t \to \infty$.

Proof. Define

$$V(t, y(\cdot)) = |y| + \int_0^t \int_t^\infty |C(u,s)|\, du\, |y(s)|\, ds$$

and differentiate V along a solution of (1.5.3) obtaining

$$V'(t, y(\cdot)) \le A(t)|y| + \int_0^t |C(t,s)|\, |y(s)|\, ds + |p(t)|$$
$$+ \int_t^\infty |C(u,t)|\, du\, |y| - \int_0^t |C(t,s)|\, |y(s)|\, ds$$
$$\le -\alpha|y| + |p(t)|$$

so that we have the pair

(1.5.15)
$$V(t, y(\cdot)) = |y| + \int_0^t \int_t^\infty |C(u,s)|\, du\, |y(s)|\, ds,$$
$$V'(t, y(\cdot)) \le -\alpha|y| + |p(t)|.$$

When $p(t) \equiv 0$, (1.5.3) becomes (1.5.1) and we have $V'(t, z(\cdot)) \le -\alpha|z|$ so that $0 \le V(t, z(\cdot)) \le V(0, z(0)) - \alpha \int_0^t |z(s)|\, ds$. We then see that $Z \in L^1[0,\infty)$ and, because of (1.5.9) and (1.5.15), it follows that Z and Z' are bounded; thus $Z(t) \to 0$ as $t \to \infty$.

Problem 1.5.0. System (1.5.15) is an example of one of the central problems in the present theory. One wishes to show that (1.5.15) implies solutions of (1.5.3) are bounded when $\alpha > 0$ and p is bounded.

Exercise 1.5.0. Do Example 1.5.0 for systems. Let A be a constant $n \times n$ matrix all of whose characteristic roots have negative real parts. Find B with $A^T B + BA = -I$. Use

$$V(t, y(\cdot)) = [y^T By]^{1/2} + K \int_0^t \int_t^\infty |C(u, s)|\, du\, |y(s)|\, ds.$$

Example 1.5.1. Let the conditions of Example 1.5.0 hold and let

$$|C(t, s)| \geq \lambda \int_t^\infty |C(u, s)|\, du$$

for some $\lambda > 0$ and $0 \leq s \leq t < \infty$. Then each solution of (1.5.3) is bounded and (1.5.4) has a T-periodic solution.

Proof. Let $K = 1 + \varepsilon$, $\varepsilon > 0$, and define

$$V(t, y(\cdot)) = |y| + K \int_0^t \int_t^\infty |C(u, s)|\, du\, |y(s)|\, ds.$$

For ε small enough we obtain

$$V'(t, y(\cdot)) \leq \left[A(t) + K \int_t^\infty |C(u, t)|\, du \right] |y|$$

$$- \varepsilon \int_0^t |C(t, s)|\, |y(s)|\, ds + |p(t)|$$

$$\leq -(\alpha/2)|y| - \varepsilon\lambda \int_0^t \int_t^\infty |C(u, s)|\, du\, |y(s)|\, ds + |p(t)|$$

$$\leq -\mu V(t, y(\cdot)) + |p(t)|$$

for some $\mu > 0$. We then have the system

(1.5.16)
$$V(t, y(\cdot)) = |y| + K \int_0^t \int_t^\infty |C(u, s)|\, du\, |y(s)|\, ds,$$
$$V'(t, y(\cdot)) \leq -\mu V(t, y(\cdot)) + |p(t)|$$

so that

$$|y(t)| \leq V(t, y(\cdot)) \leq V(0, y(0))e^{-\mu t} + \int_0^t e^{-\mu(t-s)}|p(s)|\, ds$$

$$\leq V(0, y(0))e^{-\mu t} + [\sup |p(t)|]/\mu$$

$$\to \sup |p(t)|/\mu \quad \text{as} \quad t \to \infty,$$

completing the proof.

The condition $|C(t,s)| \geq \lambda \int_t^\infty |C(u,s)|\, du$ is very strong. Essentially, C decreases exponentially.

Exercise 1.5.1. Do Example 1.5.1 for systems.

The next example is taken from Burton, Huang, Mahfoud (1985).

Example 1.5.2. Let the conditions with Example 1.5.0 hold and suppose there is a continuous function $\Phi : [0,\infty) \to [0,\infty)$, $\Phi'(t) \leq 0$, $\Phi \in L^1[0,\infty)$, and $\Phi'(t-s)+|C(t,s)| \leq 0$ for $0 \leq s \leq t < \infty$. If $\Phi(0)+A(t) \leq -\alpha < 0$ then all solutions of (1.5.3) are bounded and (1.5.4) has a T-periodic solution.

Proof. Define

$$V(t, y(\cdot)) = |y| + \int_0^t \Phi(t-s)|y(s)|\, ds$$

and obtain

$$V'(t, y(\cdot)) \leq A(t)|y| + \int_0^t |C(t,s)|\, |y(s)|\, ds + |p(t)|$$

$$+ \Phi(0)|y| + \int_0^t \Phi'(t-s)|y(s)|\, ds \leq -\alpha|y| + |p(t)|$$

$$\leq -\alpha|y| + M, \qquad M > 0.$$

In summary,

(1.5.17)
$$V(t, y(\cdot)) = |y| + \int_0^t \Phi(t-s)|y(s)|\, ds,$$
$$V'(t, y(\cdot)) \leq -\alpha|y| + M.$$

We will use Laplace transforms to obtain a variation of parameters inequality. For brevity we write

(1.5.17)
$$V(t) = |y| + \int_0^t \Phi(t-s)|y(s)|\, ds,$$
$$V'(t) \leq -\alpha|y| + M.$$

and consider the "homogeneous" system

(1.5.18)
$$H(t) = \beta(t) + \int_0^t \Phi(t-s)\beta(s)\, ds,$$
$$H'(t) = -\alpha\beta(t), \qquad \alpha > 0.$$

Then $H(t) = H(0) - \alpha \int_0^t \beta(s)\, ds$ so that

(1.5.19)
$$\beta(t) + \int_0^t [\alpha + \Phi(t-s)]\beta(s)\, ds = H(0)$$

and we take $\beta(0) = H(0) = 1$. It is easy to show that $\beta(t) \in L^1[0, \infty)$, $\beta(t) \geq 0$, and $\beta(t) \to 0$ as $t \to \infty$. The Laplace transform of (1.5.19) is

$$L(\beta) + L(\alpha)L(\beta) + L(\Phi)L(\beta) = L(1),$$

$$L(\beta)[1 + (\alpha/s) + L(\Phi)] = 1/s,$$

or

(1.5.20) $$L(\beta) = 1/[1 + L(\Phi) + (\alpha/s)]s.$$

In (1.5.17) define a positive continuous function η by $V'(t) = -\alpha|y| + M - \eta(t)$ and take the Laplace transform of (1.5.17). We have

$$sL(V) = V(0) - \alpha L(|y|) + (M/s) - L(\eta),$$

$$L(V) = [V(0) - \alpha L(|y|) + (M/s) - L(\eta)]/s,$$

and

$$L(V) = L(|y|) + L(\Phi)L(|y|)$$

so that

$$L(|y|) + L(\Phi)L(|y|) = [V(0) - \alpha L(|y|) + (M/s) - L(\eta)]/s$$

or

(1.5.20) $$L(|y|) = [V(0) + (M/s) - L(\eta)]/[1 + L(\Phi) + (\alpha/s)]s.$$

Thus,

$$L(|y|) = V(0)L(\beta) + (M/s)L(\beta) - L(\eta)L(\beta)$$
$$= L(\beta V(0)) + L(M * \beta) - L(\eta * \beta)$$
$$= L\left(\beta V(0) + \int_0^t M\beta(s)\, ds - \int_0^t \eta(t-s)\beta(s)\, ds\right).$$

Hence,

$$|y(t)| = V(0)\beta(t) + \int_0^t M\beta(s)\, ds - \int_0^t \eta(t-s)\beta(s)\, ds$$

so that

(1.5.21) $$|y(t)| \leq |y(0)|\beta(t) + M\int_0^t \beta(s)\, ds,$$

a variation of parameters inequality. This completes the proof.

Corollary. *Let* $\Phi : [0, \infty) \to [0, \infty)$, $\Phi'(t) \leq 0$, $\Phi \in L^1[0, \infty)$, $\alpha > 0$, *and* $M > 0$. *If*

$$V(t) = |y| + \int_0^t \Phi(t - s)|y(s)| \, ds,$$
$$V'(t) \leq -\alpha|y| + M,$$

and if

$$\beta(t) + \int_0^t [\alpha + \Phi(t - s)]\beta(s) \, ds = 1,$$

then

$$|y(t)| \leq |y(0)|\beta(t) + M \int_0^t \beta(s) \, ds;$$

moreover, $\beta \in L^1[0, \infty)$ *and* $\beta(t) \to 0$ *as* $t \to \infty$.

Exercise 1.5.3. Do Example 1.5.2 for systems.

We next point out that care must be taken in our deductions concerning pairs such as (1.5.17).

Example 1.5.3. Consider the pair

(1.5.22)
$$V(t) \leq \alpha(t) + \int_0^t \alpha(s) \, ds,$$
$$V'(t) = -\alpha(t) + 1$$

and let $\eta(t) \geq 0$ with $\int_0^\infty \eta(s) \, ds \leq \dfrac{1}{2}$ and $\eta(n) = n$, n any integer. Let $\alpha(t) = 1 + \eta(t)$ and $V(t) = 1 - \int_0^t \eta(s) \, ds$. Then (1.5.22) is satisfied and $\alpha(t)$ is unbounded.

Example 1.5.4. Let the conditions with Example 1.5.0 hold and suppose there is a function $\Phi : [0, \infty) \to [0, \infty)$ with $\Phi'(t) \leq 0$, $\Phi \in L^1[0, \infty)$, and $\Phi(t - s) \geq \int_t^\infty |C(u, s)| \, du$ for $0 \leq s \leq t < \infty$. Then solutions of (1.5.3) are bounded and (1.5.4) has a T-periodic solution. (Note that we have not asked $\Phi(0) + \int_t^\infty |C(u, t)| \, du \leq -\alpha < 0$.)

Proof. We have

$$V(t, y(\cdot)) \leq |y| + \int_0^t \Phi(t - s)|y(s)| \, ds$$

and

$$V'(t, y(\cdot)) \leq -\alpha|y| + M, \qquad \alpha > 0, \qquad M > 0.$$

Let $V(t) = V(t, y(\cdot))$ and consider any $t > 0$ for which $V(s) \leq V(t)$ if $0 \leq s \leq t$ and write $[V'(s) \leq -\alpha|y(s)| + M]\Phi(t - s)$. An integration yields

$$\int_0^t V'(s)\Phi(t - s)\, ds \leq -\alpha \int_0^t \Phi(t - s)|y(s)|\, ds + \int_0^t M\Phi(s)\, ds.$$

Let $\int_0^\infty \Phi(s)\, ds = K$, rewrite the last inequality, and integrate by parts to obtain

$$\alpha \int_0^t \Phi(t - s)|y(s)|\, ds \leq MK - \int_0^t V'(s)\Phi(t - s)\, ds$$

$$\leq MK - V(t)\Phi(0) + V(0)\Phi(t) - \int_0^t V(s)\Phi'(t - s)\, ds$$

$$\leq MK - V(t)\Phi(0) + V(0)\Phi(t) - V(t)\int_0^t \Phi'(t - s)\, ds$$

$$\leq MK - V(t)\Phi(0) + V(0)\Phi(t) + V(t)[\Phi(0) - \Phi(t)]$$

$$\leq MK.$$

Hence, at such a maximum

$$\int_0^t \Phi(t - s)|y(s)|\, ds \leq MK/\alpha.$$

Moreover, at such a maximum $V'(t) \geq 0$ so that $|y(t)| \leq M/\alpha$. This means that $|y(t)| \leq V(t) \leq (M/\alpha) + (MK/\alpha)$ is the largest $V(t)$ can become past $t = 0$. Hence,

$$(1.5.23) \qquad |y(t)| \leq \max\{[M(1 + K)/\alpha], |y(0)|\},$$

as required.

Problem 1.5.1. Under the assumptions of Example 1.5.4 show that there exists $B = B(M) > 0$ so that $|y(t, 0, y_0)| < B$ for t sufficiently large. Then show that for each $H > 0$ there exists S such that $|y_0| \leq H$ and $t \geq S$ imply $|y(t, 0, y_0)| < B$.

Remark 1.5.3. The great deficiency of the theory is that the variation of parameters formula (1.5.5) contains $R(t, s)$ instead of $Z(t - s)$, a deficiency also appearing in ordinary differential equations with variable coefficients. This means that $Z \in L^1[0, \infty)$ does not seem to imply $y(t)$ bounded for bounded p. Notice that in Exercise 1.2.2 if $\int_0^\infty |P(t)e^{Jt}|\, dt < \infty$, then $\int_0^t |P(t)e^{J(t-s)}P^{-1}(s)|\, ds \leq M$ for some M and all t.

Problem 1.5.2. Let A be T-periodic and $C(t+T, s+T) = C(t, s)$. Prove or give a counterexample that if $\int_0^\infty |Z(t)|\, dt < \infty$, then there is an $M > 0$ with $\int_0^t |R(t, s)|\, ds \le M$ for $0 \le t < \infty$. A positive result would bring the variation of parameters formula for equations

$$x' = Ax + p(t), \qquad\qquad\qquad\qquad A \text{ constant,}$$

$$y' = Ay + \int_0^t C(t-s)y(s)\, ds + p(t), \qquad\qquad A \text{ constant,}$$

$$x' = A(t)x + p(t), \qquad\qquad\qquad\qquad A(t+T) = A(t),$$

and

$$y' = A(t)y + \int_0^t C(t, s)y(s)\, ds + p(t)$$

onto the same level. Integrability of the principal solution matrix of the unforced equation and boundedness of p would imply boundedness of y. Thus, integrability of the principal matrix solution and integrability of C would imply a periodic solution for the equation with infinite delay when $p(t+T) = p(t)$. The fact that $R(t, s)$ is the resolvent must be brought into play for a positive result. Langenhop has constructed the example

$$R(t, s) = \{[\cos^2(t-s)]/[1 + (t-s)^2]\} + [1 - \cos^2(t-s)]\sin^2[2\pi s/T]$$

satisfying $R(t, t) = 1$ and $R(t, 0) \in L^1[0, \infty)$, but $\int_0^t |R(t, s)|\, ds$ unbounded. One might try substituting $R(t, s)$ into (1.5.1), integrating, and interchanging the order of integration. The fact that $R(t+T, s+T) = R(t, s)$ may then play some role.

Remark 1.5.4. The natural way to consider (1.5.5) is to say that if Z is bounded and if $\int_0^t |R(t, s)|\, ds \le M$ for $t \ge 0$ then for p bounded we have $y(t, 0, y_0)$ bounded. But Perron (1930) viewed things from the opposite direction for ordinary differential equations and obtained a beautiful and very useful result. The proof here is C. E. Langenhop's variation of Hale's (1969) formulation.

Theorem.   Perron   *Let $R(t, s)$ be an $n \times n$ matrix of continuous real valued functions on $0 \le s \le t < \infty$. Let $B = \{f : [0, \infty) \to R^n \mid f$ is continuous and bounded on $[0, \infty)\}$. If $\int_0^t R(t, s)f(s)\, ds \in B$ for each $f \in B$ then there exists M such that $\int_0^t |R(t, s)|\, ds < M$ for $0 \le t < \infty$.*

Proof. For $f \in B$ define

$$T_t f = \int_0^t R(t, s)f(s)\, ds\,.$$

If we use $\|f\| = \sup_{t \geq 0} |f(t)|$, then B is a Banach space. Consider the family $\{T_t\}$ as a family of operators on B to R^n. For each fixed t it follows that $T_t : B \to R^n$ is linear and bounded since

$$|T_t f| \leq \int_0^t |R(t,s)| \, |f(s)| \, ds \leq t K_t \|f\|$$

where $K_t = \max_{0 \leq s \leq t} |R(t,s)|$. Also, the family $\{T_t\}$ is pointwise bounded; that is, for each $f \in B$ there exists K_f such that $|T_t f| \leq K_f$. (This is true since $T_t f \in B$.) Hence, by the uniform boundedness principle there exists K such that for all $t \geq 0$ and for all $f \in B$ we have $|T_t f| \leq K \|f\|$; that is,

$$(1.5.24) \qquad \left| \int_0^t R(t,s) f(s) \, ds \right| \leq K \|f\|$$

for $t \geq 0$ and $f \in B$.

Now fix t, i and j ($1 \leq i, j \leq n$) and define $f_t(s)$ to be the $n \times 1$ vector function whose entries are all zero except for the jth which is taken as $\operatorname{sgn} R_{ij}(t,s)$, where $R(t,s) = (R_{ij}(t,s))$. Take $f_t(s) = 0$ for $s > t$. Now define (for $r > 0$) the function $f_{t,r}(s) = r \int_s^{s+(1/r)} f_t(\sigma) \, d\sigma$, $s \geq 0$. Since $|f_t(\sigma)| \leq 1$ for all σ we have $|f_{t,r}(s)| \leq 1$ for all s. Moreover, $f_{t,r}(s)$ is continuous on $s \geq 0$ and $\lim_{r \downarrow 0} f_{t,r}(s) = f_t(s)$ at all points s of continuity of $f_t(s)$. The points of discontinuity are at most points s where $R_{ij}(t,s)$ changes sign and has Legesgue measure zero. Since

$$|R(t,s) f_{t,r}(s)| \leq |R(t,s)|$$

and $\int_0^t |R(t,s)| \, ds < \infty$, we can apply the Lebesgue dominated convergence theorem and conclude that

$$\lim_{r \downarrow 0} \int_0^t R(t,s) f_{t,r}(s) \, ds = \int_0^t R(t,s) f_t(s) \, ds$$

$$= \int_0^t \begin{bmatrix} R_{1j}(t,s) & \operatorname{sgn} & R_{ij}(t,s) \\ & \vdots & \\ R_{nj}(t,s) & \operatorname{sgn} & R_{ij}(t,s) \end{bmatrix} ds.$$

If we use $|x| = \max_{1 \leq i \leq n} |x_i|$, then the preceeding implies that

$$\int_0^t |R_{ij}(t,s)| \, ds = \left| i\text{th component of } \int_0^t R(t,s) f_t(s) \, ds \right|$$

$$\leq \left| \int_0^t R(t,s) f_t(s) \, ds \right|$$

$$= \lim_{r \downarrow 0} \left| \int_0^t R(t,s) f_{t,r}(s) \, ds \right|.$$

But $f_{t,r} \in B$ and $\|f_{t,r}\| \leq 1$ so we have by (1.5.24) that

$$(1.5.25) \qquad \int_0^t |R_{ij}(t,s)|\, ds \leq \lim_{r \downarrow 0} K \cdot 1 = K\,.$$

But $t \geq 0$ is arbitrary and so are i and j. For $|x|$ as defined we have

$$|R(t,s)| \leq \sum_i \sum_j |R_{ij}(t,s)|\,.$$

By (1.5.25) it follows that

$$\int_0^t |R(t,s)|\, ds \leq n^2 K \quad \text{for all } t \geq 0\,.$$

This completes the proof.

Example 1.5.5. Let the conditions of Example 1.5.4 hold. Then there is an $M > 0$ such that if $|p(t)| \leq p_0$, then for any $L > 0$ there exists S such that $|y_0| \leq L$ and $t \geq S$ implies $|y(t, 0, y_0)| \leq Mp_0 + 1$.

Proof. Example 1.5.4 and inequality (1.5.23) show that $|y(t, 0, y_0)|$ is bounded for every bounded p.   Thus, Eq. (1.5.5) shows that $|\int_0^t R(t,s)p(s)\, ds|$ is bounded for every bounded p. By Perron's theorem there exists M with $\int_0^t |R(t,s)|\, ds \leq M$. By (1.5.5) the result follows since $Z(t) \to 0$ as $t \to \infty$.

The conclusion of Example 1.5.5 is a type of ultimate boundedness. It plays a central role in establishing the existence of periodic solutions using fixed-point theorems.

We now give an example using a hybrid Liapunov-Razumikhin technique. This result was formulated by Grimmer and Siefert (1975), but this is a different proof.

Example 1.5.6. Let A be a constant with all characteristic roots having negative real parts. Let $A^T B + BA = -I$ with $B = B^T$ and with α^2 and β^2 being the smallest and largest characteristic roots of B. If $\int_0^t |BC(t,s)|\, ds \leq M$ for $0 \leq t < \infty$ and $2\beta M/\alpha < 1$, then all solutions of (1.5.3) are bounded.

Proof. If the result is false, there is a solution $y(t)$ with $\limsup_{t \to \infty} y^T(t)By(t) = +\infty$. Thus, there are values of t with $|y(t)|$ as large as we please and $[y^T(t)By(t)]' \geq 0$, say at $t = S$, and $y^T(t)By(t) \leq y^T(S)By(S)$ if $t \leq S$. Hence, at $t = S$ we have

$$[y^T(t)By(t)]' = -y^T(t)y(t)$$
$$+ \int_0^t 2y^T(s)C^T(t,s)By(t)\, ds + 2p^T(t)By(t) \geq 0$$

or

$$y^T(S)y(S) \leq \int_0^S 2y^T(s)C^T(S,s)By(S)\,ds + 2p^T(S)By(S)$$

$$\leq 2|y(S)| \int_0^S |BC(S,s)|\,|y(s)|\,ds + 2y^T(S)Bp(S)$$

$$\leq 2|y(S)| \int_0^S |BC(S,s)| \left[(y^T(s)By(s))^{1/2}/\alpha \right] ds + 2y^T(S)Bp(S)$$

$$\leq (2/\alpha)|y(S)|(y^T(S)By(S))^{1/2} \int_0^S |BC(S,s)|\,ds + 2y^T(S)Bp(S)$$

$$\leq (2/\alpha)|y(S)|\,\beta\,|y(S)|\,M + 2y^T(S)Bp(S)$$

$$= (2\beta M/\alpha)|y(S)|^2 + 2y^T(S)Bp(S)\,.$$

This is a contradiction for large $y(S)$, and the proof is complete.

Our examples to this point have tended to obtain boundedness from the part $y' = A(t)y$, treating $\int_0^t C(t,s)y(s)\,ds$ as a perturbation. This is only for purposes of simple illustrations. In later sections we use the integral of C to establish boundedness.

Example 1.5.7. Let (1.5.3) be a scalar equation and suppose there is an $\alpha : [0,\infty) \to R$ and a positive constant b with $A(t) + \alpha(t) \leq -b$. Suppose also that there is a continuous scalar function $\Phi(t,s) \geq 0$ for $0 \leq s \leq t < \infty$ with

$$\alpha(s)e^{-b(t-s)} - \int_s^t e^{-b(t-u)}\,|C(u,s)|\,du \geq -\Phi(t-s)$$

for $0 \leq s \leq t < \infty$ and that $\int_0^t \Phi(t,s)\,ds \leq P < 1$ for $0 \leq t < \infty$. Then all solutions of (1.5.3) are bounded for p bounded and, under the periodicity assumptions, (1.5.4) has a T-periodic solution.

Proof. Briefly, define

$$V(t,y(\cdot)) = |y| + \int_0^t \left[\alpha(s)e^{-b(t-s)} \right.$$

$$\left. - \int_s^t e^{-b(t-u)}\,|C(u,s)|\,du \right] |y(s)|\,ds$$

and obtain $V'(t, y(\cdot)) \leq -bV + |p(t)|$. Now, let $y(t)$ be a solution of (1.5.3) with the property that $|y(S)| \geq |y(t)|$ for $0 \leq t \leq S$. Then

$$|y(S)|[1 - P] = |y(S)| - P|y(S)|$$

$$\leq |y(S)| - \int_0^S \Phi(S, s)\, |y(S)|\, ds$$

$$\leq |y(S)| - \int_0^S \Phi(S, s)\, |y(s)|\, ds$$

$$\leq |y(S)| + \int_0^S \left[\alpha(s)e^{-b(S-s)} \right.$$

$$\left. - \int_s^S e^{-b(S-u)}\, |C(u, s)|\, du \right] |y(s)|\, ds$$

$$\leq V(S, y(\cdot))$$

$$\leq V(0, y(0))e^{-bS} + \int_s^S e^{-b(S-s)}\, |p(s)|\, ds\,,$$

from which the result follows.

1.6   Stability and Boundedness

This section is devoted to the study of stability and boundedness properties of solutions of

$$(1.6.1) \qquad\qquad x' = A(t)x + \int_0^t C(t, s)x(s)\, ds$$

in which A is an $n \times n$ matrix of functions continuous for $0 \leq t < \infty$ and $C(t, s)$ is an $n \times n$ matrix of functions continuous for $0 \leq s \leq t < \infty$.

For any $t_0 \geq 0$ and any continuous function $\phi : [0, t_0] \to R^n$, a solution of (1.6.1) is a function $x : [0, \infty) \to R^n$ satisfying (1.6.1) for $t \geq t_0$ and such that $x(t) = \phi(t)$ for $t \in [0, t_0]$. Under our stated conditions (1.6.1) has a unique solution $x(t, t_0, \phi)$ defined on $[t_0, \infty)$.

Definition 1.6.1. *The solution $x = 0$ of* (1.6.1) *is*

- (i) stable *if for every $\varepsilon > 0$ and every $t_0 \geq 0$, there exists a $\delta = \delta(\varepsilon, t_0) > 0$ such that $|\phi(t)| < \delta$ on $[0, t_0]$ implies $|x(t, t_0, \phi)| < \varepsilon$ for $t \geq t_0$;*

- (ii) uniformly stable *if it is stable and the δ in the definition of stability is independent of t_0;*

- (iii) asymptotically stable *if it is stable and for each $t_0 \geq 0$ there is a $\beta = \beta(t_0) > 0$ such that $|\phi(t)| < \beta$ on $[0, t_0]$ implies $x(t, t_0, \phi) \to 0$ as $t \to \infty$;*

(iv) uniformly asymptotically stable *if it is uniformly stable, the β in the definition of asymptotic stability is independent of t_0, and for each $\eta > 0$ there is a $T = T(\eta) > 0$ such that $|\phi(t)| < \beta$ on $[0, t_0]$ implies $|x(t, t_0, \phi)| < \eta$ for $t \geq t_0 + T$.*

An $n \times n$ matrix is said to be *stable* if all of its characteristic roots have negative real parts. Also, when a function is written without its argument, then it is understood that the argument is t.

In Section 1.3 we mentioned that when $C(t, s) = C(t-s)$, $C \in L^1[0, \infty)$, and A is constant then Grossman and Miller (1973) show that the principal matrix solution $Z(t)$ of (1.6.1) is $L^1[0, \infty)$ if and only if

$$\det [Is - A - C^*(s)] \neq 0 \quad \text{for} \quad \text{Re } s \geq 0$$

where C^* is the Laplace transform of C. Miller (1971) (cf. Burton, 1983a, p. 47) also shows that, under these conditions, $Z \in L^1[0, \infty)$ if and only if the zero solution is uniformly asymptotically stable. In view of Theorem 1.4.1 we are very much interested in uniform asymptotic stability.

The results on boundedness and stability of Section 1.3 rest primarily on the assumption that $x' = A(t)x$ is strongly stable and that $C(t, s)$ is small in some sense so that the integral in (1.6.1) is a small perturbation of $x' = A(t)x$. We begin the same way here and gradually work around to a full use of $C(t, s)$ to stabilize the equation.

It should be understood that the determinant condition of Grossman and Miller is a transcendental equation, that there is no Routh–Hurwitz criterion for locating the roots, and that it holds only for A constant and C of convolution type. Our discussion of stability relies on construction of Liapunov functionals.

Before getting down to the main stability results we indicate one way in which Liapunov functionals may be derived. As we discuss in Chapter 4, given an ordinary differential equation $x' = F(x)$, if we can find a first integral which is positive definite, then it serves as a Liapunov function. Many of the most useful Liapunov functions are the result of small perturbations of a first integral Liapunov function. We now present a parallel illustration for (1.6.1). A rough "algorithm" can be given as follows;

(a) Integrate (1.6.1) from 0 to t.

(b) Interchange the order of integration in

$$\int_0^t \int_0^u C(u, s)x(s) \, ds \, du.$$

(c) Take norms of the result in (a) and (b).

We now give the details. Let $x(t)$ be a solution of (1.6.1) on $[0, \infty]$ and obtain

$$x(t) = x(0) + \int_0^t A(s)x(s)\,ds + \int_0^t \int_s^t C(u,s)\,du\,x(s)\,ds.$$

Thus, the functional

$$(1.6.2) \qquad h(t, x(\cdot)) = x(t) + \int_0^t \left[-A(s) - \int_s^t C(u,s)\,du \right] x(s)\,ds$$

is identically constant and so its derivative is zero. We can call $h(t, x(\cdot))$ a first integral of (1.6.1).

Now h may serve as a suitable Liapunov functional as it stands, or as in the case of ordinary differential equations, with certain fairly obvious changes it may be converted into an excellent Liapunov functional. As a beginning, we suppose (1.6.1) is scalar,

$$A(t) \leq 0, \qquad C(t,s) \geq 0, \quad \text{and} \qquad -A(s) - \int_s^t C(u,s)\,du \geq 0$$

for $0 \leq s \leq t < \infty$. Note that if $x(0) > 0$ then $x(t) > 0$ on $[0, \infty)$ and that if $x(t)$ is a solution so is $-x(t)$; we therefore suppose $x(t) > 0$ on $[0, \infty)$. Hence, $h(t, x(\cdot))$ is a positive definite functional which can be written as

$$(1.6.3) \qquad H(t, x(\cdot)) = |x(t)| + \int_0^t \left[|A(s)| - \int_s^t |C(u,s)|\,du \right] |x(s)|\,ds$$

and its derivative is zero. Moreover, if $|A(s)| - \int_s^t |C(u,s)|\,du \geq \alpha > 0$, then $H'(t, x(\cdot)) \leq 0$ implies H bounded so that $\int_0^t \alpha |x(s)|\,ds$ is bounded; thus $x \in L^1[0, \infty)$.

Having arrived at H, we can now drop the assumption that $C(t,s) \geq 0$, that x is a solution on $[0, \infty)$ instead of an arbitrary interval $[t_0, \infty)$, and that (1.6.1) is scalar. We formalize our findings in the next theorem. Since we reduce the conditions on $C(t, s)$, $x(t)$ might vanish and so $|x(t)|$ becomes nondifferentiable. In that case the derivative of H is interpreted as the upper right-hand derivative and is discussed quite fully in Section 4.2 for a general functional which is locally Lipschitz. But for now we simply consider the separate cases of $x(t)$ positive or negative.

Theorem 1.6.1. *Let* (1.6.1) *be a scalar equation with* $A(s) \leq 0$ *and*

$$|A(s)| - \int_s^t |C(u,s)|\,du \geq 0 \quad \text{for} \quad 0 \leq s \leq t < \infty.$$

Then the zero solution of (1.6.1) is stable. If, in addition, there is a $t_2 \geq 0$ and an $\alpha > 0$ with $|A(s)| - \int_s^t |C(u,s)|\, du \geq \alpha$ for $t_2 \leq s \leq t < \infty$, and if both $\int_0^t |C(t,s)|\, ds$ and $A(t)$ are bounded, then $x = 0$ is asymptotically stable.

Proof. Let $x(t) = x(t, t_0, x_0)$ be a solution of (1.6.1) on $[t_0, \infty)$. Then

$$H'_{(1.6.1)}(t, x(\cdot)) \leq A(t)|x| + \int_0^t |C(t,s)|\, |x(s)|\, ds$$

$$+ |A(t)|\, |x| - \int_0^t |C(t,s)|\, |x(s)|\, ds \equiv 0.$$

Given $\varepsilon > 0$ and $t_0 \geq 0$, let $\phi : [0, t_0] \to R$ be continuous with $|\phi(t)| < \delta$ and $\delta > 0$ to be determined. Then

$$|x(t)| \leq H(t, x(\cdot)) \leq H(t_0, \phi(\cdot))$$

$$\leq |\phi(t_0)| + \int_0^{t_0} \left[|A(s)| - \int_s^{t_0} |C(u,s)|\, du \right] |\phi(s)|\, ds$$

$$\leq \delta \left\{ 1 + \int_0^{t_0} \left[|A(s)| - \int_s^{t_0} |C(u,s)|\, du \right] ds \right\}$$

$$< \varepsilon$$

if $\delta = \beta(\varepsilon, t_0)$ is small enough.

If t_2 and α exist, then $H' \leq 0$ implies that

$$|x(t)| + \int_{t_2}^t \alpha |x(s)|\, ds \leq H(t, x(\cdot)) \leq H(t_0, \phi(\cdot))$$

so that $x \in L^1[0, \infty)$. We also have $x'(t)$ bounded so $x(t) \to 0$ as $t \to \infty$. This completes the proof.

The form of H leads us naturally to some very general functionals. When x is a vector we may interpret $|x|$ in many different ways. Our treatment of the linear constant coefficient system

$$x' = Ax$$

by way of

$$V(x) = x^T B x$$

with

$$A^T B + B A = -I$$

suggests that $|x|$ in $H(t, x(\cdot))$ be interpreted as

$$[x^T B x]^{1/2}.$$

To be definite, let A be a constant $n \times n$ matrix all of whose characteristic roots have negative real parts. Find $B = B^T$ with

$$(1.6.4) \qquad\qquad A^T B + BA = -I.$$

Write $H(t, x(\cdot))$ as

$$(1.6.5) \qquad V(t, x(\cdot)) = [x^T Bx]^{1/2} + \int_0^t \Phi(t, s)|x(s)| \, ds$$

where Φ is to be determined and where $|x|$ is the Euclidean norm of x. There are then positive constants k, K, and r with

$$(1.6.6) \qquad\begin{aligned} |x| &\geq 2k[x^T Bx]^{1/2}, \qquad |Bx| \leq K[x^T Bx]^{1/2}, \\ r|x| &\leq [x^T Bx]^{1/2}. \end{aligned}$$

Taking the derivative of V along solutions of (1.6.1) yields

$$\begin{aligned} V'_{(1.6.1)}&(t, x(\cdot)) \\ &= \left[-x^T x + 2 \int_0^t x^T(s) C^T(t, s) \, ds \, Bx \right] \Big/ 2[x^T Bx]^{1/2} \\ &\quad + \Phi(t, t)|x| + \int_0^t [\partial \Phi(t, s)/\partial t] \, |x(s)| \, ds \\ &\leq [-k + \Phi(t, t)] \, |x| + \int_0^t [K \, |C(t, s)| + \Phi_t(t, s)] \, |x(s)| \, ds. \end{aligned}$$

We would like to make

$$\Phi_t(t, s) + |C(t, s)| \, K < 0 \, ;$$

and this can be done if we choose $\bar{K} > K$ and

$$(1.6.7) \qquad\qquad \Phi(t, s) = \bar{K} \int_t^\infty |C(u, s)| \, du \, ,$$

provided the integral converges. We then have

$$(1.6.8) \qquad V(t, x(\cdot)) = [x^T Bx]^{1/2} + \bar{K} \int_0^t \int_t^\infty |C(u, s)| \, du \, |x(s)| \, ds$$

with

$$\begin{aligned} V'_{(1.6.1)}(t, x(\cdot)) &\leq \left[-k + \bar{K} \int_t^\infty |C(u, t)| \, du \right] |x| \\ &\quad - (\bar{K} - K) \int_0^t |C(t, s)| \, |x(s)| \, ds \, . \end{aligned}$$

Theorem 1.6.2. *In* (1.6.1) *let A be constant with all characteristic roots having negative real parts, $B = B^T$ positive definite with $A^T B + BA = -I$, and let* (1.6.6) *hold. Suppose there is a $\bar{K} \geq K$ and $\bar{k} \geq 0$ with*

$$\bar{k} \leq k - \bar{K} \int_0^\infty |C(u,t)| \, du \quad for \quad 0 \leq t < \infty.$$

(a) *Then the zero solution of* (1.6.1) *is stable.*

(b) *If $\bar{K} > K$ and $\bar{k} > 0$, then $x = 0$ is asymptotically stable.*

(c) *If $\int_0^t \int_t^\infty |C(u,s)| \, du \, ds$ is bounded for $0 \leq t < \infty$, then $x = 0$ is uniformly stable.*

(d) *Suppose* (b) *and* (c) *hold. If for each $\rho > 0$ there exists $S > 0$ such that $P \geq S$ and $t \geq 0$ imply $\int_0^t \int_{t+P}^\infty |C(u,s)| \, du \, ds < \rho$, then $x = 0$ is uniformly asymptotically stable.*

Proof of (a) *and* (c). Let $\varepsilon > 0$ and $t_0 \geq 0$ be given. We must find $\delta > 0$ such that $[|\phi(t)| < \delta$ on $[0, t_0]$ and $t \geq t_0]$ imply that $|x(t, t_0, \phi)| < \varepsilon$. Since $V' \leq 0$ we have

$$r|x(t)| \leq V(t, x(\cdot)) \leq V(t_0, \phi(\cdot))$$
$$= [\phi(t_0)^T B \phi(t_0)]^{1/2} + \bar{K} \int_0^{t_0} \int_{t_0}^\infty |C(u,s) \, du \, |\phi(s)| \, ds$$
$$\leq (\delta/2k) + \delta \int_0^{t_0} \int_{t_0}^\infty \bar{K}|C(u,s)| \, du \, ds < \varepsilon r$$

if $\delta = \delta(t_0, \varepsilon)$ is small enough. Moreover, if the integral in (c) is bounded, say by M, then

$$\delta = \varepsilon r / [(1/2k) + \bar{K}M]$$

is the relation for uniform stability. This completes the proof of (a) and (c).

Proof of (b). By standard manipulation we find that

(1.6.9) $$V'(t, x(\cdot)) \leq -\mu[|x| + |x'|]$$

for some $\mu > 0$. This turns out to be a fundamental type of relation seen throughout Chapter 4. Since $V(t, x(\cdot)) \geq 0$, it follows that $x \in L^1[0, \infty)$ and that x has finite Euclidean length. This implies that $x(t) \to 0$ as $t \to \infty$ and the proof of (b) is complete.

The proof of (d) is long, can be found in Burton (1983a; pp. 40-41), and the types of arguments are repeated several times in Chapter 4; thus, we do not give it here.

These two theorems have depended on $x' = A(t)x$ being stable and have treated $\int_0^t C(t,s)x(s)\,ds$ as a perturbation. But stability can be derived from C by variants of the following technique. Write (1.6.1) as

$$(1.6.10) \qquad x' = A(t)x - G(t,t)x + (d/dt)\int_0^t G(t,s)x(s)\,ds$$

where

$$(1.6.11) \qquad \partial G(t,s)/\partial t = C(t,s)\,.$$

Integrate (1.6.10) from 0 to t obtaining

$$x(t) - \int_0^t G(t,s)x(s)\,ds = x(0) + \int_0^t [A(s) - G(s,s)]x(s)\,ds$$

or

$$h(t,x(\cdot)) = x(t) - \int_0^t G(t,s)x(s)\,ds + \int_0^t [-A(s) + G(s,s)]x(s)\,ds$$

whose derivative is identically zero. If we proceed as we did in the course of obtaining Theorems 1.6.1 and 1.6.2, then we will obtain stability and instability results using $A(t)$ or $C(t,s)$. Such work was introduced in Burton (1983a, Chapter 5) and was refined in Burton and Mahfoud (1983, 1984). The next few results give the flavor of that work.

Theorem 1.6.3. *Let* (1.6.1) *be scalar and* $\int_t^\infty |C(u,t)|\,du$ *be continuous for* $t \geq 0$. *Suppose there is an* $\alpha > 0$ *with*

$$(1.6.12) \qquad \int_0^t |C(t,s)|\,ds + \int_t^\infty |C(u,t)|\,du - 2|A(t)| \leq -\alpha\,.$$

Then (1.6.1) *is stable if and only if* $A(t) < 0$.

Proof. Suppose $A(t) < 0$ and consider the functional

$$V(t,x(\cdot)) = x^2 + \int_0^t \int_t^\infty |C(u,s)|\,du\,x^2(s)\,ds\,.$$

The derivative of $V(t, x(\cdot))$ along a solution $x(t)$ of (1.6.1) satisfies

$$V'_{(1.6.1)}(t, x(\cdot)) \leq 2Ax^2 + 2\int_0^t |C(t,s)|\,|x(s)|\,|x|\,ds$$
$$+ \int_t^\infty |C(u,t)|\,du\,x^2 - \int_0^t |C(t,s)|\,x^2(s)\,ds$$
$$\leq 2Ax^2 + \int_0^t |C(t,s)|\,(x^2(s) + x^2)\,ds$$
$$+ \int_t^\infty |C(u,t)|\,du\,x^2 - \int_0^t |C(t,s)|\,x^2(s)\,ds$$
$$= \left[2A + \int_0^t |C(t,s)|\,ds + \int_t^\infty |C(u,t)|\,du\right]x^2 \leq -\alpha x^2\,.$$

As V is positive definite and $V' \leq 0$ it follows that $x = 0$ is stable. Suppose that $A(t) > 0$ and consider the functional

$$W(t, x(\cdot)) = x^2 - \int_0^t \int_t^\infty |C(u,s)|\,du\,x^2(s)\,ds$$

so that

$$W'_{(1.6.1)}(t, x(\cdot)) \geq 2Ax^2 - 2\int_0^t |C(t,s)|\,|x(s)|\,|x|\,ds$$
$$- \int_t^\infty |C(u,t)|\,du\,x^2 + \int_0^t |C(t,s)|\,x^2(s)\,ds$$
$$\geq 2Ax^2 - \int_0^t |C(t,s)|\,(x^2(s) + x^2)\,ds$$
$$- \int_t^\infty |C(u,t)|\,du\,x^2 + \int_0^t |C(t,s)|\,x^2(s)\,ds$$
$$= \left[2A - \left(\int_0^t |C(t,s)|\,ds + \int_t^\infty |C(u,t)|\,du\right)\right]x^2 \geq \alpha x^2\,.$$

Now, given any $t_0 \geq 0$ and any $\delta > 0$, we can find a continuous function $\phi : [0, t_0] \to R$ with $|\phi(t)| < \delta$ and $W(t_0, \phi(\cdot)) > 0$ so that if $x(t) = x(t, t_0, \phi)$ is a solution of (1.6.1), then we have

$$x^2(t) \geq W(t, x(\cdot)) \geq W(t_0, \phi(\cdot)) + \alpha \int_{t_0}^t x^2(s)\,ds$$

(1.6.13)
$$\geq W(t_0, \phi(\cdot)) + \alpha \int_{t_0}^t W(t_0, \phi(\cdot))\,ds$$
$$= W(t_0, \phi(\cdot)) + \alpha W(t_0, \phi(\cdot))\,(t - t_0)\,.$$

As $t \to \infty$, $|x(t)| \to \infty$. This completes the proof.

Corollary 1. *If* (1.6.12) *holds and* $A(t) < 0$ *and bounded, then the zero solution of* (1.6.1) *is asymptotically stable.*

Proof. We showed in the proof of Theorem 1.6.3 that $V'_{(1.6.1)}(t, x(\cdot)) \leq -\alpha x^2$. This implies that $x^2(t)$ is in $L^1[0, \infty)$ and $x^2(t)$ is bounded. It follows from (1.6.12) and (1.6.1) that $x'(t)$ is bounded. Thus, $x(t) \to 0$ as $t \to \infty$. The proof is now complete

Corollary 2. *If* (1.6.12) *holds and* $A(t) > 0$, *then the zero solution of* (1.6.1) *is completely unstable. Furthermore, for any* $t_0 \geq 0$ *and any* $\delta > 0$ *there is a continuous function* $\phi : [0, t_0] \to R$ *and a solution* $x(t, t_0, \phi)$ *with* $|\phi(t)| < \delta$ *and*

$$|x(t, t_0, \phi)| \geq [c_1 + c_2(t - t_0)]^{1/2}$$

where c_1 *and* c_2 *are positive constants depending on* t_0 *and* ϕ.

Proof. This is an immediate consequence of (1.6.13).

We select a continuous function $G(t, s)$ with

(1.6.14) $\partial G(t, s)/\partial t = C(t, s)$

and let

$$Q(t) = A(t) - G(t, t),$$

so that (1.6.1) may be written as

(1.6.1)′ $x' = Q(t)x + (d/dt) \int_0^t G(t, s)x(s)\,ds\,.$

Theorem 1.6.4. *Suppose* (1.6.14) *holds, that* (1.6.1) *is a scalar equation, and there are constants* Q_1, Q_2, J, *and* R *with* $R < 2$ *such that*

(i) $0 < Q_1 \leq |Q(t)| \leq Q_2,$

(ii) $\int_0^t |G(t, s)|\,ds \leq J < 1,$ *and*

(iii) $\int_0^t |G(t, s)|\,ds + \int_t^\infty |G(u, t)|\,du \leq RQ_1/Q_2$

for $0 \leq t < \infty$. *Furthermore, suppose there is a continuous function* $h : [0, \infty) \to [0, \infty)$ *with* $|G(t, s)| \leq h(t - s)$ *and* $h(u) \to 0$ *as* $u \to \infty$. *Then the zero solution is stable if and only if* $Q(t) < 0$.

Proof. Suppose $Q(t) < 0$ and consider the functional

$$V(t, x(\cdot)) = \left(x - \int_0^t G(t, s) x(s) \, ds \right)^2$$
$$+ Q_2 \int_0^t \int_t^\infty |G(u, s)| \, du \, x^2(s) \, ds \, .$$

The derivative of V along a solution $x(t)$ of $(1.6.1)'$ satisfies

$$V'_{(1.6.1)'}(t, x(\cdot)) = 2 \left(x - \int_0^t G(t, s) x(s) \, ds \right) Q(t) x$$
$$+ Q_2 \int_t^\infty |G(u, t)| \, du \, x^2 - Q_2 \int_0^t |G(t, s)| x^2(s) \, ds$$
$$\leq 2Q x^2 + Q_2 \int_0^t |G(t, s)| \left(x^2(s) + x^2 \right) ds$$
$$+ Q_2 \int_t^\infty |G(u, t)| \, du \, x^2 - Q_2 \int_0^t |G(t, s)| x^2(s) \, ds$$
$$= \left[2Q + Q_2 \left(\int_0^t |G(t, s)| \, ds + \int_t^\infty |G(u, t)| \, du \right) \right] x^2$$
$$\leq [2Q + RQ_1] x^2 \leq [-2Q_1 + RQ_1] x^2 = -\beta x^2, \quad \beta > 0 \, .$$

Let $\varepsilon > 0$ and $t_0 \geq 0$ be given. We propose to find $\delta > 0$ so that if $|\phi(t)| < \delta$ on $[0, t_0]$, then $|x(t, t_0, \phi)| < \varepsilon$ for all $t \geq t_0$.

Since $V'_{(1.6.1)'}(t, x(\cdot)) \leq 0$ for $t \geq t_0$, then

$$V(t, x(\cdot)) \leq V(t_0, \phi(\cdot))$$
$$= \left| \phi(t_0) - \int_0^{t_0} G(t_0, s) \phi(s) \, ds \right|^2 + Q_2 \int_0^{t_0} \int_{t_0}^\infty |G(u, s)| \, du \, \phi^2(s) \, ds$$
$$\leq \delta^2 \left[1 + \int_0^{t_0} |G(t_0, s)| \, ds \right]^2 + Q_2 \delta^2 \int_0^{t_0} \int_{t_0}^\infty |G(u, s)| \, du \, ds \leq \delta^2 N^2$$

where

$$N^2 = (1 + RQ_1/Q_2)^2 + Q_2 \int_0^{t_0} \int_{t_0}^\infty |G(u, s)| \, du \, ds \, .$$

Now,

$$V(t, x(\cdot)) \geq \left(x(t) - \int_0^t G(t, s) x(s) \, ds \right)^2$$
$$\geq \left(|x(t)| - \left| \int_0^t G(t, s) x(s) \, ds \right| \right)^2 \, .$$

Thus

$$\left| |x(t)| - \left| \int_0^t G(t,s)x(s)\,ds \right| \right| \leq \delta N$$

or

$$|x(t)| \leq \delta N + \int_0^t |G(t,s)|\,|x(s)|\,ds\,.$$

Now, so long as $|x(t)| < \varepsilon$, we have

$$|x(t)| < \delta N + \varepsilon \int_0^t |G(t,s)|\,ds \leq \delta N + J\varepsilon < \varepsilon\,,$$

for all $t \geq t_0$ provided $\delta < \varepsilon(1 - J)/N$. Since (1.6.1) and (1.6.1)' are the same equation, then the zero solution of (1.6.1) is stable.

Suppose now that $Q(t) > 0$ and consider the functional

$$W(t,x(\cdot)) = \left(x - \int_0^t G(t,s)x(s)\,ds \right)^2 - Q_2 \int_0^t \int_t^\infty |G(u,s)|\,du\,x^2(s)\,ds\,.$$

Then

$$\begin{aligned}
W'_{(1.6.1)'}(t,x(\cdot)) &= 2\left(x - \int_0^t G(t,s)x(s)\,ds \right) Q(t)x \\
&\quad - Q_2 \int_t^\infty |G(u,t)|\,du\,x^2 + Q_2 \int_0^t |G(t,s)|x^2(s)\,ds \\
&\geq 2Qx^2 - Q_2 \int_0^t |G(t,s)|\,(x^2(s) + x^2)\,ds \\
&\quad - Q_2 \int_t^\infty |G(u,t)|\,du\,x^2 + Q_2 \int_0^t |G(t,s)|x^2(s)\,ds \\
&= \left[2Q - Q_2 \left(\int_0^t |G(t,s)|\,ds + \int_t^\infty |G(u,t)|\,du \right) \right] x^2 \\
&\geq [2Q_1 - RQ_1]x^2 = \gamma x^2
\end{aligned}$$

where $\gamma = 2Q_1 - RQ_1$.

Now, given any t_0 and $\delta > 0$, we can find a continuous function $\phi : [0,t_0] \to R$ with $|\phi(t)| < \delta$ and $W(t_0,\phi(\cdot)) > 0$ so that if $x(t) = x(t,t_0,\phi)$ is a solution of (1.6.1), then we have

$$\left(x(t) - \int_0^t G(t,s)x(s)\,ds \right)^2 \geq W(t,x(\cdot))$$

$$\geq W(t_0,\phi(\cdot)) + \gamma \int_{t_0}^t x^2(s)\,ds\,.$$

We shall show that $x(t)$ is unbounded. If $x(t)$ is not unbounded, then as $\int_0^t |G(t,s)| \, ds$ is bounded, we have $\int_0^t G(t,s)x(s) \, ds$ bounded and hence $x^2(t)$ is in $L^1[0,\infty)$.

Using the Schwartz inequality, we have

$$\left(\int_0^t |G(t,s)| \, |x(s)| \, ds \right)^2 = \left(\int_0^t |G(t,s)|^{1/2} \, |G(t,s)|^{1/2} \, |x(s)| \, ds \right)^2$$

$$\leq \int_0^t |G(t,s)| \, ds \int_0^t |G(t,s)| x^2(s) \, ds$$

$$\leq \int_0^t |G(t,s)| \, ds \int_0^t h(t-s) x^2(s) \, ds \,.$$

The last integral is the convolution of an L^1 function with a function tending to zero. Thus the integral tends to zero as $t \to \infty$ and hence

$$\int_0^t G(t,s)x(s) \, ds \to 0 \quad \text{as } t \to \infty \,.$$

Since

$$\left| x - \int_0^t G(t,s)x(s) \, ds \right| \geq \left[W(t_0, \phi(\cdot)) \right]^{1/2} \,,$$

then for sufficiently large T, $|x(t)| \geq \alpha$ for some $\alpha > 0$ and all $t > T$. This contradicts $x^2(t)$ being in $L^1[0,\infty)$. Thus, $x(t)$ is unbounded and the zero solution of (1.6.1) is unstable. This completes the proof

Now, we would like to extend Theorem 1.6.3 to the system

$$(1.6.15) \qquad\qquad x' = Ax + \int_0^t C(t,s)x(s) \, ds$$

where A and C are $n \times n$ matrices, A is constant, and C is continuous for $0 \leq s \leq t < \infty$.

To this end, we are interested in finding an $n \times n$ symmetric matrix B which satisfies

$$(1.6.16) \qquad\qquad A^T B + BA = -I \,.$$

If all characteristic roots of A have negative real parts, a unique positive definite matrix B can be found. If all characteristic roots of A have positive real parts then a unique negative definite matrix B can also be found.

Moreover, according to Barbashin (1968, p. 1100), if $\gamma_1, \ldots, \gamma_k$ are the characteristic roots of A, then (1.6.16) has a unique symmetric solution provided that $\gamma_i + \gamma_j \neq 0$ for any i and j.

It is easily shown that if A has a characteristic root with zero real part then (1.6.16) has no solution.

One may construct examples of A having $\gamma_i + \gamma_j = 0$ and (1.6.16) having a solution, but not a unique one. Moreover, there are examples in which $\gamma_i + \gamma_j = 0$ and (1.6.16) has no solution; however, in Burton (1983a, Chapter 5) we show that there is a positive definite matrix D such that

$$(1.6.16)' \qquad\qquad A^T B + BA = -D$$

can be solved for B. In the following theorems equation $(1.6.16)'$ is a satisfactory substitute for (1.6.16).

Theorem 1.6.5. *Suppose* (1.6.16) *holds for some symmetric matrix B and that there is a constant $M > 0$ such that*

$$(1.6.17) \qquad |B| \left(\int_0^t |C(t,s)|\, ds + \int_t^\infty |C(u,t)|\, du \right) \le M < 1\,.$$

Then the zero solution of (1.6.15) *is stable if and only if B is positive definite.*

Proof. We consider the functional

$$V(t, x(\cdot)) = x^T B x + |B| \int_0^t \int_t^\infty |C(u,s)|\, du\, |x(s)|^2\, ds\,.$$

Differentiate V along a solution $x(t)$ of (1.6.15) to obtain

$$V'_{(1.6.15)}(t, x(\cdot))$$

$$= \left[x^T A^T + \int_0^t x^T(s) C^T(t,s)\, ds \right] Bx + x^T B \left[Ax + \int_0^t C(t,s) x(s)\, ds \right]$$

$$\quad + |B| \int_t^\infty |C(u,t)|\, du\, |x|^2 - |B| \int_0^t |C(t,s)|\, |x(s)|^2\, ds$$

$$= -|x|^2 + 2x^T B \int_0^t C(t,s) x(s)\, ds$$

$$\quad + |B| \int_t^\infty |C(u,t)|\, du\, |x|^2 - |B| \int_0^t |C(t,s)|\, |x(s)|^2\, ds$$

$$\le -|x|^2 + |B| \int_0^t |C(t,s)|\, \left(|x|^2 + |x(s)|^2 \right) ds$$

$$\quad + |B| \int_t^\infty |C(u,t)|\, du\, |x|^2 - |B| \int_0^t |C(t,s)|\, |x(s)|^2\, ds$$

$$= \left[-1 + |B| \left(\int_0^t |C(t,s)|\, ds + \int_0^\infty |C(u,s)|\, du \right) \right] |x|^2$$

$$\le [-1 + M]\, |x|^2 = -\alpha |x|^2$$

where $\alpha = 1 - M > 0$.

Now, if B is positive definite, then $x^T B x > 0$ for all $x \neq 0$ and hence $V(t, x(\cdot))$ is positive definite with $V'_{(1.6.15)}(t, x(\cdot))$ negative definite. Thus $x = 0$ is stable.

Suppose that $x = 0$ is stable but B is not positive definite. Then there is an $x_0 \neq 0$ with $x_0^T B x_0 \leq 0$. If $x_0^T B x_0 = 0$, then along the solution $x(t, 0, x_0)$, $V(0, x_0) = x_0^T B x_0 = 0$ and $V'(t, x(\cdot)) \leq -\alpha |x|^2$ so that for some $t_1 > 0$ we have $V(t_1, x(\cdot)) < 0$. Thus $x^T(t_1) B x(t_1) < 0$.

Hence, if $x^T B x$ is not always positive for $x \neq 0$, we may suppose there is an $x_0 \neq 0$ with $x_0^T B x_0 < 0$.

Let $\varepsilon = 1$ and $t_0 = 0$. Since $x = 0$ is stable, there is a $\delta > 0$ such that $|x_0| < \delta$ implies $|x(t, 0, x_0)| < 1$ for $t \geq 0$. We may choose x_0 so that $|x_0| < \delta$ and $x_0^T B x_0 < 0$. Letting $x(t) = x(t, 0, x_0)$, we have

$$x^T(t) B x(t) \leq V(t, x(\cdot)) \leq V(0, x_0) - \alpha \int_0^t |x(s)|^2 \, ds$$

$$\leq x_0^T B x_0 - \alpha \int_0^t |x(s)|^2 \, ds \,.$$

We show that $x(t)$ is bounded away from zero. Suppose not; then there is a sequence $\{t_n\}$ tending to infinity monotonically such that $x(t_n) \to 0$. Hence, $x^T(t_n) B x(t_n) \to 0$, a contradiction to $x^T(t) \cdot B x(t) \leq x_0^T B x_0 < 0$.

Thus there is a $\gamma > 0$ with $|x(t)|^2 \geq \gamma$ so that $x^T(t) B x(t) \leq x_0^T B x_0 - \alpha \gamma t$, implying that $|x(t)| \to \infty$ as $t \to \infty$. This contradicts $|x(t)| < 1$ and completes the proof.

We now suppose that C is of convolution type and use C to stabilize the equation. Consider the system

$$(1.6.18) \qquad x' = Ax + \int_0^t C(t - s) x(s) \, ds$$

with A constant and C continuous on $[0, \infty)$.

Select an $n \times n$ matrix $G(t)$ with

$$(1.6.19) \qquad dG(t)/dt = C(t)$$

and set

$$(1.6.20) \qquad Q = A - G(0) \,,$$

so that (1.6.18) takes the form

$$(1.6.21) \qquad x' = Qx + (d/dt) \int_0^t G(t - s) x(s) \, ds \,.$$

Let D be a symmetric matrix satisfying

(1.6.22) $Q^T D + DQ = -I$.

If D is any positive definite matrix, then there is a positive constant k such that

(1.6.23) $k|x|^2 \leq x^T Dx$ for all x .

Theorem 1.6.9. *Suppose* (1.6.19)–(1.6.22) *hold. Let*

(i) $2|DQ| \int_0^\infty |G(v)| \, dv < 1$ *and*

(ii) $G(t) \to 0$ *as* $t \to \infty$.

Then the zero solution of (1.6.18) *is stable if and only if* D *is positive definite.*

Proof. Consider the functional

$$V(t, x(\cdot)) = \left(x - \int_0^t G(t - s)x(s) \, ds \right)^T D \left(x - \int_0^t G(t - s)x(s) \, ds \right)$$
$$+ |DQ| \int_0^t \int_t^\infty |G(u - s)| \, du \, |x(s)|^2 \, ds \, .$$

so that along a solution $x(t)$ of (1.6.21) we have

$$V'_{(1.6.21)}(t, x(\cdot)) = x^T Q^T D \left(x - \int_0^t G(t - s)x(s) \, ds \right)$$
$$+ \left(x - \int_0^t G(t - s)x(s) \, ds \right)^T DQx$$
$$+ |DQ| \int_t^\infty |G(u - t)| \, du \, |x|^2$$
$$- |DQ| \int_0^t |G(t - s)| \, |x(s)|^2 \, ds$$
$$\leq -|x|^2 + |DQ| \int_0^t |G(t - s)| \left(|x|^2 + |x(s)|^2 \right) ds$$
$$+ |DQ| \int_t^\infty |G(u - t)| \, du \, |x|^2$$
$$- |DQ| \int_0^t |G(t - s)| \, |x(s)|^2 \, ds$$
$$= \left[-1 + |DQ| \left(\int_0^t |G(t - s)| \, ds + \int_t^\infty |G(u - t)| \, du \right) \right] |x|^2$$
$$\leq \left[-1 + 2|DQ| \int_0^\infty |G(v)| \, dv \right] |x|^2 \stackrel{\text{def}}{=} -\mu |x|^2 \, .$$

Suppose D is positive definite and let $\varepsilon > 0$ and $t_0 \geq 0$ be given. We must find $\delta > 0$ so that if $|\phi(t)| < \delta$ on $[0, t_0]$, then $|x(t, t_0, \phi)| < \varepsilon$ for $t \geq t_0$. As $V'_{(1.6.21)}(t, x(\cdot)) \leq 0$, then

$$V(t, x(\cdot)) \leq V(t_0, \phi(\cdot))$$

$$\leq |D| \left(|\phi(t_0)| + \int_0^{t_0} |G(t_0 - s)|\, |\phi(s)|\, ds \right)^2$$

$$+ |DQ| \int_0^{t_0} \int_{t_0}^{\infty} |G(u - s)|\, du\, |\phi(s)|^2\, ds \leq \delta^2 N^2$$

for some $N > 0$.

Using (1.6.23) we have

$$V(t, x(\cdot)) \geq \left(x - \int_0^t G(t - s)x(s)\, ds \right)^T D \left(x - \int_0^t G(t - s)x(s)\, ds \right)$$

$$\geq k^2 \left(|x| - \left| \int_0^t G(t - s)x(s)\, ds \right| \right)^2, \quad k \neq 0.$$

Thus,

$$|x(t)| \leq (\delta N / k) + \int_0^t |G(t - s)|\, |x(s)|\, ds.$$

So long as $|x(t)| < \varepsilon$, we have $|x(t)| < (\delta N / k) + \varepsilon \int_0^{\infty} |G(v)|\, dv < \varepsilon$ for all $t \geq t_0$, provided that $\delta < (k/N)(1 - \int_0^{\infty} |G(v)|\, dv)\varepsilon$. By (i) and the fact that $|2DQ| \geq 1$, the right-hand side of the above inequality is positive. Hence, $x = 0$ is stable.

Now suppose that $x = 0$ is stable but D is not positive definite. Then it can be shown that there is an $x_0 \neq 0$ with $x_0^T D x_0 < 0$ and $|x_0| < \delta$ for any given $\delta > 0$. As $x = 0$ is stable, we may choose δ so that $|x_0| < \delta$ implies $|x(t, 0, x_0)| < 1$ for all $t \geq 0$.

Letting $x(t) = x(t, 0, x_0)$, we have

$$V(t, x(\cdot)) \leq V(0, x_0) - \mu \int_0^t |x(s)|^2\, ds = -\eta - \mu \int_0^t |x(s)|^2\, ds$$

where $\eta = -x_0^T D x_0 > 0$. Thus

(1.6.24)
$$\left(x(t) - \int_0^t G(t - s)x(s)\, ds \right)^T D \left(x(t) - \int_0^t G(t - s)x(s)\, ds \right)$$

$$\leq -\eta - \mu \int_0^t |x(s)|^2\, ds.$$

Using the Schwartz inequality as in the proof of Theorem 1.6.4 we conclude that

$$\left(\int_0^t |G(t-s)|\,|x(s)|\,ds \right)^2 \leq \int_0^t |G(t-s)|ds \int_0^t |G(t-s)|\,|x(s)|^2\,ds .$$

As $|x(t)| < 1$ and $\int_0^t |G(t-s)|\,ds$ is bounded, then $\int_0^t G(t-s)x(s)\,ds$ is bounded. By (1.6.24) it follows that

$$\eta + \mu \int_0^t |x(s)|^2\,ds \leq |D| \left(|x(t)| + \left| \int_0^t G(t-s)x(s)\,ds \right| \right) \leq K$$

for some constant K. Thus, $|x(t)|^2$ is in $L^1[0,\infty)$. Now, $G(t) \to 0$ as $t \to \infty$ and $|x(t)|^2$ in L^1 imply that $\int_0^t |G(t-s)|\,|x(s)|^2\,ds \to 0$ as $t \to \infty$. Thus, by the Schwartz inequality argument, $\int_0^t G(t-s)x(s)\,ds \to 0$ as $t \to \infty$.

By (1.6.24) we see that for large t then $x^T(t)Dx(t) \leq -\eta/2$. Moreover, as $x \to 0$ we have $x^T Dx \to 0$. Hence, we conclude that $|x(t)|^2 \geq \gamma$ for some $\gamma > 0$ and all t sufficiently large. Thus, $\int_0^t |x(t)|^2\,dt \to \infty$ as $t \to \infty$, contradicting $|x(t)|^2$ being in $L^1[0,\infty)$. Therefore, the assumption that D is not positive definite is false and the proof is complete.

If

(1.6.25) $$\left| \int_0^\infty C(v)\,dv \right| < \infty ,$$

we may define $G(t)$ by

(1.6.26) $$G(t) = - \int_t^\infty C(v)\,dv .$$

Thus,

(1.6.27) $$Q = A - G(0) = A + \int_0^\infty C(v)\,dv .$$

Corollary 3. *Suppose (1.6.22) and (1.6.25)–(1.6.27) hold. Let*

(1.6.28) $$2|DQ| \int_0^\infty \left| \int_t^\infty C(v)\,dv \right| dt < 1 .$$

Then the zero solution of (1.6.18) is stable if and only if D is positive definite.

Theorem 1.6.10. *Suppose* (1.6.22) *and* (1.6.25)–(1.6.27) *hold. If*

(i) $A + \int_0^\infty C(v)\, dv$ *is a stable matrix and*

(ii) $2|DQ| \int_0^\infty |\int_t^\infty C(v)\, dv|\, dt < 1$ *then the zero solution of* (1.6.18) *is stable and, furthermore, all solutions of* (1.6.18) *are in* $L^2[0, \infty)$ *and bounded.*

If, in addition,

(iii) $\int_0^\infty |C(v)|^2\, dv < \infty$ *or* $\int_0^\infty |C(v)|\, dv < \infty$, *then all solutions of* (1.6.18) *tend to zero as* $t \to \infty$ *and, hence, the zero solution of* (1.6.18) *is asymptotically stable.*

If, in addition

(iv) $\int_0^\infty \int_t^\infty |C(v)|\, dv\, dt < \infty$, *then all solutions of* (1.6.18) *are in* $L^1[0, \infty)$ *and the zero solution of* (1.6.18) *is uniformly asymptotically stable.*

Proof. Let (i) and (ii) hold. Then D is positive definite and by Corollary 3, $x = 0$ is stable; hence, all solutions are bounded.

Using the functional $V(t, x(\cdot))$ in the proof of Theorem 1.6.9 we have

$$V'_{(1.6.21)}(t, x(\cdot)) \leq -\mu|x(t)|^2$$

for $\mu > 0$ and $t \geq t_0$. Thus, $|x(t)|^2$ is in $L^1[0, \infty)$.

If in addition, (iii) holds, then (1.6.18) yields

$$|x'(t)| \leq |A|\,|x(t)| + \int_0^t |C(t - s)|\,|x(s)|\, ds$$

$$\leq |A|\,|x(t)| + \frac{1}{2}\int_0^t |C(t - s)|^2\, ds + \frac{1}{2}\int_0^t |x(s)|^2\, ds\,.$$

Thus, $|x'(t)|$ is bounded and so is $(|x(t)|^2)'$; for,

$$(|x(t)|^2)' = (x^T(t)x(t))' \leq 2|x(t)|\,|x'(t)|\,.$$

Hence, $x(t) \to 0$ as $t \to \infty$ and $x = 0$ is asymptotically stable.

Suppose (iv) holds. Since all solutions of (1.6.18) tend to zero as $t \to \infty$, we may apply Theorem 7 of Burton and Mahfoud (1983) to conclude that the zero solution of (1.6.18) is uniformly asymptotically stable and all solutions of (1.6.18) are in $L^1[0, \infty)$.

Corollary 4. *Suppose* (1.6.22) *and* (1.6.27) *hold. If*

(i) $A + \int_0^\infty C(v)\,dv$ *is a stable matrix and*

(ii) $2|DQ| \int_0^\infty \int_t^\infty |C(v)|\,dv\,dt < 1,$

then the zero solution of (1.6.18) *is uniformly asymptotically stable.*

Proof. This is an immediate consequence of Theorem 1.6.10 as, by (i) and (ii), all conditions of the theorem are satisfied.

We have used $A(t)$ alone to stabilize (1.6.1) and we have used all of $C(t, s)$ to stabilize (1.6.1). We now show that we can select part of $C(t, s)$ to help stabilize (1.6.1).

Consider the scalar equation

$$(1.6.29) \qquad x' = L(t)x + \int_0^t C_1(t, s)x(s)\,ds + \frac{d}{dt}\int_0^t H(t, s)x(s)\,ds$$

where $L(t)$ is continuous for $0 \le t < \infty$, and $C_1(t, s)$ and $H(t, s)$ are continuous for $0 \le s \le t < \infty$. Here L, C, H, and x are all scalars.

Let

$$(1.6.30) \qquad\qquad P(t) = \int_0^t |C_1(t, s)|\,ds\,,$$

$$(1.6.31) \qquad\qquad J(t) = \int_0^t |H(t, s)|\,ds\,,$$

and

$$(1.6.32) \qquad \Phi(t, s) = \int_t^\infty \big[(1 + J(u))\,|C_1(u, s)| \\ + (|L(u)| + P(u))\,|H(u, s)|\big]\,du$$

assuming, of course, that $\Phi(t, s)$ exists for $0 \le s \le t < \infty$.

Let

$$(1.6.33) \qquad V(t, x(\cdot)) = \left(x - \int_0^t H(t, s)x(s)\,ds\right)^2 + v\int_0^t \Phi(t, s)x^2(s)\,ds$$

where v is an arbitrary constant. Thus, if $x(t) = x(t, t_0, \phi)$ is a solution of (1.6.29), then the derivative $V'_{(1.6.29)}(t, x(\cdot))$ of $V(t, x(\cdot))$ along $x(t)$ satisfies

$$V'_{(1.6.29)}(t, x(\cdot)) = 2 \left(x - \int_0^t H(t, s)x(s)\, ds \right) \left(L(t)x + \int_0^t C_1(t, s)x(s)\, ds \right)$$

$$+ v \frac{d}{dt} \int_0^t \Phi(t, s)x^2(s)\, ds$$

$$= 2L(t)x^2 + 2x \int_0^t C_1(t, s)x(s)\, ds - 2L(t)x \int_0^t H(t, s)x(s)\, ds$$

$$- 2 \int_0^t H(t, s)x(s)\, ds \int_0^t C_1(t, s)x(s)\, ds + v \frac{d}{dt} \int_0^t \Phi(t, s)x^2(s)\, ds .$$

Thus,

$$\left| V'_{(1.6.29)}(t, x(\cdot)) - 2L(t)x^2 - v \frac{d}{dt} \int_0^t \Phi(t, s)x^2(s)\, ds \right|$$

$$\leq 2 \int_0^t |C_1(t, s)|\, |x|\, |x(s)|\, ds + 2|L(t)| \int_0^t |H(t, s)|\, |x|\, |x(s)|\, ds$$

$$+ 2 \int_0^t |H(t, s)|\, |x(s)|\, ds \int_0^t |C_1(t, s)|\, |x(s)|\, ds .$$

Using the Schwartz inequality, we may write

$$\int_0^t |H(t, s)|\, |x(s)|\, ds = \int_0^t |H(t, s)|^{1/2}|H(t, s)|^{1/2}|x(s)|\, ds$$

$$\leq \left[\int_0^t |H(t, s)|\, ds \int_0^t |H(t, s)|x^2(s)\, ds \right]^{1/2}$$

$$= \left[J(t) \int_0^t |H(t, s)|x^2(s)\, ds \right]^{1/2}$$

so that

$$2 \int_0^t |H(t, s)|\, |x(s)|\, ds \int_0^t |C_1(t, s)|\, |x(s)|\, ds$$

$$\leq 2 \left[J(t) \int_0^t |H(t, s)|x^2(s)\, ds \right]^{1/2} \left[P(t) \int_0^t |C_1(t, s)|x^2(s)\, ds \right]^{1/2}$$

$$= 2 \left[P(t) \int_0^t |H(t, s)|x^2(s)\, ds \right]^{1/2} \left[J(t) \int_0^t |C_1(t, s)|x^2(s)\, ds \right]^{1/2}$$

$$\leq P(t) \int_0^t |H(t, s)|x^2(s)\, ds + J(t) \int_0^t |C_1(t, s)|x^2(s)\, ds .$$

Thus,

$$\left| V'_{(1.6.29)}(t, x(\cdot)) - 2L(t)x^2 - v\frac{d}{dt}\int_0^t \Phi(t,s)x^2(s)\,ds \right|$$

$$\leq \int_0^t |C_1(t,s)|\,(x^2 + x^2(s))\,ds + |L(t)|\int_0^t |H(t,s)|\,(x^2 + x^2(s))\,ds$$

$$+ P(t)\int_0^t |H(t,s)|x^2(s)\,ds + J(t)\int_0^t |C_1(t,s)|x^2(s)\,ds$$

$$= \left[P(t) + |L(t)|J(t) \right]x^2 + \int_0^t \left[(1 + J(t))\,|C_1(t,s)| \right.$$

$$+ \left. (|L(t)| + P(t))\,|H(t,s)| \right]x^2(s)\,ds.$$

Using (1.6.32), we obtain

(1.6.34)
$$\left| V'_{(1.6.29)}(t, x(\cdot)) - 2L(t)x^2 - v\frac{d}{dt}\int_0^t \Phi(t,s)x^2(s)\,ds \right|$$

$$\leq \left[P(t) + |L(t)|J(t) \right]x^2 - \int_0^t \frac{\partial\Phi(t,s)}{\partial t}x^2(s)\,ds\,.$$

Theorem 1.6.11. *Let P, J and Φ be defined by* (1.6.30)–(1.6.32). *If $L(t) < 0$,*

(i) $\sup\limits_{t\geq 0} J(t) < 1,$

and

(ii) $J(t)|L(t)| + P(t) + \Phi(t,t) \leq 2|L(t)|\,,$

then the zero solution of (1.6.29) *is stable.*

Proof. Taking $v = 1$ in (1.6.33) and observing that

(1.6.35)
$$\frac{d}{dt}\int_0^t \Phi(t,s)x^2(s)\,ds = \Phi(t,t)x^2 + \int_0^t \frac{\partial\Phi(t,s)}{\partial t}\,x^2(s)\,ds\,,$$

we obtain from (1.6.34) and (ii) that $V'_{(1.6.29)}(t, x(\cdot)) \leq 0$. Since $V(t, x(\cdot))$ is not positive definite, we still need to prove stability. Let $t_0 \geq 0$ and $\varepsilon > 0$ be given; we must find $\delta > 0$ so that if $\phi[0, t_0] \to R$ is continuous with $|\phi(t)| < \delta$, then $|x(t, t_0, \phi)| < \varepsilon$ for all $t \geq t_0$. As $V'_{(1.6.29)}(t, x(\cdot)) \leq 0$

for $t \geq t_0$, we have

$$V(t, x(\cdot)) \leq V(t_0, \phi(\cdot)) = \left(\phi(t_0) - \int_0^{t_0} H(t_0, s)\phi(s) \, ds \right)^2 + \int_0^{t_0} \Phi(t_0, s)\phi^2(t) \, ds$$

$$\leq \delta^2 \left[(1 + J(t_0))^2 + \int_0^{t_0} \Phi(t_0, s) \, ds \right]$$

$$\leq \delta^2 \left[4 + \int_0^{t_0} \Phi(t_0, s) \, ds \right] \stackrel{\text{def}}{=} \delta^2 N^2.$$

On the other hand,

$$V(t, x(\cdot)) \geq \left(x(t) - \int_0^t H(t, s)x(s) \, ds \right)^2$$

$$\geq \left(|x(t)| - \int_0^t |H(t, s)| \, |x(s)| \, ds \right)^2.$$

Thus

$$|x(t)| \leq \delta N + \int_0^t |H(t, s)| \, |x(s)| \, ds.$$

So long as $|x(t)| < \varepsilon$, we have

$$|x(t)| \leq \delta N + \varepsilon \sup_{t \geq 0} J(t) \stackrel{\text{def}}{=} \delta N + \varepsilon \beta < \varepsilon$$

for all $t \geq t_0$ provided that $\delta < \varepsilon(1 - \beta)/N$. Thus, the solution $x = 0$ is stable.

The foregoing is but a brief introduction to the subject of stability of Volterra equations using Liapunov functionals. Many other Liapunov functionals are to be found in Burton and Mahfoud (1983, 1984), in Burton (1983a, Chapters 2 and 5–8), and in Gopalsamy (1983).

Chapter 2

History, Motivation, Examples

In this chapter we introduce a large number of problems, both old and new, which are treated using the general theory of diffcrential equations. We attempt to give sufficient description concerning the derivation, solution, and properties of solutions so that the reader will be able to appreciate some of the flavor of the problem. In none of the cases do we give a complete treatment of the problem, but offer references for further study.

2.1 Classical Second-Order Equations

2.1.1 Positive Damping, History, Philosophy

The results of the nonlinear theory treated in this book can be traced, in large measure, to problems suggested by the second-order equation

$$(2.1.1) \qquad x'' + f(t, x, x')x' + g(x) = p(t).$$

In mechanical problems, f usually represents a damping or friction term, g represents a restoring force, and p is an externally applied force. The discussion is divided into two parts, depending on whether f slows the motion or in some way causes motion.

Equation (2.1.1) may arise from a simple statement of Newton's law of motion: If a rigid body moves in a straight line then the rate of change of momentum equals the sum of the forces acting on the body in the direction of motion. There are well-known corresponding statements concerning electrical RLC circuits and chemistry problems. When we denote the position of the body by $x = x(t)$, then $x'(t)$ is the velocity, $x''(t)$ is the acceleration, and mx' is the momentum when m is the mass of the body.

The classical elementary textbook problem is formulated as follows. We attach a coil spring to a solid beam, add a mass to the spring, and affix a dasher to the mass with the dasher in a container of liquid. This is called a spring-mass-dashpot system. See Fig. 2.1. The front suspension system on an automobile is essentially such a system with the shock absorber being the dashpot.

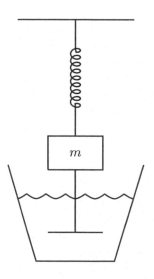

Fig. 2.1

To derive the equation of motion for the mass, it is usually assumed that the spring obeys Hooke's law so that, for small displacements, the spring force is proportional to the elongation (or amount of compression). It is also supposed that the dashpot exerts a force proportional to the velocity. If $p(t)$ is an external applied force, then there results the linear constant coefficient equation

$$(2.1.2) \qquad (mx')' = -cx' - kx + p(t)$$

where m is the mass, c is the positive constant of proportionality for the dashpot, and k is the positive spring constant. The external force p may be applied in a variety of ways. Some investigators have suggested driving the automobile at a constant velocity over a "washboard" road to produce $p(t) = A \cos \omega t$. For later contrast, we stress that the damping slows the mass and the restoring force always acts to bring the mass to the equilibrium position.

There is a direct analog of (2.1.2) in electrical circuit theory. The equation of a simple RLC circuit is

$$(2.1.3) \qquad LQ'' + RQ' + (1/C)Q = E(t)$$

where L is the inductance, R is the resistance, C is the capacitance, $E(t)$ is the impressed voltage, and Q is the charge on the capacitor. A sinusoidal or a constant input for $E(t)$ is very common. The reader is referred to standard introductory texts on differential equations for more detail and examples. [See, for example, Boyce and DiPrima (1969).]

While (2.1.2) is very simple and is, at best, a first approximation to the actual description of a real-world system, its solutions display a qualitative pattern that is central in the study of n-dimensional nonlinear systems. Two properties are prominent and they form the basis for our study in Chapter 4.

I. First, if $m > 0$, $c > 0$, $k > 0$, and $p(t)$ is bounded and continuous, then all solutions are bounded in the future and converge to a bounded function. If $p(t)$ is periodic, then all solutions converge to a unique periodic solution. In Chapter 4 we formalize this type of boundedness and call it uniform boundedness and uniform ultimate boundedness.

II. Next, if $m > 0$, $c > 0$, $k > 0$, and $p(t) \equiv 0$, then all solutions converge to zero as $t \to \infty$. Moreover, the zero solution attracts all other solutions so strongly that if certain error terms are added into the equation, solutions still tend to zero. Such behavior is formalized as uniform asymptotic stability.

Every engineering student studies (2.1.2) and draws these conclusions. But the interesting part is that, in fact, (2.1.2) is not even faintly accurate as a description of the physical processes usually involved in its derivation. One must instead go to the very nonlinear equation (2.1.1) for a more accurate description.

One of the big projects of investigators between the years 1920 and 1970 was to show that, under generous conditions, conclusions I and II for (2.1.2) are largely true for (2.1.1) so that the intuitive notions derived from (2.1.2) are mainly valid. Detailed accounts of the results of these studies are found in Sansone and Conti (1964), Reissig et al. (1963), Graef (1972), and Burton and Townsend (1968, 1971). There is an enormous literature concerning periodic solutions of special forms of (2.1.1) obtained by mathematicians from the People's Republic of China. An introduction to some such work, with references, is found in Shi (1980) and Yanqian (1982) in fairly accessible journals.

The work on (2.1.1) had spectacular consequences. Boundedness work with (2.1.1) and the associated work in relaxation oscillation theory (the topic of Section 2.2) led Levinson (1944) to introduce transformation theory which ultimately expanded to delay equations of both finite and infinite type and inspired a dozen sophisticated fixed-point theorems. Levinson's work seems to have generated the definitions of uniform boundedness and uniform ultimate boundedness. With these properties in mind, Browder (1959) proved his main asymptotic fixed-point theorem which allowed Yoshizawa (1966) to prove the existence of a periodic solution of an ordinary differential equation. At the same time, delay differential equations were being investigated with a view to proving the existence of periodic solutions under the uniform boundedness assumptions. Yoshizawa (1966) extended his periodic result to these delay equations in the case in which the period is as large as the delay. In a series of three papers, Jones (1963, 1964, 1965) proved fixed-point theorems useful in the search for periodic solutions. Browder (1970) continued with his research into asymptotic fixed-point theorems and cited many examples in which investigators had used his asymptotic theory for differential equations. But it was Horn (1970) who ultimately proved the asymptotic fixed-point theorem which has enabled investigators during the last few months to bring the periodic theory of equations with both finite and infinite delay on the same level as that for ordinary differential equations. Chapter 4 of this book is devoted in some measure to showing how that is done.

It is easy to see, in cases like the shock absorber problem, why we are interested in proving that solutions tend to zero. The front wheel of the automobile drops into a hole and sets the spring-mass-dashpot system into motion. A good design should cause the oscillations to die out quickly. And it is proper to ask now just why one is really so interested in the existence of periodic solutions.

A. First, when there is a periodic solution, and especially when all solutions approach a periodic solution, then there emerges orderliness in what may otherwise seem to be a chaotic situation.

B. Among all qualitative properties of solutions of differential equations, none is quite so satisfying aesthetically as that of periodicity.

C. Items A and B refer to our perceptions and desires. Thus, it is more to the point that periodicity is a property inherent in the structure of the solution space of a given differential equation even when there is nothing even slightly suggestive of periodicity in the structure of the equation itself. It is a fundamental property which imposes itself, perhaps when we least expect it. Consider a pair of scalar equations

$$(2.1.4) \qquad x' = P(x, y), \qquad y' = Q(x, y)$$

in which P and Q satisfy a local Lipschitz condition in both x and y. With nothing more said, it is unclear that (2.1.4) possesses periodicity properties; yet, such properties are there.

Theorem. Poincaré-Bendixson (cf. Bendixson, 1901) *Let $\phi(t)$ be a solution of (2.1.4) which is bounded for $t \geq 0$. Either*

(i) *ϕ is periodic,*

(ii) *ϕ converges to a periodic solution, or*

(iii) *there is a sequence $\{t_n\} \to \infty$ such that $\phi(t_n) \to (x_0, y_0)$ and $P(x_0, y_0) = Q(x_0, y_0) = 0$.*

Note that, in the last case $x(t) \equiv x_0$ and $y(t) \equiv y_0$ is a periodic solution. Thus, we conclude that whenever (2.1.4) has a bounded solution, then it has a periodic solution. It is important to add that (2.1.4) may have nontrivial periodic solutions of every real period T. Thus, (2.1.4) contains no hint of the possible periods of its periodic solutions.

D. Physical systems are frequently subjected to periodic disturbances even when the forces acting on the system are in no manner periodic. The resulting differential equation may, nevertheless, contain a periodic forcing function. A classic example of this occurs in the case of an elastic object (such as a steel bridge span) being acted upon by a wind of constant force. As the body of moving air strikes the object, it parts and moves around the object setting up "streets" of vortices behind the object. The net effect is a force on the object acting perpendicular to the direction of the wind with a magnitude of $c_0 \cos \omega t$. Such a force was the cause of the vertical motion of the bridge across the narrows at Tacoma, Washington, causing a "galloping" of the span and its ultimate collapse in 1940. In the same way, air striking the wing of an airplane results in a periodic force causing the wing to flutter up and down.

A classroom movie of the collapse of the Tacoma bridge is available in many physics departments in universities. Discussions and newspaper accounts of such events are found in Dickinson (1972) and in Braun (1975).

E. Periodic disturbances are cause for great concern, as indicated in D. The tremendous response to a (possibly small) periodic force is known as resonance, a condition occurring when the natural frequency of vibration coincides with the frequency of the disturbance. The response can be a periodic solution, or it can be disaster.

F. There are obvious periodic forces, actions, and events throughout the natural and man made environment. Examples include planetary motion, heartbeat, physiological cycles, and the ubiquitous electrical and petroleum engines.

Finally, having faulted the linear model (2.1.2) one should certainly supply some reason for it. Let us consider the front suspension of an automobile as a spring-mass-dashpot system. Hooke's law refers to small elongations.

Are the elongations in our example small? Seldom. One may note that there are triangular-shaped pieces of hard rubber in the suspension system to absorb some of the shock when the spring is totally compressed. And, even with good shock absorbers and a smooth road, these rubber bumpers do their work during fast stops. In other words, not only is the elongation not small, but when the spring is totally compressed the equation itself is not describing the motion.

Even when springs are being used within their proper range of operation, they are seldom linear springs as Hooke's law requires. It is interesting to observe the design of nonlinear springs. For example, look at the suspension of the rear axle of a very large freight truck. A *leaf spring* joins the truck chassis to the axle. This leaf spring consists of a bundle of strips (leaves) of spring steel of varying lengths. Only the longest leaf is compressed when the truck is empty; but, as weight is added and the body of the truck descends, more leaves bend, thereby exerting more upward pressure on the body. The graph of this spring force is, roughly, piecewise linear and is sharply concave upward. The nonlinearity is further accented by the second set of leaf springs (called overload springs) which only come into contact between the chassis and the axle after a great amount of weight is added. A picture of this could easily be shown here, but the actual object is close at hand and is well worth seeing.

2.1.2 Negative Damping, Relaxation Oscillations

From the constant coefficient system (2.1.2) with $p(t)$ not zero, it was easy to obtain an illustration of a nontrivial periodic solution to which all other solutions converge. When $p(t)$ is zero, it is not possible to find constants m, c, and k for which (2.1.2) has a nontrivial solution to which all others converge. But, starting around 1920 there began appearing a great variety of nonlinear problems with this property. Many of these periodic solutions were of a steep or zigzag character and were called relaxation oscillations.

The most common such equation was Liénard's equation

$$(2.1.5) \qquad x'' + f(x)x' + g(x) = 0$$

with $f(0) < 0$, $f(x) > 0$ if $|x| > a > 0$, and $xg(x) > 0$ if $x \neq 0$, of which van der Pol's equation

$$(2.1.6) \qquad x'' + \varepsilon(x^2 - 1)x' + x = 0, \qquad \varepsilon > 0,$$

is a typical example. These equations are purported to describe heartbeat (van der Pol and van der Mark, 1928), electrical circuits (Liénard, 1928; van der Pol, 1927; cf. Bellman, 1961; Minorsky, 1962; Haag, 1962), belt-driven oscillators (Haag, 1962), communication equipment (Cartwright, 1950), and many other phenomena. In these problems investigators speak of "negative damping." In some way the damping or friction acts to produce motion, rather than reduce the motion as in the classical theory. While there is frequently no external applied force $p(t)$, all the models show some type of energy being fed into the system so that the oscillations really are "forced vibrations" as opposed to their usual classification as "free vibrations."

We first display a physical model illustrating the general type of behavior being discussed for (2.1.5). No attempt is made to relate the model to (2.1.5).

Example 2.1.1. Imagine a child's seesaw with a water container on each side of the fulcrum with a hole in each container near the bottom on the side of the container farthest from the fulcrum; in fact, that whole side can be removed. Water pours down from a source directly above the fulcrum. It fills one container, whose weight then rotates the seesaw, putting the other container in line for the water. The first container empties because its side is missing. There results an oscillatory motion of the seesaw. See Fig. 2.2. This model allows us to see how oscillatory motion can be induced, but does not illustrate a change in friction or damping. The oscillations are caused by a change in the center of gravity, much as occurs in a playground swing. Notice how irregular the shape of the periodic motion would be.

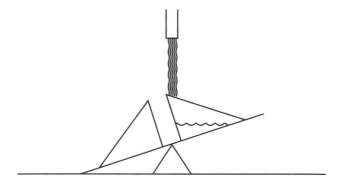

Fig. 2.2

The next example, from Haag (1962, p. 165), shows how friction can produce a "restoring force" and contribute to oscillations.

Example 2.1.2. Consider a mechanical system consisting of a movable block B of mass m resting on frictionless guides on a table. The block is attached by a coil spring to a solid vertical support on the table. At each end of the table is a belt pulley supporting an endless belt, the top side of which rests on the movable block. See Fig. 2.3.

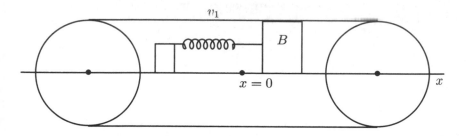

Fig. 2.3

It is assumed that the spring satisfies Hooke's law so that

$$F_s = -kx$$

for some $k > 0$. Now the top of the belt moves to the right with a constant velocity v_1 and exerts a force on B because of friction. There are two possible cases: if the belt slides over the block, then there is a force F of sliding friction, which we can write as

$$F = ka, \qquad a > 0;$$

if the belt is pulling B along at velocity v_1, then there is a force of static friction,

$$\alpha F = \alpha ka, \qquad \alpha > 1.$$

If $x' = v < v_1$ the sliding velocity $v_1 - v$ is positive and the algebraic value of the friction is ka so that the forces acting on the block yield

(2.1.7) $$(mx')' = -kx + ka.$$

If we let $y = x - a$, we obtain

(2.1.8) $$y'' + \omega^2 y = 0$$

with $\omega^2 = k/m$ and general solution

$$y = A \cos \omega t + B \sin \omega t,$$

which is harmonic motion induced in part by the friction which is "negative."

If $x' = v = v_1$, then this equation will hold until the spring elongates enough so that

$$F_s = -kx$$

can overcome the force due to friction between the block and belt. Thus, if $x' = v = v_1$, this holds for

$$-\alpha a < x < \alpha a,$$

so that x increases linearly. When $x > \alpha a$, then the belt slides and (2.1.8) holds.

The complete analysis, given by Haag (1962), shows the harmonic motion for small amplitudes. The other solutions behave as shown in Fig. 2.4. Here, all solutions starting outside the larger circle eventually reach it and are periodic. Haag (1962, pp. 165–167) also considers the case in which the guides are not frictionless.

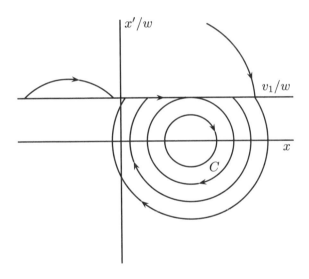

Fig. 2.4

In many electrical circuits a potential builds up gradually without current flowing through a certain part of the circuit. Then, when the potential reaches a threshold value, current flows. Cronin (cf. Burton, 1981, pp. 33–45) discusses this property in nerve conduction. Much of this type of work was initiated by B. van der Pol in the early 1920s in journals which are difficult to find. Discussions are found in Minorsky (1962), Haag (1962), and Andronow and Chaikin (1949).

Example 2.1.3. Haag (1962, p. 169) considers an electrical circuit consisting of a battery powering a circuit with a resistor, a condenser, and a neon tube shunted across the condenser. See Fig. 2.5.

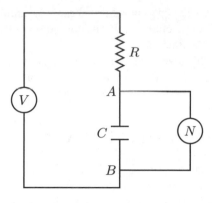

Fig. 2.5

When the potential difference between A and B,

$$v = V - Ri,$$

is less than a threshold voltage v_1, no current flows in the tube N. But the condenser is charged up by the relation

$$(*) \qquad \int_0^t i \, dt = C(V - Ri),$$

or, since V and C are constant,

$$(**) \qquad i + CR\,i' = 0.$$

At $t = 0$, $(*)$ yields

$$i_0 = V/R,$$

so a solution of the linear first-order equation $(**)$ is

$$i = (V/R)e^{-(1/CR)t},$$

yielding

$$v = V(1 - e^{-(1/CR)t}).$$

When

$$t = CR\ln[V/(V - v_1)]$$

then

$$v = v_1$$

and the condenser discharges suddenly across N and the process starts again, yielding a periodic charge on the condenser. It is a zigzag curve, typical of relaxation oscillations.

Haag (1962, p. 1970) also considers a circuit containing a triode having a descriptive equation equivalent to a van der Pol equation. That is an interesting example, but will not be considered here.

2.2   Problems with a Delay

In Section 2.1 we discussed problems in which different types of motions were caused by variations in the friction or damping. Frequently, behavior differs because of variations in the restoring force. A spring attached to a mass acts instantaneously. But when man devises a machine to run a machine there is always a delay in response.

Let us illustrate the case of a delayed response. Consider first the alert and experienced driver of a good automobile on a good highway. The driver quickly observes any straying from the proper lane and moves the wheel so slightly to correct the path of the vehicle that neither an observer nor a passenger notices the controlling influence. By contrast, the drinking driver notices too late that the car is on the shoulder of the road, yanks the steering wheel causing the car to swerve across the road, and continues to over correct until the highway patrolman mercifully puts the driver away for a time. This is a case of too much delay and too much response.

2.2.1   Controlling a Ship

Minorsky (1962) designed an automatic steering device for the battleship *New Mexico*. The following is a sketch of the problem.

Let the rudder of the ship have angular position $x(t)$ and suppose there is a friction force proportional to the velocity, say $-cx'(t)$. There is a direction indicating instrument which points in the actual direction of motion and there is an instrument pointing in the desired direction. These two are connected by a device which activates an electric motor producing a certain force to move the rudder so as to bring the ship onto the desired course. There is a time lag of amount $h > 0$ between the time the ship gets off course and the time the electric motor activates the restoring force. The equation for $x(t)$ is

(2.2.1) $$x''(t) + cx'(t) + g(x(t - h)) = 0$$

where $xg(x) > 0$ if $x \neq 0$ and c is a positive constant. The object is to give conditions ensuring that $x(t)$ will stay near zero so that the ship closely follows its proper course. Such equations are discussed in Chapter 4.

Minorsky (1962, pp. 53–57, 534–537) also considers the problem of stabilizing the rolling of a ship by the "activated tanks method" in which ballast water is pumped from one position to another by means of a propeller pump whose blade angles are controlled by electronic instruments giving the angular motion of the ship. Without the pumping, the rolling is idealized by the ordinary differential equation

$$x'' + kx' + \omega_0^2 x = a \sin \omega t$$

with k and ω_0 positive constants. The stabilizing equipment "quenches" the rolling according to the equation

$$x'' + (k + K)x' + \omega_0^2 x = a \sin \omega t,$$

where K is the coefficient of damping produced by the pumping. When the pump becomes overworked, a delay in the damping results and we have

(2.2.2) $$x'' + kx' + Kx'(t - h) + \omega_0^2 x = a \sin \omega t.$$

Such equations are also treated in Chapter 4, although one may note that this is linear. The best possible behavior is a small globally stable periodic solution.

We remarked at the start of this chapter that there are electrical and chemical analogs of the mechanical systems. In fact, a mechanical system is frequently replaced by an electrical analog which is easier to construct, modify, and observe. One modifies the electrical system to improve performance and then makes the same corrections on the mechanical system.

Minorsky's ship rolling problem was modeled by an electronic circuit (cf. Minorsky, 1962, p. 536) given in Fig. 2.6.

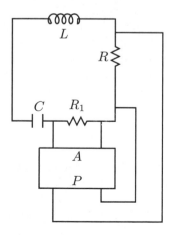

Fig. 2.6

The figure is self-explanatory except that A is a linear amplifier with a phase-shifting network P producing a constant time delay between the input and output of P. The resulting equation is

$$L(di/dt) + (R + R_1)i + \lambda R_1 i(t - h) + (1/C) \int_0^t i(s)\,ds = 0.$$

Upon differentiation we have

(2.2.3) $\qquad Li'' + (R + R_1)i' + \lambda R_1 i'(t - h) + (1/C)i = 0.$

The parameters can be matched to those in (2.2.2) in the unforced case. Evidently, a sinusoidal input can be placed in the circuit.

2.2.2   The Sunflower Equation

Somolinos (1978) has considered the equation

$$x'' + (a/r)x' + (b/r)\sin x(t - r) = 0$$

and has obtained interesting results on the existence of periodic solutions. The study of this problem goes back to the early 1800s and has attracted much attention. It involves the motion of a sunflower plant.

The tip of the plant is observed to move from side to side in a periodic fashion. Investigators believe that a growth hormone, Auxin, causes the motion in the following way. Suppose we designate sides of the plant as left and right. While the plant is bent to the left, gravity causes Auxin to gather on the left side; this makes the left side grow faster than the right side, so the plant bends to the right. It then oscillates from left to right in a T-periodic fashion. In this particular experiment, $T = 20$ minutes.

To model the process we suppose the rate of growth is proportional to the concentration of Auxin. Let L_1 and L_2 be the lengths of the left and right sides of the stem and C_1 and C_2 the corresponding concentrations of Auxin. Then

(2.2.4) $\qquad dL_i/dt = K_1 C_i, \qquad i = 1, 2.$

The difference in concentration of Auxin is proportional to $\sin\alpha$, where α is the angle between stem and the vertical. Then

(2.2.5) $\qquad \Delta C = C_1 - C_2 = K_2 \sin\alpha.$

In an increment of time Δt, if d is the diameter of the stem then

$$d(\Delta\alpha) = \Delta L_1 - \Delta L_2$$

so that

$$\frac{d\alpha}{dt} = \frac{1}{d}\left(\frac{dL_1}{dt} - \frac{dL_2}{dt}\right)$$

which, with (2.2.4) yields

(2.2.6) $d\alpha/dt = (1/d)K_1(C_1 - C_2) = (-K_1 d)(C_2 - C_1).$

Now, to substitute (2.2.5) into (2.2.6) would be to assume that the influence of Auxin is instantaneous. It is not; rather, it takes an amount r of time before it influences the bending. Moreover, we assume that past concentrations of Auxin have an exponentially decaying memory (or effect). Thus, we decide to write

(2.2.7) $$C_2(t) - C_1(t) = K_2 \int_1^\infty e^{-a(s-1)} \sin \alpha(t - sr)\, ds$$

and combine (2.2.6) and (2.2.7) to get

(2.2.8) $$\alpha' = -b \int_1^\infty e^{-a(s-1)} \sin \alpha(t - sr)\, ds$$

with $b = K_1 K_2/d$. Let $\omega = sr - t$ to obtain

$$\alpha' = -\frac{b}{r} \int_{r-t}^\infty \exp\left[-a\left(\frac{\omega + t - r}{r}\right)\right] \sin \alpha(-\omega)\, d\omega.$$

Upon differentiation we obtain

(2.2.9) $$\alpha'' + (a/r)\alpha' + (b/r)\sin \alpha(t - r) = 0,$$

the "sunflower equation."

Somolinos (1978) shows that if $a < br$, plus other mild conditions, then there is a periodic solution.

We encountered Liapunov functionals in Chapter 1. Here, one may define

(2.2.10) $$V(t, x, y) = \frac{y^2}{2} + \frac{b}{r}(1 - \cos x) + \frac{b}{r}\int_{-r}^0 \int_{t+s}^t y^2(u)\, du\, ds$$

(here $\alpha = x$, $x' = y$) and obtains the derivative along a solution as

$$V'(t, x, y) \leq (-(a/r) + 2b)y^2,$$

after some labor of the type we saw in Chapter 1. It is then possible to show that if $a > 2br$, then all solutions approach constants as $t \to \infty$. Thus, under that condition there would be no periodic solution.

To see that solutions tend to constants, we note first that

$$y' = -(a/r)y - (b/r)\sin x(t - r)$$

and the last term is a bounded function; thus, the variation of parameters formula yields $y(t)$ bounded and, hence, $y'(t)$ is bounded. Thus, $(y^2)'$ is bounded. Also, $V' \leq -\mu y^2$, $\mu > 0$, implies $y^2 \in L^1[0, \infty)$. But $y^2 \in L^1$ and $(y^2)'$ bounded imply that $y^2(t) \to 0$ as $t \to \infty$.

Next, by the Cauchy criterion for convergence of $\int_0^\infty y^2(t)\, dt$, it follows that

$$\int_{-r}^0 \int_{t+s}^t y^2(u)\, du\, ds \to 0 \qquad \text{as} \quad t \to \infty.$$

Since $V'(t, x, y) \leq 0$ and V is bounded from below, V approaches a constant. Thus,

$$V(t, x, y) \to 0^2 + (b/r)(1 - \cos x(t)) + 0$$

which approaches a constant; hence, $\cos x(t)$ approaches a constant and so $x(t)$ approaches a constant.

Thus, depending on a, b, and r (all positive) sometimes (2.2.9) has a nontrivial periodic solution, and sometimes it does not.

2.2.3   Some Models of War and Peace

L. F. Richardson (1881–1953), a British Quaker, observed two world wars and was concerned about them (cf. Richardson, 1960; Jacobson, 1984). He speculated that wars begin where arms races end and he felt that international dynamics could be modeled mathematically because of human motivations. He claimed that men are guided by "their traditions, which are fixed, and their instincts which are mechanical"; thus, on a grand scale they are incapable of good and evil. He sought to develop a theory of international dynamics to guide statesmen with domestic and foreign policy, much as dynamics guides machine design.

Let X and Y be nations suspicious of each other. Suppose X and Y create stocks of arms x and y, respectively; more generally, x and y represent "threats minus cooperation" so that negative values have meaning. At least three things affect the arms buildup of X;

(a) economic burden;

(b) terror at the sight of $y(t)$ (or national pride);

(c) grievances and suspicions of Y.

the same will, of course, apply to Y.

Richardson assumed that each side had complete and instantaneous knowledge of the arms of the other side and that each side could react instantaneously. He reasoned from (a) that

$$dx/dt = -a_1 x$$

because the burden is proportional to the size x, and he argued from (b) that

$$dx/dt = -a_1 x + b_1 y$$

because the terror is proportional to the size y. Finally, Richardson assumed constant standing grievances, say g_i, so that the complete system is

(2.2.11)
$$x' = -a_1 x + b_1 y + g_1,$$
$$y' = -a_2 y + b_2 x + g_2$$

with a_i, b_i, and g_i being positive constants. Domestic and foreign policy will set the a_i and b_i, although Richardson maintained a more mechanical view.

Remark 2.2.1. This model is a first approximation and the assumptions on time response are certainly naive; yet, the model suggests several very interesting ideas which would likely never occur to an observer without having a model. And this is, of course, the object of a mathematical model. Surely, we would all reason that the economically stronger nation could win an arms race. But this model, and its generalizations, suggest that the stronger nation could set policy in such a way that an arms race could be avoided, even if the weaker national desired an arms race. Moreover, this model and its generalizations suggest that unilateral disarmament is potential disaster; if one nation agrees to "forgive and forget" ($g_1 = b_1 = 0$), then $x(t) \to 0$, but $y(t)$ does not tend to zero.

To understand (2.2.11) better, write it as $P' = AP + G$ where $P = (x, y)^T$, $G = (g_1, g_2)^T$, and

$$A = \begin{pmatrix} -a_1 & b_1 \\ b_2 & -a_2 \end{pmatrix}.$$

Assume $\det A \neq 0$ and transform (2.2.11) by $P = Q + B$, where $B = -A^{-1}G$, into

$$Q' = AQ.$$

Then A has characteristic equation

$$\lambda^2 + (a_1 + a_2)\lambda + a_1 a_2 - b_1 b_2 = 0$$

and (2.2.11) has the equilibrium point

$$P_0 = -A^{-1}G.$$

Theorem 2.2.1. *Solutions of* (2.2.11) *approach the equilibrium point* P_0 *exponentially provided that* $a_1a_2 > b_1b_2$.

Thus, an arms race does not occur if the economic burdens are greater than the fears. Faulty as such models are, they have attracted enormous attention as may be seen from the number of papers citing Richardson's work in the Science Citation Index. We now describe two recent generalizations.

Hill (1978) recognized deficiencies in Richardson's model. He reasoned that it takes time to respond to an observed situation and, therefore, proposed the model

$$(2.2.12) \qquad \begin{aligned} x' &= -a_1x(t-T) + b_1y(t-T) + g_1, \\ y' &= b_2x(t-T) - a_2y(t-T) + g_2, \end{aligned}$$

where T is a positive constant. As before, one can transform this system to a system

$$Q' = AQ(t-T)$$

and try for a solution $Q = Q_0e^{\lambda t}$. This results in the characteristic quasipolynomial

$$(2.2.13) \qquad \lambda^2 + (a_1 + a_2)e^{-\lambda T}\lambda + (a_1a_2 - b_1b_2)e^{-2\lambda T} = 0.$$

That is a transcendental equation and the following is Hill's sufficient condition for Re $\lambda < 0$.

Theorem 2.2.2. *Solutions of* (2.2.12) *approach the equilibrium exponentially if*

$$a_1a_2 > b_1b_2, \qquad (3\pi/T) \geq a_1 + a_2,$$

and

$$b_1 \leq [a_1a_2/b_2] - [3\pi(a_1+a_2)/2Tb_2] + [9\pi^2/4b_2\,T^2].$$

While the Hill model is certainly a step in the right direction, it can be criticized for all the same reasons applied to (2.2.11). Surely, all the delays are not the same; the reactions are not based only on an observation at time $t - T$, but on a long period of observation; grievances are not constant.

Gopalsamy (1981) constructed certain models which overcome some of the criticism. He reasoned that X remembers the distant past dimly, the recent past quite well, but cannot react instantly to observed threats. His model is

$$(2.2.14) \qquad \begin{aligned} x' &= -a_1x + b_1 \int_{-\infty}^{t} C(t-s)y(s)\,ds + g_1, \\ y' &= -a_2y + b_2x + g_2 \end{aligned}$$

for $a_i > 0$, $b_i > 0$, $g_i > 0$ and all are constant, while $C(t) \geq 0$ and

$$\int_0^\infty C(t)\,dt < \infty.$$

Because of the difficulty in analysis he requires g_i constant and $C(t) = t^n e^{-\alpha t}$ with n a positive integer and α a positive constant. Under those restrictions it is possible to reduce (2.2.14) to a system of ordinary differential equations (as we discussed in Chapter 1).

But if a reasonable model is to be obtained then it needs to be based on the observed characteristics of the problem and not on mathematical convenience. Consider the question of X observing and reacting to $y(t)$. Suppose that it takes α-time units for X to begin to react to $y(t)$ and that a complete response to $y(t)$ takes $(\beta - \alpha)$-time units to be accomplished, with events occurring more than β-time units ago tending to fade in the eyes of X. We then begin with an idealized $C(t)$ of the form given in Fig. 2.7 where $\int_0^\infty C(t)\,dt = 1$.

Then for fixed $t > 0$, $C(t - s)$ has a graph as indicated in Fig. 2.8. We also graph $y(s)$.

We then have in Fig. 2.9 the graph of the product $C(t - s)y(s)$.

Fig. 2.7

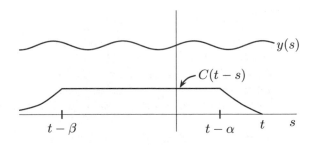

Fig. 2.8

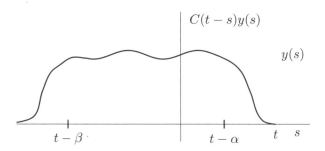

Fig. 2.9

At a particular value of t, the function in Fig. 2.9 is the one being integrated. The integral in

$$x' = -a_1 x + b_1 \int_{-\infty}^{t} C(t - s) y(s) \, ds + g_1$$

introduces a selective memory of $y(t)$.

Obviously, determination of C and the a_i, b_i, and g_i in a really precise manner is impossible, although Richardson tries very hard to do so. The object is to analyze such problems in a qualitative manner to minimize the need for exact determination of the constants and C.

Example 2.2.1. The Sputnik episode was, in a general way, an arms race and it provides a clear example of the response to a perceived threat. Let X be the United States and Y be the Soviet Union. Although X and Y were allies in the Second World War, by 1950 there were certainly tensions between them. The fact that Y had finally constructed and tested an atomic bomb caused these tensions to increase, but there was no general alarm.

In 1957 the world observed an orbiting satellite named Sputnik which was launched by Y. This meant that Y had the bomb and Y had the capacity to drop it on any city in the world. The perception of X concerning $y(t)$ can be qualitatively illustrated in Fig. 2.10.

One of Richardson's basic assumptions was that both X and Y have immediate and exact knowledge of both x and y. Certainly, knowledge will not be immediate, nor will it ever be exact. The models can only reflect perception.

In 1957 the scientific and technical gaps between X and Y were too large to be grasped by even the best informed observers. On the eve of the launching of Sputnik, one of the most successful university texts on differential equations (by Spiegel, 1958) was going to press in America. In

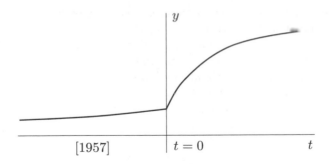

Fig. 2.10

that book the author discusses in a philosophical fashion whether or not it might someday be possible to launch a satellite. He notes that the "escape velocity" from the earth's field is approximately 6.9 miles per second and then states: "At present this speed is well out of attainment. The "Big Bertha" cannon of World War I had a muzzle velocity of 1 mile per second, so perhaps there may still be a chance." That "chance" was circling the earth before he had received his galley proofs.

While X did launch a satellite in 1958, it was 1959 before X could substantially respond to $y(t)$; thus, we take $\alpha = 2$ years. In Fig. 2.11 we have the graph of $C(2 - s)$ and the response graph of $C(2 - s)y(s)$.

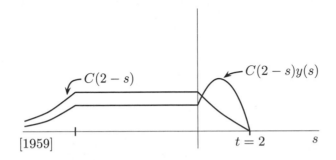

Fig. 2.11

By 1961 there was significant response by X which recognized her basic scientific weaknesses. All school programs were being bolstered; the NSF and NDEA fellowships for graduate study were being awarded; research contracts were given to professors in areas even remotely concerned with a response to Sputnik. Initially, industrial companies were given contracts to produce rockets capable of launching satellites; the rockets were produced; when fired, those rockets exploded on their launch pads. Newspaper accounts of those disasters ultimately claimed that the problem was that engineers were using linear differential equations to approximate nonlinear relations, a problem discussed in Section 2.1 of this chapter.

But by 1964 the full force of response by X to $y(t)$ was felt, so one might take $\beta - \alpha = 5$ years. In Fig. 2.12 we have graphs of $C(7 - s)$ and $C(7 - s)y(s)$. Recall that at each t, $x(t)$ is increasing by

$$\int_{-\infty}^{t} C(t - s)y(s)\, ds.$$

A first approximation to a good linear model is

(2.2.15)
$$x' = -a_1 x + b_1 \int_{-\infty}^{t} C_1(t - s)y(s)\, ds + g_1(t),$$
$$y' = -a_2 y + b_2 \int_{-\infty}^{t} C_2(t - s)x(s)\, ds + g_2(t),$$

with $a_i > 0$, $b_i > 0$, g_i bounded and continuous, $C_i \geq 0$, C_i continuous, and $\int_0^{\infty} C_i(t)\, dt = 1$, $i = 1, 2$.

Such a model does not have an equilibrium point. Since the g_i are variable, the most one can hope for is a bounded set into which solutions all enter with increasing time.

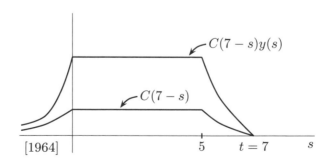

Fig. 2.12

Theorem 2.2.3. *If $a_1 > b_2$ and $a_2 > b_1$, then there is a ball*

$$B^* = \{(x,y)|x^2 + y^2 \leq B, \quad B > 0\}$$

such that each solution of (2.2.15) enters and remains inside B^*.

The proof utilizes a Liapunov functional

$$V(t, x(\cdot), y(\cdot)) = |x| + |y| + b_1 \int_0^t \int_{t-s}^\infty C_1(u)\, du\, |y(s)|\, ds$$

$$+ b_2 \int_0^t \int_{t-s}^\infty C_2(u)\, du\, |x(s)|\, ds,$$

applied to (2.2.15) with integrals from 0 to t. The rest of the integral contains the initial function and is treated as a perturbation term in a variation of parameters formula.

It is also possible to analyze some very general nonlinear models using the Liapunov technique. While the models are descriptive, rather than based on first principles, they offer a way to organize material and suggest answers to a very complex problem which threatens the very existence of life on this planet. At the time of this writing, the president of X argues for "peace through strength" and seeks funds for improved weapons systems. Others argue for unilateral disarmament, clutching at the straw of hope that all others will follow suit, tightly closing their eyes to the invasion by Y of Afghanistan, for example. Each of these is a position based on vague philosophical notions. A good and comprehensive model may indicate a course of action which will lead to arms reduction and some chance of global safety. But the model must reflect reality and not mathematical convenience if it is to be effective.

Other discussions of war models are found in Braun (1975) and Coleman (1976).

2.3 Biology, Economics, and Epidemics

Cooke (1976), Cooke and Yorke (1973), Haddock and Terjécki (1983), Kaplan *et al.* (1979), and many others have considered a delay problem related to biological growth, epidemics, and economic capital growth. They frequently concentrate on problems in which all constant functions are solutions. They show that if a solution stays above a constant for a time, it will stay above the constant forever. This can be good economic news, but a dreadful prediction about epidemics.

Example 2.3.1. Early in this century A.J. Lotka (1907a, b) formulated basic principles for the mathematical theory of population growth. He assumed that:

(a) individuals belong to different classes and the relative proportion of members of each class during any time period is constant,

(b) the life span of an individual in a class is independent of the number of members of that class and independent of the age distribution in the class, and

(c) the life support conditions remain constant.

On the basis of these assumptions Lotka derived the equation

$$(2.3.1) \qquad\qquad x' = B_t - D_t$$

where $x(t)$ is the population, B_t is the number of births per unit time, and D_t the deaths. He generalized the model to the renewal equation (cf. Bellman and Cooke, 1963; Feller, 1941)

$$(2.3.2) \qquad\qquad B(t) = G(t) + \int_0^t B(t-s)P(s)m(s)\,ds$$

where $B(t)$ is the number of births between 0 and t, $G(t)$ is the number of births at time t to parents surviving from an initial population at $t = 0$, $P(s)$ is the probability density of survival to age at least s, and $m(s)$ is the probability density for a parent of age s giving birth.

Example 2.3.2. In the work of Cooke and Yorke (1973) the Lotka assumption is changed so that

(c) the number of births per unit time is a function only of the population size, not of the age distribution.

Under this assumption, we let $x(t)$ be the population size and let the number of births be $B(t) = g(x(t))$. Assume each individual has life span L so that the number of deaths per unit time is $g(x(t - L))$. Then the population size is described by

$$(2.3.3) \qquad\qquad x'(t) = g(x(t)) - g(x(t-L)),$$

where g is some differentiable function. We note that every constant function is a solution of (2.3.3).

Cooke and Yorke (1973) prove that: if $t_0 \geq 0$ and if $x(t)$ is a solution of (2.3.3) satisfying $g(x(t_0)) \geq g(x(t_0 - s))$ for $s \in [0, L]$, then $x(t) \geq x(t_0)$ for all $t \geq t_0$. They interpret this result as follows. If the number of births per unit time is as large or larger now (at time t_0) than at every time during the previous life span, then the future population will never be smaller than it is today. They also study the case with probability distributions in the form of integral equations.

Example 2.3.3. The following model for the spread of gonorrhea is considered by Cooke and Yorke (1973). The population is divided into two classes:

$S(t)$ = the number of susceptibles, and

$x(t)$ = the number of infectious.

The rate of new infection depends only on contacts between susceptible and infectious individuals. Since $S(t)$ equals the constant total population minus $x(t)$, the rate is some function $g(x(t))$. Assume that an exposed individual is immediately infectious and stays infectious for a period L (the time for treatment and cure). Then x also satisfies (2.3.3) and the conclusion of Example 2.3.2 holds in this case also. But that conclusion here does not fill one with optimism. They do, however, also have the result that if the infections stay low, they will not later grow.

Example 2.3.4. Cooke and Yorke (1973) also briefly consider a related economic model. Let $x(t)$ be the value of capital stock. Assume that production of new capital depends only on $x(t)$ and that the rate of production is $g(x(t))$. Also, assume that the lifetime of equipment is L and that depreciation is independent of the type of equipment; in particular, at time s after production the value of a unit of capital equipment has decreased in value to $P(s)$ times its original value. Here, $P(0) = 1$, $P(L) = 0$.

Now, at any time t, $x(t)$ equals the sum of capital produced over the period $[t - L, t]$ plus a constant c denoting the value of nondepreciating assets. Thus,

(2.3.4)
$$x(t) = \int_0^L P(s)g[x(t - s)]\,ds + c$$
$$= \int_{t-L}^t P(t - u)g[x(u)]\,du + c.$$

Cooke and Yorke obtain certain boundedness results for (2.3.4) and they pose the problem of determining conditions under which (2.3.4) has a periodic solution.

Example 2.3.5. Kolecki (cf. Minorsky, 1962, p. 516) considers an econometric model in which $I(t)$ is the rate of investment, $U(t)$ is the rate of depreciation of capital goods, and

$$W(t) = I(t) - U(t)$$

is the state of the economy. One assumes that the producers control production according to unfilled orders. Also, the time between placing the order and filling it is a fixed amount h. If $A(t)$ is the production rate, then

$$A(t) = \int_{t-h}^{t} I(s)\, ds,$$

while production is

$$P(t) = \int_{t-h}^{t} A(s)\, ds.$$

Next, if $k(t)$ is the amount of capital and $L(t)$ is the rate of delivery of the product, then

$$k'(t) = L(t) - U(t)$$

and, with $L(t) = I(t - h)$, then

$$k'(t) = I(t - h) - U(t).$$

One observes the economic situation and concludes that investment, capital, and production are related by a linear equation

$$I(t) = m[c + A(t) - nk(t)]$$

for m, n, and c being positive constants. If we let

$$u(t) = I(t) - U(t),$$

these relations result in the delay equation

$$u' - pu - qu(t - h) = 0$$

for p and q constant.

2.4   Source of Models

In an earlier book (Burton, 1983a) we devoted Chapter 4 to a parallel set of physical problems including some from biology, elasticity, dynamics, viscoelasticity, electricity, reactor dynamics, heat flow, and chemical oscillations. The books of Andronow and Chaikin (1949), Bellman and Cooke (1963), Braun (1975), Davis (1962), Minorsky (1962), and Haag (1962) are wonderful sources of problems.

Chapter 3

Fixed-Point Theory

3.1 Compactness in Metric Spaces

Questions concerning existence of solutions of differential equations and the existence of periodic solutions can be formulated in terms of fixed points of mappings. In fact, fixed-point theory was developed, in large measure, as a means of answering such questions. All but one of the fixed-point theorems which we consider here require a setting in a compact subset of a metric space. We consider a variety of differential equations and as the equations become more general it becomes increasingly difficult to find a space in which the set in question is compact. In this section we discuss six compact sets which are central to this book.

Definition 3.1.1. *A pair* (S, ρ) *is a metric space if* S *is a set and* $\rho :$ $S \times S \to [0, \infty)$ *such that when* y, z, *and* u *are in* S *then*

  (a) $\rho(y, z) \geq 0$, $\rho(y, y) = 0$, *and* $\rho(y, z) = 0$ *implies* $y = z$,

  (b) $\rho(y, z) = \rho(z, y)$, *and*

  (c) $\rho(y, z) \leq \rho(y, u) + \rho(u, z)$.

The metric space is complete *if every Cauchy sequence in* (S, ρ) *has a limit in that space. A sequence* $\{x_n\} \subset S$ *is a* Cauchy *sequence if for each* $\varepsilon > 0$ *there exists* N *such that* $n, m > N$ *imply* $\rho(x_n, x_m) < \varepsilon$.

Definition 3.1.2. *A set* L *in a metric space* (S, ρ) *is* compact *if each sequence* $\{x_n\} \subset L$ *has a subsequence with limit in* L.

Definition 3.1.3. *Let* $\{f_n\}$ *be a sequence of functions with* $f_n : [a, b] \to R$, *the reals.*

(a) $\{f_n\}$ *is* uniformly bounded *on* $[a, b]$ *if there exists* $M > 0$ *such that* $|f_n(t)| \leq M$ *for all* n *and all* $t \in [a, b]$.

(b) $\{f_n\}$ *is* equicontinuous *if for any* $\varepsilon > 0$ *there exists* $\delta > 0$ *such that* $t_1, t_2 \in [a, b]$ *and* $|t_1 - t_2| < \delta$ *imply* $|f_n(t_1) - f_n(t_2)| < \varepsilon$ *for all* n.

The first result gives the main method of proving compactness in the spaces in which we are interested.

Theorem 3.1.1. Ascoli-Arzela   *If* $\{f_n(t)\}$ *is a uniformly bounded and equicontinuous sequence of real functions on an interval* $[a, b]$, *then there is a subsequence which converges uniformly on* $[a, b]$ *to a continuous function.*

Proof. As the rational numbers are countable, we may let $t_1, t_2, \ldots$ be a sequence of all rational numbers on $[a, b]$ taken in any fixed order. Consider the sequence $\{f_n(t_1)\}$. This sequence is bounded so it contains a convergent subsequence, say $\{f_n^1(t_1)\}$, with limit $\phi(t_1)$. The sequence $\{f_n^1(t_2)\}$ also has a convergent subsequence, say $\{f_n^2(t_2)\}$, with limit $\phi(t_2)$. If we continue in this way we obtain a sequence of sequences (there will be one sequence for each value of m):

$$f_n^m(t), \qquad m = 1, 2, \ldots; \quad n = 1, 2, \ldots,$$

each of which is a subsequence of all the preceding ones, and such that for each m we have

$$f_n^m(t_m) \to \phi(t_m) \quad \text{as} \quad n \to \infty.$$

We select the diagonal. That is, consider the sequence of functions

$$F_k(t) = f_k^k(t).$$

It is a subsequence of the given sequence and is, in fact, a subsequence of each of the sequences $\{f_n^m(t)\}$, for n large. As $f_n^m(t_m) \to \phi(t_m)$, it follows that $F_k(t_m) \to \phi(t_m)$ as $k \to \infty$ for each m.

We now show that $\{F_k(t)\}$ converges uniformly on $[a, b]$. Let $\varepsilon_1 > 0$ be given, and let $\varepsilon = \varepsilon_1/3$. Denote by δ the number with the property described in the definition of equicontinuity for the number ε. Now, divide the interval $[a, b]$ into p equal parts, where p is any integer larger than $(b - a)/\delta$. Let ξ_j be a rational number in the jth part $(j = 1, \ldots, p)$; then $\{F_k(t)\}$ converges at each of these points. Hence, for each j there exists an integer M_j such that $|F_r(\xi_j) - F_s(\xi_j)| < \varepsilon$ if $r > M_j$ and $s > M_j$. Let M be the largest of the numbers M_j.

If t is in the interval $[a, b]$, it is in one of the p parts, say the jth; so $|t - \xi_j| < \delta$, and $|F_k(t) - F_k(\xi_j)| < \varepsilon$ for every k. Also, if $r > M \geq M_j$ and $s > M$, then $|F_r(\xi_j) - F_s(\xi_j)| < \varepsilon$. Hence, if $r > M$ and $s > M$ then

$$|F_r(t) - F_s(t)| = |(F_r(t) - F_r(\xi_j)) + (F_r(\xi_j) - F_s(\xi_j)) - (F_s(t) - F_s(\xi_j))|$$

$$\leq |F_r(t) - F_r(\xi_j)| + |F_r(\xi_j) - F_s(\xi_j)| + |F_s(t) - F_s(\xi_j)|$$

$$< 3\varepsilon = \varepsilon_1.$$

By the Cauchy criterion for uniform convergence, the sequence $\{F_k(t)\}$ converges uniformly to some function $\phi(t)$. As each $F_k(t)$ is continuous, so is $\phi(t)$. This completes the proof.

An important complete metric space is a Banach space which we now define in several steps.

Definition 3.1.4. *A triple* $(V, +, \cdot)$ *is said to be a* linear *(or* vector*) space over a field* F *if* V *is a set and for* $x, y, z, w, \ldots \in V$ *then*

   I. *For* $x, y \in V$ *there is a unique* $z \in V$ *with* $z = x + y$ *such that*

      (1) $x + y = y + x$,

      (2) $x + (y + w) = (x + y) + w$,

      (3) *there is a unique* $0 \in V$ *with* $0 + x = x$ *for all* $x \in V$, *and*

      (4) *for each* $x \in V$, *there is a unique* $-x \in V$ *with* $x + (-x) = 0$.

  II. *For any* $\alpha, \beta \in F$ *and* $x \in V$, *then* $\alpha \cdot x = \alpha x$ *is defined and*

      (1) $\alpha(\beta x) = (\alpha \beta) x$, *and*

      (2) $1 \cdot x = x$.

 III. *For any* $\alpha, \beta \in F$ *and* $x, y \in V$ *then*

      (1) $(\alpha + \beta) x = \alpha x + \beta x$, *and*

      (2) $\alpha(x + y) = \alpha x + \alpha y$.

The only field used here will be the reals R.

Definition 3.1.5. *A linear space* $(V, +, \cdot)$ *is a* normed space *if for each* $x \in V$ *there is a nonnegative real number* $\|x\|$, *called the* norm *of* x, *such that*

   (1) $\|x\| = 0$ *if and only if* $x = 0$,

   (2) $\|\alpha x\| = |\alpha| \|x\|$ *for each* $\alpha \in R$, *and*

   (3) $\|x + y\| \leq \|x\| + \|y\|$.

Note. A normed space is a vector space and it is a metric space with $\rho(x, y) = \|x - y\|$. But a vector space with a metric is not always a normed space.

Definition 3.1.6. *A* Banach space *is a complete normed space.*

We often say a Banach space is a complete normed vector space.

Example 3.1.1. (a) The space R^n over the field R is a vector space and there are many suitable norms for it. For example, if $x = (x_1, \ldots, x_n)$ then

(1) $\|x\| = \max_i |x_i|$,

(2) $\|x\| = \left[\sum_{i=1}^{n} x_i^2 \right]^{1/2}$, or

(3) $\|x\| = \sum_{i=1}^{n} |x_i|$

are all suitable norms. Norm (2) is the Euclidean norm. Notice that the square root is required in order that $\|\alpha x\| = |\alpha| \|x\|$.

(b) With any of these norms, $(R^n, \|\cdot\|)$ is a Banach space. It is complete because the real numbers are complete.

(c) A set L in $(R^n, \|\cdot\|)$ is compact if and only if it is closed and bounded, as is seen in any text on advanced calculus.

Example 3.1.2. (a) The space $C([a, b], R^n)$ consisting of all continuous functions $f : [a, b] \to R^n$ is a vector space over the reals.

(b) If $\|f\| = \max_{a \leq t \leq b} |f(t)|$, where $|\cdot|$ is a norm in R^n, then it is a Banach space.

(c) For a given pair of positive constants M and K, the set

$$L = \left\{ f \in C([a, b], R^n) \mid \|f\| \leq M, \ |f(u) - f(v)| \leq K|u - v| \right\}$$

is compact. To see this, note first that Ascoli's theorem is also true for vector sequences; apply it to each component successively. If $\{f_n\}$ is any sequence in L, then it is uniformly bounded and equicontinuous. By Ascoli's theorem it has a subsequence converging uniformly to a continuous function $f : [a, b] \to R^n$. But $|f_n(t)| \leq M$ for any fixed t, so $\|f\| \leq M$. Moreover, if we denote the subsequence by $\{f_n\}$ again, then for fixed u and v there exist $\varepsilon_n > 0$ and $\delta_n > 0$ with

$$|f(u) - f(v)| \leq |f(u) - f_n(u)| + |f_n(u) - f_n(v)| + |f_n(v) - f(v)|$$

$$\overset{\text{def}}{=} \varepsilon_n + |f_n(u) - f_n(v)| + \delta_n$$

$$\leq \varepsilon_n + \delta_n + K|u - v| \to K|u - v|$$

as $n \to \infty$. Hence, $f \in L$ and L is compact.

Example 3.1.3. (a) Let $\phi : [a, b] \to R^n$ be continuous and let $\mathcal{S}$ be the set of continuous functions $f : [a, c] \to R^n$ with $c > b$ and with $f(t) = \phi(t)$ for $a \leq t \leq b$. Define $\rho(f, g) = \sup_{a \leq t \leq c} |f(t) - g(t)|$ for $f, g \in \mathcal{S}$.

(b) Then $(\mathcal{S}, \rho)$ is a complete metric space but not a Banach space because $f + g$ is not in $\mathcal{S}$.

Example 3.1.4. (a) Let $(\mathcal{S}, \rho)$ denote the space of bounded continuous functions $f : (-\infty, 0] \to R^n$ with $\rho(\phi, \psi) = \|\phi - \psi\| = \sup_{-\infty < s \leq 0} |\phi(s) - \psi(s)|$ where $|\cdot|$ is the Euclidean norm in R^n.

(b) Then $(\mathcal{S}, \rho)$ is a Banach space.

(c) The set

$$L = \{f \in \mathcal{S} \mid \|f\| \leq 1,\ |f(u) - f(v)| \leq |u - v|\}$$

is not compact in $(\mathcal{S}, \rho)$. To see this, consider the sequence of functions $\{f_n\}$ from $(-\infty, 0]$ into $[0, 1]$ with $f_n(t) = 0$ for $t \leq -n$, $f_n(t)$ is the straight line between the points $(-n, 0)$ and $(0, 1)$. Any subsequence of $\{f_n\}$ converges pointwise to $f \equiv 1$. But $\rho(f_n, 1) = 1$ for all n. Thus, there is no subsequence of $\{f_n\}$ with a limit in $(\mathcal{S}, \rho)$.

Example 3.1.5. (a) Let $(\mathcal{S}, \rho)$ denote the space of continuous functions $f : (-\infty, 0] \to R^n$ with

$$\rho(f, g) = \sum_{n=1}^{\infty} 2^{-n} \rho_n(f, g)/\{1 + \rho_n(f, g)\}$$

where

$$\rho_n(f, g) = \max_{-n \leq s \leq 0} |f(s) - g(s)|$$

and $|\cdot|$ is the Euclidean norm on R^n.

(b) Then $(\mathcal{S}, \rho)$ is a complete metric space. The distance between all functions is bounded by 1.

(c) And $(\mathcal{S}, +, \cdot)$ is a vector space over R.

(d) But $(\mathcal{S}, \rho)$ is not a Banach space because ρ does not define a norm; $\rho(x, 0) = \|x\|$ will not satisfy $\|\alpha x\| = |\alpha| \|x\|$.

(e) The space $(\mathcal{S}, \rho)$ is a locally convex topological vector space. For details and properties see DeVito (1978, pp. 55–69), Reed and Simon (1972), or Dunford and Schwarz (1964). The reason we need to identify the space here is that we shall later use the Schauder-Tychonov fixed-point theorem which is applicable to this particular space.

(f) Let M and K be given positive constants. The set

$$L = \{f \in \mathcal{S} \mid |f(t)| \leq M \text{ on } (-\infty, 0],\ |f(u) - f(v)| \leq K|u - v|\}$$

is compact. To see this, let $\{f_n\}$ be a sequence in L. We must show that there is an $f \in L$ and a subsequence, say $\{f_n\}$ again, such that $\rho(f_n, f) \to 0$ as $n \to \infty$. If we examine ρ we see that $\rho(f_n, f) \to 0$ as $n \to \infty$ just in case f_n converges to f uniformly on compact subsets of $(-\infty, 0]$. Consider $\{f_n\}$ on $[-1, 0]$; it is uniformly bounded and equicontinuous so there is a subsequence, say $\{f_n^1\}$ converging uniformly to some continuous f on $[-1, 0]$. Moreover, the argument in Example 3.1.2 shows that $|f(t)| \le M$ and $|f(u) - f(v)| \le K|u - v|$. Next, consider $\{f_n^1\}$ on $[-2, 0]$; it is uniformly bounded and equicontinuous so there is a subsequence $\{f_n^2\}$ converging uniformly to a continuous function, say f again, on $[-2, 0]$. Continue in this way and have $F_n = f_n^n$ which is a subsequence of $\{f_n\}$ and it converges uniformly on compact subsets of $(-\infty, 0]$ to a function $f \in L$. Thus, L is compact.

Example 3.1.6. Let $g : (-\infty, 0] \to [1, \infty)$ be a continuous strictly decreasing function with $g(0) = 1$ and $g(r) \to \infty$ as $r \to -\infty$.

(a) Let $(\mathcal{S}, |\cdot|_g)$ be the space of continuous functions $f : (-\infty, 0] \to R^n$ for which

$$\sup_{-\infty < t \le 0} |f(t)/g(t)| \overset{\text{def}}{=} |f|_g$$

exists.

(b) Then $(\mathcal{S}, |\cdot|_g)$ is a Banach space.

(c) For positive constants M and K the set

$$L = \{f \in \mathcal{S} \mid |f(t)| \le M \text{ on } (-\infty, 0], \ |f(u) - f(v)| \le K|u - v|\}$$

is compact. Let $\{f_n\}$ be a sequence in L and construct the subsequence of Example 3.1.5 so that $\{F_n\}$ converges to $f \in L$ uniformly on compact subsets of $(-\infty, 0]$. We need to show that if

$$\delta_n = \sup_{-\infty < t \le 0} |(F_n(t) - f(t))/g(t)|$$

then $\delta_n \to 0$ as $n \to \infty$. For a given $\varepsilon > 0$ there exists $T > 0$ such that $2M/g(-T) < \varepsilon/2$. Thus,

$$\delta_n \le (\varepsilon/2) + \max_{-T \le t \le 0} |F_n(t) - f(t)|.$$

Since the convergence is uniform on $[-T, 0]$, there is an N such that $n \ge N$ implies $\max_{-T \le t \le 0} |F_n(t) - f(t)| < \varepsilon/2$.

Example 3.1.7. Let $(\mathcal{S}, |\cdot|_g)$ be as in Example 3.1.6 with (a) and (b) holding. Then the set

$$L = \{f \in \mathcal{S} \mid |f(t)| \le \sqrt{g(t)} \text{ on } (-\infty, 0], \ |f(u) - f(v)| \le |u - v|\}$$

is compact. To prove this, let $\{f_n\} \subset L$. We must show that there is $f \in L$ and a subsequence $\{f_{n_k}\}$ such that $|f_{n_k} - f|_g \to 0$ as $k \to \infty$. Use Ascoli's theorem repeatedly as before and obtain a subsequence $\{f_k^k\}$ converging uniformly to a continuous f on compact subsets of $(-\infty, 0]$ and $|f(t)| \leq \sqrt{g(t)}$. Let $\varepsilon > 0$ be given and find $K > 0$ with $2/\sqrt{g(-K)} < \varepsilon/2$. Find N such that

$$\max_{-K \leq t \leq 0} |f_k^k(t) - f(t)| < \varepsilon/2$$

if $k > N$. Then $k > N$ implies

$$|f_k^k - f|_g \leq \sup_{-\infty < t \leq -K} |(f_k^k(t) - f(t))/g(t)| + \max_{-K \leq t \leq 0} |(f_k^k(t) - f(t))/g(t)|$$

$$\leq \left[\sup_{-\infty < t \leq -K} 2\sqrt{g(t)}/g(t) \right] + \max_{-K \leq t \leq 0} |f_k^k(t) - f(t)|$$

$$\leq \varepsilon.$$

It is readily established that

$$|f(u) - f(v)| \leq |u - v|.$$

Exercise 3.1.1. Return to Example 3.1.5 and suppose there is a continuous function $r : (-\infty, 0] \to [0, \infty)$ and a positive constant K with

$$L = \{f \in \mathcal{S} \mid |f(t)| \leq r(t) \text{ on } (-\infty, 0], \ |f(u) - f(v)| \leq K|u - v|\}.$$

Show that L is compact in $(\mathcal{S}, \rho)$. This will show that $(\mathcal{S}, \rho)$ is much richer in compact sets than is $(\mathcal{S}, |\cdot|_g)$.

3.2 Contraction Mappings

Recall from Chapter 1 that an initial value problem

(3.2.1) $$x' = f(t, x), \qquad x(t_0) = x_0$$

can be expressed as an integral equation

(3.2.2) $$x(t) = x_0 + \int_{t_0}^t f(s, x(s)) \, ds$$

from which a sequence of functions $\{x_n\}$ may be inductively defined by

$$x_0(t) = x_0, \qquad x_1(t) = x_0 + \int_{t_0}^t f(s, x_0) \, ds,$$

and, in general,

$$(3.2.3) \qquad x_{n+1}(t) = x_0 + \int_{t_0}^{t} f(s, x_n(s)) \, ds.$$

This is called Picard's method of successive approximations and, under liberal conditions on f, one can show that $\{x_n\}$ converges uniformly on some interval $|t - t_0| \le k$ to some continuous function, say $x(t)$. Taking the limit in the equation defining $x_{n+1}(t)$, we pass the limit through the integral and have

$$x(t) = x_0 + \int_{t_0}^{t} f(s, x(s)) \, ds$$

so that $x(t_0) = x_0$ and, upon differentiation, we obtain $x'(t) = f(t, x(t))$. Thus, $x(t)$ is a solution of the initial value problem

Banach realized that this was actually a fixed-point theorem with wide application. For if we define an operator P on a complete metric space $([t_0, t_0 + k], R)$ with the supremum norm $\| \cdot \|$ (see Example 3.1.2) by $x \in C$ implies

$$(3.2.4) \qquad (Px)(t) = x_0 + \int_{t_0}^{t} f(s, x(s)) \, ds$$

then a fixed point of P, say $P\phi = \phi$, is a solution of the initial value problem.

The idea had two outstanding features. First, it had application to problems in every area of mathematics which used complete metric spaces. And it was clean. For example, the standard muddy and shaky proofs of implicit function theorems became clear and solid using the fixed-point theory. We will use it here to prove existence of solutions of various kinds of differential equations.

Definition 3.2.1. *Let* (S, ρ) *be a complete metric space and* $P : S \to S$. *The operator* P *is a* contraction operator *if there is an* $\alpha \in (0, 1)$ *such that* $x, y \in S$ *imply*

$$\rho(Px, Py) \le \alpha \rho(x, y).$$

Theorem 3.2.1. Contraction Mapping Principle *Let* (S, ρ) *be a complete metric space and* $P : S \to S$ *a contraction operator. Then there is a unique* $x \in S$ *with* $Px = x$. *Furthermore, if* $y \in S$ *and if* $\{y_n\}$ *is defined inductively by* $y_1 = Py$ *and* $y_{n+1} = Py_n$, *then* $y_n \to x$, *the unique fixed point. In particular, the equation* $Px = x$ *has one and only one solution.*

Proof. Let $x_0 \in S$ and define a sequence $\{x_n\}$ in S by $x_1 = Px_0$, $x_2 = Px_1 = P^2 x_0, \ldots, x_n = Px_{n-1} = P^n x_0$. To see that $\{x_n\}$ is a Cauchy sequence, note that if $m > n$ then

$$\rho(x_n, x_m) = \rho(P^n x_0, P^m x_0)$$
$$\leq \alpha\rho(P^{n-1}x_0, P^{m-1}x_0)$$
$$\vdots$$
$$\leq \alpha^n \rho(x_0, x_{m-n})$$
$$\leq \alpha^n \{\rho(x_0, x_1) + \rho(x_1, x_2) + \cdots + \rho(x_{m-n-1}, x_{m-n})\}$$
$$\leq \alpha^n \{\rho(x_0, x_1) + \alpha\rho(x_0, x_1) + \cdots + \alpha^{m-n-1}\rho(x_0, x_1)\}$$
$$= \alpha^n \rho(x_0, x_1)\{1 + \alpha + \cdots + \alpha^{m-n-1}\}$$
$$\leq \alpha^n \rho(x_0, x_1)\{1/(1-\alpha)\}.$$

Because $\alpha < 1$, the right side tends to zero as $n \to \infty$. Thus, $\{x_n\}$ is a Cauchy sequence and $(\mathcal{S}, \rho)$ is complete so it has a limit $x \in \mathcal{S}$. Now P is certainly continuous so

$$Px = P\left(\lim_{n\to\infty} x_n\right) = \lim_{n\to\infty}(Px_n) = \lim_{n\to\infty} x_{n+1} = x$$

and x is a fixed point. To see that x is the unique fixed point, let $Px = x$ and $Py = y$. Then

$$\rho(x, y) = \rho(Px, Py) \leq \alpha\rho(x, y)$$

and, because $\alpha < 1$, we conclude that $\rho(x, y) = 0$ so that $x = y$. This completes the proof.

In applying this result to (3.2.1), a distressing event occurred which we now briefly describe. Assume that f is continuous and satisfies a global Lipschitz condition in x, say

$$|f(t, x_1) - f(t, x_2)| \leq L|x_1 - x_2|$$

for $t \in R$ and $x_1, x_2 \in R^n$. Then by (3.2.4) we obtain (for $t \geq t_0$)

$$|Px_1(t) - Px_2(t)| = \left| \int_{t_0}^t [f(s, x_1(s)) - f(s, x_2(s))] \, ds \right|$$
$$\leq \int_{t_0}^t L|x_1(s) - x_2(s)| \, ds$$

so that if $\| \cdot \|$ is the sup norm on continuous functions on $[t_0, t_0 + k]$, then

$$\|Px_1 - Px_2\| \leq Lk\|x_1 - x_2\|.$$

This is a contraction if $Lk = \alpha < 1$. Now L is fixed and we take k small enough that $Lk < 1$. This gives a fixed point which is a solution of (3.2.1) on $[t_0, t_0 + k]$.

But the distressing part is that this interval is shorter than the one given by the results of Picard's successive approximations. While this can be satisfactorily dealt with in most cases of interest, it is upsetting. Fortunately there are two ways to cure it. Hale (1969) adopts a different metric which resolves the discrepancy. A different way is through use of asymptotic fixed-point theorems. We shall see two other asymptotic fixed-point theorems, Browder's and Horn's, in addition to the following one.

Theorem 3.2.2. *Let* $(\mathcal{S}, \rho)$ *be a complete metric space and suppose that* $P : \mathcal{S} \to \mathcal{S}$ *such that* P^m *is a contraction for some fixed positive integer* m. *Then* P *has a fixed point in* $\mathcal{S}$.

Proof. Let x be the unique fixed point of P^m, $P^m x = x$. Then $PP^m x = Px$ and $PP^m x = P^m Px$ so $P^m Px = Px$. Thus, Px is also a fixed point of P^m and so, by uniqueness, $Px = x$. Thus, x is a fixed point of P. Moreover, it is unique because if $Py = y$, then $P^m y = y$ so $x = y$. This completes the proof.

The term "contraction" is used in several different ways in the literature. Our use is sometimes denoted by "strict contraction." The property $\rho(Px, Py) \leq \rho(x, y)$ is sometimes called "contraction" but it has limited use in fixed-point theory. A concept in between these two which is frequently useful is portrayed in the next result.

Theorem 3.2.3. *Let* $(\mathcal{S}, \rho)$ *be a compact nonempty metric space,*

$$P : \mathcal{S} \to \mathcal{S}, \quad and \quad \rho(Px, Py) < \rho(x, y)$$

for $x \neq y$. *Then* P *has a unique fixed point.*

Proof. We have

$$\rho(x, Px) \leq \rho(x, y) + \rho(y, Px) \leq \rho(x, y) + \rho(y, Py) + \rho(Py, Px)$$

and since $\rho(Py, Px) \leq \rho(x, y)$ we conclude

$$\rho(x, Px) - \rho(y, Py) \leq 2\rho(x, y).$$

Interchanging x and y yields

$$|\rho(x, Px) - \rho(y, Py)| \leq 2\rho(x, y).$$

Thus the function $B : \mathcal{S} \to [0, \infty)$ defined by $B(x) = \rho(x, Px)$ is continuous on $\mathcal{S}$. The compactness of $\mathcal{S}$ yields $z \in \mathcal{S}$ with $\rho(z, Pz) = \rho(Pz, z) =$

$\inf_{x \in S} \rho(x, Px)$. If $\rho(Pz, z) \neq 0$ then $0 \leq \rho(P(Pz), Pz) < \rho(Pz, z)$ contradicting the infimum property. Thus $\rho(Pz, z) = 0$ and $Pz = z$. If there is another distinct fixed point, say $Py = y$, then $\rho(y, z) = \rho(Py, Pz) < \rho(y, z)$, a contradiction for $y \neq z$. This completes the proof.

Notice that the successive approximations are constructive in spirit. At least in theory one may begin with $x_0 \in S$, compute $x_1, \ldots, x_n$. Frequently one is interested in determining just how near x_0 and x_n are to that unique fixed point x. The next result gives an approximation.

Theorem 3.2.4. *If (S, ρ) is a complete metric space and $P : S \to S$ is a contraction operator with fixed point x, then for any $y \in S$ we have*

(a) $\rho(x, y) \leq \rho(Py, y)/(1 - \alpha)$

and

(b) $\rho(P^n y, x) \leq \alpha^n \rho(Py, y)/(1 - \alpha)$.

Proof. To prove (a) we note that

$$\rho(y, x) \leq \rho(y, Py) + \rho(Py, Px) \leq \rho(y, Py) + \alpha\rho(y, x)$$

so that

$$\rho(y, x)(1 - \alpha) \leq \rho(y, Py).$$

For (b), recall that in the proof of Theorem 3.2.1 we had

$$\rho(P^n y, P^m y) \leq \alpha^n \rho(y, Py)/(1 - \alpha).$$

As $m \to \infty$, $P^m y \to x$ so that

$$\rho(P^n y, x) \leq \alpha^n \rho(y, Py)/(1 - \alpha).$$

This completes the proof.

3.3 Existence Theorems for Linear Equations

This section is an introduction to the application of fixed-point theorems. In Chapter 1 we obtained two existence theorems. The following results will offer alternate, and generally more streamlined, proofs. We are interested in periodic solutions on the whole real axis and so our equations are defined for all real t. Thus, let $A(t)$ be an $n \times n$ matrix of functions continuous on $(-\infty, \infty)$, $C(t, s)$ be an $n \times n$ matrix of functions continuous on $(-\infty, \infty) \times (-\infty, \infty)$, and $f : (-\infty, \infty) \to R^n$ be continuous. Then

(3.3.1) $$x' = A(t)x + f(t)$$

is a system of linear differential equations,

$$(3.3.2) \qquad x' = A(t)x + \int_0^t C(t, s)x(s)\, ds + f(t)$$

is a system of linear integrodifferential equations, and

$$(3.3.3) \qquad x(t) = f(t) + \int_0^t C(t, s)x(s)\, ds$$

is a system of linear integral equations.

Theorem 3.3.1. *Let* $(t_0, x_0) \in R^{n+1}$ *be given. Then there is a unique function* $\psi : (-\infty, \infty) \to R^n$ *with*

$$\psi'(t) = A(t)\psi(t) + f(t), \qquad -\infty < t < \infty,$$

and $\psi(t_0) = x_0$.

Proof. From (3.3.1) and $x(t_0) = x_0$ we obtain

$$x(t) = x_0 + \int_{t_0}^t A(s)x(s)\, ds + \int_{t_0}^t f(s)\, ds.$$

Let $S > 0$ be given and find $M > 0$ with $|A(t)| \le M$ for $|t_0 - t| \le S$. Let $(\mathcal{S}, \|\cdot\|)$ be the Banach space of continuous functions $g : [t_0 - S, t_0 + S] \to R^n$ with the supremum norm. Define $P : \mathcal{S} \to \mathcal{S}$ by $g \in \mathcal{S}$ implies

$$(Pg)(t) = x_0 + \int_{t_0}^t A(s)g(s)\, ds + \int_{t_0}^t f(s)\, ds.$$

Then for $g, h \in \mathcal{S}$ we have

$$|(Pg)(t) - (Ph)(t)| \le \left| \int_{t_0}^t |A(s)[g(s) - h(s)]|\, ds \right| \le MS\|g - h\|$$

so that

$$\|Pg - Ph\| \le MS\|g - h\|.$$

Thus, if $MS = \alpha < 1$, then P is a contraction and there is a unique $\phi \in \mathcal{S}$ with $P\phi = \phi$, a solution of (3.3.1) on $|t - t_0| \le S$ when $MS < 1$. But we have claimed the result for any $S > 0$. Thus, we will let S be arbitrary and find an integer m so that P^m is a contraction. In fact,

$$|(P^m g)(t) - (P^m h)(t)| \le M^m |t - t_0|^m \|g - h\|/m!,$$

as we now show by induction. The result is already proved for $m = 1$. Assume

$$|(P^k g)(t) - (P^k h)(t)| \le M^k |t - t_0|^k \|g - h\|/k!.$$

Then

$$|(P^{k+1}g)(t) - (P^{k+1}h)(t)| \leq \left| \int_{t_0}^{t} |A(s)||(P^k g)(s) - (P^k h)(s)| \, ds \right|$$

$$\leq M^{k+1} \|g - h\| \left| \int_{t_0}^{t} |s - t_0|^k \, ds/k! \right|$$

$$\leq M^{k+1} \|g - h\| |t - t_0|^{k+1}/(k+1)!.$$

Thus,

$$\|P^m g - P^m h\| \leq M^m S^m \|g - h\|/m!$$

and for large m we have $(MS)^m/m! < 1$, completing the proof.

While one frequently considers (3.3.1) for $t < 0$, it is most unusual to consider (3.3.2) for $t < 0$. The latter is an equation with memory; for $t > 0$ it remembers its past, namely $x(t)$ for $0 \leq s < t$. But if $t < 0$ and t is time, then $x(t)$ seems to remember its future. Nevertheless there are good reasons to consider (3.3.2) for $t < 0$ as well as $t > 0$.

If $t_0 = 0$, then (3.3.2) may be treated in a manner almost identically to that for (3.3.1). But if $t_0 \neq 0$ then, in order to obtain a solution of (3.3.2), we must specify $x(t)$ between $t = 0$ and $t = t_0$ so that (3.3.2) will make sense.

Thus, as we saw in Chapter 1 an *initial value problem* for (3.3.2) consists of a $t_0 \in R$, say $t_0 > 0$, and a continuous *initial function* $\phi : [0, t_0] \to R^n$. We then seek a function $\psi : [0, \infty) \to R^n$ with $\psi(t) = \phi(t)$ on $[0, t_0]$ and for $t > t_0$,

$$\psi'(t) = A(t)\psi(t) + \int_0^t C(t, s)\psi(s) \, ds + f(t).$$

In Chapter 1 we made a change of variable resulting in ϕ being combined with f so that the initial value problem for (3.3.2) becomes almost identical for (3.3.1); however, the fixed-point theorem is more interesting if we retain ϕ.

Theorem 3.3.2. *For each $t_0 > 0$ and each continuous function $\phi : [0, t_0] \to R^n$, there is one and only one continuous function $\psi : [0, \infty) \to R^n$ with $\psi(t) = \phi(t)$ on $[0, t_0]$ and satisfying (3.3.2) on (t_0, ∞). We denote the solution by $x(t, t_0, \phi)$.*

Proof. Integrate (3.3.2) from t_0 to t obtaining

$$x(t) = x(t_0) + \int_{t_0}^{t} \left[A(s)x(s) + \int_0^s C(s, u)x(u) \, du \right] ds + \int_{t_0}^{t} f(s) \, ds.$$

Let $K > t_0$ be given and let $(\mathcal{S}, \rho)$ be the complete metric space of continuous functions $g : [0, K] \to R^n$ with $g(t) = \phi(t)$ on $[0, t_0]$ and $g, h \in \mathcal{S}$ implies

$$\rho(g, h) = \max_{t_0 \leq s \leq K} |g(s) - h(s)|$$

and $|\cdot|$ is the Euclidean norm. (Of course, $(\mathcal{S}, \rho)$ is not a Banach space.) Define $P : \mathcal{S} \to \mathcal{S}$ by $g \in \mathcal{S}$ implies

$$(Pg)(t) = \phi(t_0) + \int_{t_0}^{t} \left[A(s)g(s) + \int_{0}^{s} C(s, u)g(u)\, du \right] ds + \int_{t_0}^{t} f(s)\, ds$$

for $t_0 \leq t \leq K$, while $(Pg)(t) = \phi(t)$ on $[0, t_0]$. Then for $t_0 \leq t \leq K$ and $g, h \in \mathcal{S}$ we have

$$|(Pg)(t) - (Ph)(t)|$$

$$\leq \left| \int_{t_0}^{t} \left\{ A(s)[g(s) - h(s)] + \int_{0}^{s} C(s, u)[g(u) - h(u)]\, du \right\} ds \right|$$

$$\leq \rho(g, h) \int_{t_0}^{t} \left[|A(s)| + \int_{0}^{t} |C(s, u)|\, du \right] ds$$

$$\leq \rho(g, h)[K - t_0]M$$

where

$$M = \max_{t_0 \leq s \leq K} \left[|A(s)| + \int_{0}^{s} |C(s, u)|\, du \right].$$

If $M[K - t_0] = \alpha < 1$, then P is a contraction and there is a unique fixed point. In any case, one may argue exactly as in the proof of Theorem 3.3.1 that P^m is a contraction for some m so that P has a unique fixed point. That will complete the proof.

Remark. Since $K > t_0$ is arbitrary the theorem gives a unique local and global solution. If $y(t, t_0, \phi)$ is any solution of (3.3.2) on any interval past t_0, it agrees with $x(t, t_0, \phi)$ so long as it is defined.

We want to clearly establish that when the initial condition for (3.3.2) is $x(0) = x_0$, then there is a function satisfying (3.3.2) on $(-\infty, \infty)$. The proof will also establish a unique solution of (3.3.3) on $(-\infty, \infty)$.

Theorem 3.3.3. *For each $x_0 \in R^n$ there is one and only one function $\psi : (-\infty, \infty) \to R^n$ with $\psi(0) = x_0$ and satisfying (3.3.2) on $(-\infty, \infty)$. We denote ψ by $x(t, 0, x_0)$.*

Proof. Integrate (3.3.2) from 0 to t obtaining

$$x(t) = x_0 + \int_0^t A(s)x(s)\,ds + \int_0^t \int_0^u C(u,s)x(s)\,ds\,du + \int_0^t f(s)\,ds$$

$$= x_0 + \int_0^t A(s)x(s)\,ds + \int_0^t \int_s^t C(u,s)\,du\,x(s)\,ds + \int_0^t f(s)\,ds$$

$$= x_0 + \int_0^t \left[A(s) + \int_s^t C(u,s)\,du\right]x(s)\,ds + \int_0^t f(s)\,ds.$$

Note that this equation is now of the form of (3.3.3). Let K be an arbitrary positive number and consider the Banach space $(\mathcal{S}, \|\cdot\|)$ of continuous functions $g : [-K, K] \to R^n$ with the supremum norm. Define $P : \mathcal{S} \to \mathcal{S}$ by $g \in \mathcal{S}$ implies

$$(Pg)(t) = x_0 + \int_0^t \left[A(s) + \int_s^t C(u,s)\,du\right]g(s)\,ds + \int_0^t f(s)\,ds.$$

Then for

$$M = \max_{\substack{-K \le s \le K \\ -K \le t \le K}} \left[|A(s)| + \left|\int_s^t |C(u,s)|\,du\right|\right],$$

we have for $g, h \in \mathcal{S}$ that

$$|(Pg)(t) - (Ph)(t)| = \left|\int_0^t \left[A(s) + \int_s^t C(u,s)\,du\right](g(s) - h(s))\,ds\right|$$

$$\le M\|g - h\||t|.$$

For $MK < 1$, P is a contraction. For arbitrary K, if m is large enough, P^m is a contraction. Thus P has a unique fixed point, and the proof is complete.

Corollary. *There is one and only one function $x(t)$ satisfying (3.3.3) on* $(-\infty, \infty)$.

3.4 Schauder's Fixed-Point Theorem

In this section we give a brief development of fixed-point theory as it relates to nonlinear differential equations. The focal point is Schauder's theorem, with the important generalizations of Browder and Horn for Banach spaces and Tychonov for locally convex spaces. A proof of Schauder's theorem may be obtained from the finite-dimensional Brouwer's theorem which also has great application in differential equations. In this connection, one frequently sees the Brouwer theorem applied to sets which clearly do not satisfy the hypotheses of Brouwer's theorem; the discrepancy is resolved through the theory of retracts. The sketch given here closely follows Smart (1980, pp. 9–11), and the reader will find the details given there to be most rewarding.

Definition 3.4.1. *A topological space X has the* fixed-point property *if whenever $P : X \to X$ is continuous, then P has a fixed point.*

Definition 3.4.2. *Suppose that X and Y are topological spaces and $f : X \to Y$ with $f(X) = Y$, with f continuous and one to one, and with f^{-1} continuous. Then f is a* homeomorphism.

Theorem 3.4.1. *If X is homeomorphic to Y and if X has the fixed-point property, so does Y.*

Proof. Let $g : Y \to Y$ be continuous and let $f : X \to Y$ be a homeomorphism. Define a function $P : X \to X$ by $x \in X$ implies $P(x) = f^{-1}(g(f(x)))$, which is continuous by the composite function theorem. Hence, there is a point x with $P(x) = x$ or $x = f^{-1}(g(f(x)))$ so that $f(x) = g(f(x))$ or g has a fixed point, $f(x)$. This completes the proof.

Definition 3.4.3. *A topological space X is a* retract *of Y if $X \subset Y$ and there exists a continuous function $r : Y \to X$ such that r is the identity on X.*

Definition 3.4.4. *A set S in a vector space is* convex *if $x, y \in S$ and $0 \le k \le 1$ imply $kx + (1 - k)y \in S$.*

In this discussion, E^n denotes R^n with the Euclidean norm.

Example 3.4.1. A closed convex nonempty subset X of E^n is a retract of any $Y \supset X$.

To see this, define $r : Y \to X$ by $r(x) = x$ if $x \in X$ and, if $y \notin X$, then the closedness and convexity of X implies there is a unique point $x \in X$ closest to y, so define $r(y) = x$. We leave it as an exercise to show that r is continuous (cf. Langenhop, 1973).

Theorem 3.4.2. *If Y has the fixed-point property and X is a retract of Y then X has the fixed-point property.*

Proof. Let $r : Y \to X$ be a retraction map. If $T : X \to X$ is continuous, then for $y \in X$ the map $T(r(y))$ maps $Y \to X$. Since $Tr : Y \to Y$ is continuous, it has a fixed point; but, since $Tr : Y \to X$, the fixed point is in X. Thus, there is an $x \in X$ with $T(r(x)) = x$; since $x \in X$, $r(x) = x$, so $T(x) = x$. This completes the proof.

Definition 3.4.5. *A topological space X is contractible to a point $x_0 \in X$ if there is a continuous function $f(x,t)$ with $f : X \times [0,1] \to X$ such that $f(x,0) = x$ and $f(x,1) = x_0$.*

Notation. In Euclidean space E^n,

(i) $B^n = \{x | x_1^2 + \cdots + x_n^2 \leq 1\}$ is the closed n-ball, and

(ii) $S^{n-1} = \{x | x_1^2 + \cdots + x_n^2 = 1\}$ is the $(n-1)$-sphere.

The following ideas form a sketch of the proof of Brouwer's fixed-point theorem. For detail and references see Smart (1980, pp. 10–11).

(a) There is a retraction map of R^n onto B^n.

(b) For $n \geq 0$, S^n is not contractible. This is a result from homology theory.

(c) If Y is contractible then any retract of Y is contractible.

(d) For $n \geq 1$, S^{n-1} is not a retract of B^n. (B^n is contractible, but not S^{n-1}.)

BROUWER'S FIXED-POINT THEOREM.

(i) B^n *has the fixed-point property.*

(ii) *Every compact convex nonempty subset X of E^n has the fixed-point property.*

Part (i) is argued as follows. Suppose there is a continuous function $T : B^n \to B^n$ with no fixed pont. Construct a retraction of B^n onto S^{n-1} as follows (see Fig. 3.1). For $x \in B^n$, find Tx and extend a line from Tx through x to its intersection with S^{n-1} at a point rx. The map is continuous because T is continuous and such a retraction is impossible by (d).

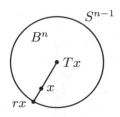

Fig. 3.1

Brouwer's theorem is for a finite-dimensional space, while the upcoming Schauder theorem is for a Banach space which can be infinite dimensional. One may pass from Brouwer's result to Schauder's by way of the following sequence of ideas.

(a) The Hilbert cube H_0 is the subset of l^2 consisting of points $x = (x_1, x_2, \ldots)$ with $|x_r| \leq 1/r$ for all r.

(b) Every compact convex subset K of a Banach space B is homeomorphic under a linear mapping to a compact convex subset of H_0.

(At this point finite dimensionality is brought in allowing the use of Brouwer's theorem.)

(c) P_n is the projection of l^2 onto an n-dimensional subspace given by

$$P_n(x_1, x_2, \ldots) = (x_1, x_2, \ldots, x_n, 0, 0, \ldots).$$

(d) The Hilbert cube H_0 has the fixed-point property.

(e) If Y is a nonempty compact convex subset of a Banach space, then by (b) we have Y homeomorphic to a subset of H_0. Since H_0 has the fixed-point property, by Theorem 3.4.1 it follows that Y does also.

SCHAUDER'S FIRST FIXED-POINT THEOREM. *Any compact convex nonempty subset Y of a Banach space has the fixed-point property.*

It is interesting to note that this has not been the preferred form of Schauder's fixed-point theorem for investigators in the area of differential equations. Most of these have used the next form, and a survey of the literature will reveal that it has caused a good bit of grief.

Definition 3.4.6. *Let M be a subset of a Banach space X and $f : M \to X$. If f is continuous and $f(M)$ is contained in a compact subset of X, then f is a compact mapping.*

SCHAUDER'S SECOND FIXED-POINT THEOREM. Cf. Smart (1980, p. 25) *Let M be a nonempty convex subset of a Banach space X and $P : M \to M$ be a compact mapping. Then P has a fixed point.*

We saw earlier that if P^m is a contraction on a complete matric space, then P itself has a fixed point. This is called an asymptotic fixed-point theorem and it is very useful, as we have already seen. Browder (1970) proved a long line of asymptotic fixed-point theorems related to Schauder's theorem. The most useful of which will now be stated. Incidentally, it too is proved by mapping sets into appropriate finite-dimensional spaces (cf. Browder, 1959).

BROWDER'S FIXED-POINT THEOREM. *Let S and S_1 be open convex subsets of the Banach space X, S_0 a closed convex subset of X, $S_0 \subset S_1 \subset S$, f a compact mapping of S into X. Suppose that for a positive integer m, f^m is well defined on S_1, $\bigcup_{0 \leq j \leq m} f^j(S_0) \subset S_1$, while $f^m(S_1) \subset S_0$. Then f has a fixed point in S_0.*

Definition 3.4.7. *Let Y be a nonempty subset of a topological space X and let $U \subset Y$. Then U is relatively open in Y if $U = V \cap Y$ for some open set V in the topology on X.*

HORN'S (1970) FIXED-POINT THEOREM. *Let $S_0 \subset S_1 \subset S_2$ be convex subsets of the Banach space X, with S_0 and S_2 compact and S_1 open relative to S_2. Let $P : S_2 \to X$ be a continuous mapping such that, for some integer $m > 0$,*

(a) $P^j(S_1) \subset S_2$, $1 \leq j \leq m - 1$,

and

(b) $P^j(S_1) \subset S_0$, $m \leq j \leq 2m - 1$.

Then P has a fixed point in S_0.

In applications to differential equations with a delay, the overriding problem has been to show the mappings compact. Thus, one searches for spaces in which compact sets are plentiful. One such space is a locally convex topological space, as discussed in Section 3.1. It is not a Banach space and the mappings on it are generally not contractions, so another fixed-point theorem is needed.

SCHAUDER-TYCHONOV FIXED-POINT THEOREM. Cf. Smart (1980, p. 15) *Let M be a compact convex nonempty subset of a locally convex topological vector space and $P : M \to M$ be continuous. Then P has a fixed point.*

3.5 Existence Theorems for Nonlinear Equations

It is possible to prove existence and uniqueness theorems for most nonlinear problems using contraction mappings if the functions satisfy a local Lipschitz condition. But the Schauder theorem yields existence from continuity alone. Moreover, once existence is proved it is sometimes possible to prove uniqueness with something less than a Lipschitz condition. In addition, the study of existence using Schauder's theorem can produce some interesting side results.

We begin with a system of nonlinear differential equations

(3.5.1) $x' = f(t, x)$

in which $f : (-\infty, \infty) \times R^n \to R^n$ and f is continuous. Two proofs of the first result are given using Schauder's first and second theorems.

Theorem 3.5.1. *For each (t_0, x_0) in R^{n+1}, there is a solution of (3.5.1), denoted by $x(t, t_0, x_0)$, on an interval $|t - t_0| < \alpha$ for some $\alpha > 0$ and satisfying $x(t_0, t_0, x_0) = x_0$.*

Proof 1. For a given $\varepsilon > 0$ there exists $K > 0$ with $|f(t, x)| \leq K$ when $|t - t_0| \leq \varepsilon$ and $|x - x_0| \leq \varepsilon$. Choose $\delta > 0$ with $\delta < \varepsilon$ and $\delta K \leq \varepsilon$.

Let $\mathcal{S}$ be the space of continuous functions $\phi : [t_0 - \delta, t_0 + \delta] \to E^n$ with the supremum norm. Define

$$M = \{\phi \in \mathcal{S} \,|\, |\phi(t) - x_0| \leq \delta K \quad \text{and} \quad |\phi(u) - \phi(v)| \leq K|u - v|\}.$$

By Ascoli's theorem M is compact. And M is certainly convex.

Define an operator $U : M \to M$ by

$$(U\phi)(t) = x_0 + \int_{t_0}^{t} f(s, \phi(s)) \, ds.$$

Then

$$|(U\phi)(t) - x_0| \leq \left| \int_{t_0}^{t} f(s, \phi(s)) \, ds \right| \leq \delta K$$

and

$$|(U\phi)(u) - (U\phi)(v)| = \left| \int_{v}^{u} f(s, \phi(s)) \, ds \right| \leq K|u - v|,$$

so $U : M \to M$. To see that U is continuous, let $\phi \in M$ and let $\mu > 0$ be given. We must find $\eta > 0$ such that $\psi \in M$ and $\|\phi - \psi\| < \eta$ imply $\|U\phi - U\psi\| \leq \mu$. Now

$$|(U\phi)(t) - (U\psi)(t)| = \left| \int_{t_0}^{t} [f(s, \phi(s)) - f(s, \psi(s))] \, ds \right|$$

and f is uniformly continuous so for the $\mu > 0$ there is an $\eta > 0$ such that $|\phi(s) - \psi(s)| < \eta$ implies $|f(s, \phi(s)) - f(s, \psi(s))| < \mu/\delta$. Thus, for $\|\phi - \psi\| < \eta$ we have

$$|(U\phi)(t) - (U\psi)(t)| \leq \delta \mu / \delta = \mu,$$

as required. By Schauder's first theorem, there is a fixed point which is, of course, a solution.

Remark 3.5.1. The second proof uses Schauder's second theorem and is interesting in that the set M lacks the Lipschitz condition. The integral $\int_{t_0}^t f(s, \phi(s))\, ds$ smooths the image of ϕ so that U becomes a compact mapping. The idea works because the image of ϕ is smoothed instantly, whereas in delay equations it takes a certain amount of time for the image of ϕ to be smoothed.

Proof 2. Proceed with all choices as in Proof 1 except we define

$$\widetilde{M} = \{\phi \in S \mid |\phi(t) - x_0| \le \delta K\}.$$

It is still true that

$$|(U\phi)(u) - (U\phi)(v)| \le K|u - v|$$

so $U : \widetilde{M} \to M$, a compact set. Thus, U will be a compact map and the proof is completed using Schauder's second theorem.

Remark 3.5.2. The classical Cauchy-Peano proof obtains an interval of existence which is usually longer than the one just obtained using Schauder's fixed-point theorem. That interval is obtained as follows. If $f(t, x)$ is continuous for $|x - x_0| \le a$ and $|t - t_0| \le b$, then $|f(t, x)| \le M$ on that set and the interval of existence is found to be $|t - t_0| \le \alpha = \min[b, a/M]$. The asymptotic fixed-point theorem for contraction mappings allowed us to improve the interval of existence for linear systems. One wonders if Browder's or Horn's asymptotic fixed-point theorem could be used to improve the interval of existence in the proof of Theorem 3.5.1. So far, no one seems to have tried to do so.

For our purposes here the length of the interval of existence obtained in Theorem 3.5.1 is not a basic problem. Given any $a > 0$ and $b > 0$ with f continuous for $|t - t_0| \le b$ and $|x - x_0| \le a$, one applies Theorem 3.5.1 and obtains a solution $x(t, t_0, x_0)$ on an interval $[t_0 - \alpha, t_0 + \alpha]$. If $\alpha < b$, then we have a new point $(t_0 + \alpha, x(t_0 + \alpha, t_0, x_0))$ and we can apply Theorem 3.5.1 to obtain a *continuation* of $x(t, t_0, x_0)$. In this way, $x(t, t_0, x_0)$ is continued to the very boundary of the region.

Now if $a = \infty$ and $b = \infty$, then we can choose successively larger finite a and b and extend the interval of existence. If the solution remains bounded, then we can obtain a solution $x(t, t_0, x_0)$ on the whole t-axis. Thus, the procedure can fail only if there is a value of t, say $T > t_0$, with $x(t, t_0, x_0)$ defined on $[t_0, T)$ and $\limsup_{t \to T^-} |x(t, t_0, x_0)| = +\infty$. It is then interesting, and an easy exercise, to show that, in fact, $\lim_{t \to T^-} |x(t, t_0, x_0)| = +\infty$; the key is that we are considering a continuous $f(t, x)$ on a compact t interval so that if $\liminf_{t \to T^-} |x(t, t_0, x_0)| = K < \infty$, then $|f(t, x)| \le M$ on the set $t_0 \le t \le T$ and $|x| \le 2K + 2$. Thus, if there are sequences $\{t_n\} \uparrow T$,

$\{T_n\} \uparrow T$ with $t_n < T_n < t_{n+1}$ and $|x(t_n)| = K + 1$, $|x(T_n)| = 2K + 2$ and $K + 1 \leq |x(t)| \leq 2K + 2$ for $t_n \leq t \leq T_n$ then

$$K \leq |x(t_n) - x(T_n)| = \left| \int_{t_n}^{T_n} f(s, x(s)) \, ds \right| \leq M|T_n - t_n|$$

so that $|T_n - t_n| \geq K/M$. It cannot then be the case that $\{t_n\} \uparrow T$. This is a fundamental property which is missing in the theory of delay equations; for such equations, one must be satisfied with $\limsup_{t \to T^-} |x(t)| = +\infty$.

For our results on periodic solutions it is necessary that solutions depend continuously on initial conditions, a property equivalent to uniqueness of solutions in the case of ordinary differential equations.

Definition 3.5.1. *Solutions of* (3.5.1) *are uniquely determined to the right by* (t_0, x_0) *if whenever* $x(t)$ *and* $y(t)$ *are two solutions of* (3.5.1) *on a common interval, say* $t_0 \leq t \leq t_0 + \alpha$ *with* $x(t_0) = y(t_0) = x_0$, *then* $x(t) = y(t)$ *for* $t_0 \leq t \leq t_0 + \alpha$.

Definition 3.5.2. *Solutions of* (3.5.1) *depend continuously on initial conditions to the right if whenever a solution* $x(t, t_0, x_0)$ *is defined on an interval* $t_0 \leq t \leq t_0 + \alpha$, *given* $\varepsilon > 0$ *there exists* $\delta > 0$ *such that* $|x_0 - x_1| < \delta$ *and* $t_0 \leq t \leq t_0 + \alpha$ *implies* $|x(t, t_0, x_0) - x(t, t_0, x_1)| < \varepsilon$.

The reader will find it an interesting exercise to prove the equivalence of these two concepts. Many different conditions are known for f which ensure uniqueness, and the reader is referred to Hartman (1964) for some of the common ones. But the simplest is the local Lipschitz condition.

Definition 3.5.3. *The function* f *of* (3.5.1) *satisfies a local Lipschitz condition in* x *if for each* (t_0, x_0) *and for each* $M > 0$, *there is a* $K > 0$ *such that* $|x - x_0| \leq M$, $|y - x_0| \leq M$, *and* $|t - t_0| \leq M$ *imply* $|f(t, x) - f(t, y)| \leq K|x - y|$.

Given the local Lipschitz condition, an existence and uniqueness theorem for (3.5.1) may be proved using contraction mappings. But, since existence is already known, Gronwall's inequality easily yields uniqueness.

Theorem 3.5.2. *Suppose that in* (3.5.1) *the function* f *satisfies a local Lipschitz condition in* x. *Then solutions of* (3.5.1) *are unique to the right and depend continuously on initial conditions to the right.*

Proof. For uniqueness, suppose $x(t, t_0, x_0)$ and $y(t, t_0, x_0)$ are two solutions of (3.5.1) on an interval $[t_0, t_0 + \alpha]$ with $x(t_0 + \alpha, t_0, x_0) \neq y(t_0 + \alpha, t_0, x_0)$.

Then for $t_0 \leq t \leq t_0 + \alpha$ we have

$$|x(t, t_0, x_0) - y(t, t_0, x_0)| \leq \int_{t_0}^{t} |f(s, x(s, t_0, x_0)) - f(s, y(s, t_0, x_0))| \, ds$$

$$\leq K \int_{t_0}^{t} |x(s, t_0, x_0) - y(s, t_0, x_0)| \, ds$$

so that by Gronwall's inequality

$$|x(t, t_0, x_0) - y(t, t_0, x_0)| \leq 0 \, e^{K(t-t_0)}.$$

This completes the proof.

We turn now to a system of functional differential equations with finite delay. The initial formulation of the problem given here closely follows that of Yoshizawa (1966, pp. 183–185). This notation has become quite standard throughout the literature. We remark that in later sections we use the symbol x_t in two different ways so the reader is advised to read the full context when encountering that symbol. (See Remark 3.5.4).

For $x \in R^n$, $|x|$ is any norm. Let $\alpha > 0$ and let C denote the Banach space of continuous functions $\phi : [-\alpha, 0] \to R^n$ with

$$\|\phi\| = \sup_{-\alpha \leq s \leq 0} |\phi(s)|.$$

Also,

$$C_H = \{\phi \in C \mid \|\phi\| \leq H\}.$$

If $x : [-\alpha, A) \to R^n$ is continuous and $A > 0$, then for fixed $t \in [0, A)$, x_t denotes the restriction of $x(u)$ to the interval $[t - \alpha, t]$ translated to $[-\alpha, 0]$; that is, $x_t \in C$ and $x_t(s) = x(t+s)$, $-\alpha \leq s \leq 0$.

Let $x'(t)$ denote the right-hand derivative and consider the system of functional differential equations

$$(3.5.2) \qquad\qquad x' = F(t, x_t)$$

with $F : [0, \infty) \times C \to R^n$ being continuous. We suppose that F takes closed bounded sets of $[0, \infty) \times C$ into bounded sets in R^n. Here, F continuous means that for each $(t_1, \phi) \in [0, \infty) \times C$ and for each $\varepsilon > 0$ there exists $\delta > 0$ such that

$$[t \geq 0, \ |t - t_1| < \delta, \ \psi \in C, \ \|\phi - \psi\| < \delta]$$

imply that $|F(t_1, \phi) - F(t, \psi)| < \varepsilon$.

Definition 3.5.4. *A function $x(t_0, \phi)$ is said to be a solution of (3.5.2) with initial function $\phi \in C$ at $t = t_0 \geq 0$ and having value $x(t, t_0, \phi)$, if there is an $A > 0$ such that $x : [t_0 - \alpha, t_0 + A) \to R^n$ with $x_t \in C$ for $t_0 \leq t \leq t_0 + A$, $x_{t_0} = \phi$, and $x(t, t_0, \phi)$ satisfies (3.5.2) on $[t_0, t_0 + A)$.*

Theorem 3.5.3. *If F is continuous on $[0, \infty) \times C$, then for each $\phi \in C$ and $t_0 \geq 0$, there is a solution $x(t_0, \phi)$ of (3.5.2) on an interval $[t_0, t_0 + A)$ for some $A > 0$ and $x(t, t_0, \phi)$ has a continuous derivative for $t > t_0$.*

Proof. Let $t_0 \geq 0$ be given and let $\phi : [t_0 - \alpha, t_0] \to R^n$ be continuous. Thus, $\phi_{t_0} \in C$ and we denote $\phi_{t_0} = \phi$. Since F is continuous there is an $\varepsilon > 0$ and a $\delta > 0$ with $|F(t_0, \phi) - F(t, \psi)| \leq 1$ if $t_0 \leq t \leq t_0 + \varepsilon$ and $\|\phi - \psi\| \leq \delta$. Thus, $|F(t, \psi)| \leq |F(t_0, \phi)| + 1$. Find $a > 0$ and $b > 0$ with $|\phi(t_0)| < a$ and $|F(t_0, \phi)| + 1 < b$. Since $\phi(t_0 + s)$ is continuous on $-\alpha \leq s \leq 0$, there is a $\delta_0 > 0$ such that for $|\tau| \leq \delta_0$, then $|\phi(t_0 + s + \tau) - \phi(t_0 + s)| < \delta/2$. Find a number A with $A \leq \varepsilon$, $A \leq \delta_0$, $bA \leq \delta/2$, $A - \alpha < 0$, and $|\phi(t_0)| + bA < a$.
Define

$$S(a, b, A) = \left\{ x : [t_0 - \alpha, t_0 + A] \to R^n \mid |x(t)| \leq a \text{ on} \right.$$

$$[t_0, t_0 + A], |x(t_1) - x(t_2)| \leq b|t_1 - t_2| \text{ on}$$

$$\left. [t_0, t_0 + A], x(t) = \phi(t) \text{ for } t_0 - \alpha \leq t \leq t_0 \right\}.$$

The set $S(a, b, A)$ is convex and, by Ascoli's theorem, it is compact. Moreover, S is contained in a Banach space. Let P be an operator defined by

$$(Px)(t) = \phi(t) \qquad\qquad \text{for } t \in [t_0 - \alpha, t_0],$$

$$(Px)(t) = \phi(t_0) + \int_{t_0}^{t} F(s, x_s)\, ds \qquad \text{for } t > t_0.$$

Then P is clearly continuous. We will show $P : S \to S$. Let $-\alpha \leq s \leq 0$ and $t_0 \leq u \leq t_0 + A$. If $s + u \leq t_0$, then

$$x(s + u) - \phi(t_0 + s) = \phi(s + u) - \phi(t_0 + s)$$

and hence $|x(s + u) - \phi(t_0 + s)| < \delta/2$ because $0 \leq u - t_0 \leq A \leq \delta_0$. If $u + s > t_0$, then

$$|x(u + s) - \phi(t_0 + s)| \leq |x(u + s) - \phi(t_0)| + |\phi(t_0) - \phi(t_0 + s)|$$

$$\leq b(s + u - t_0) + |\phi(t_0) - \phi(t_0 + s)|$$

$$\leq bA + (\delta/2) < \delta$$

because $bA \leq \delta/2$ and $0 \leq -s \leq u - t_0 \leq A \leq \delta_0$. We conclude that for any $u \in [t_0, t_0 + A]$, $\|\phi_u - x_u\| \leq \delta$. Therefore,

$$|(Px)(t)| \leq |\phi(t_0)| + b(t - t_0) \leq |\phi(t_0)| + bA < a$$

for $t \in [t_0, t_0 + A]$; and for $t_1, t_2 \in [t_0, t_0 + A]$ then $|(Px)(t_1) - (Px)(t_2)| \leq b|t_1 - t_2|$, implying $PS \subset S$. Thus, by Schauder's theorem there is a fixed point x which is a solution with

$$x(t) = \phi(t_0) + \int_{t_0}^{t} F(s, x_s)\, ds$$

so that $x'(t)$ is continuous for $t > t_0$. This completes the proof.

Remark 3.5.2. As in the case of ordinary differential equations, a solution $x(t, t_0, \phi)$ may be continued on any interval $[t_0, L]$ if it remains bounded, by repeated application of the existence theorem. Unlike ordinary differential equations, if a solution $x(t)$ exists on $[t_0, T)$ but cannot be continued past T, then

$$\limsup_{t \to T^-} |x(t)| = +\infty$$

(rather than $\lim_{t \to T^-} |x(t)| = +\infty$ as is the case for ordinary differential equations), as may be seen from examples of Herdman (1980) and Myshkis (1951). The next two propositions provide some detail in support of this remark and are adapted from Hale (1977).

Proposition 3.5.1. *Let Q be an open subset of $[0, \infty) \times C$ and F a continuous function from Q to R^n. If $x(t)$ is a solution of (3.5.2) on $[t_0 - \alpha, \beta)$, $0 \le t_0 < \beta < \infty$, which cannot be continued past β, then for any compact subset P of Q, there is a t_P such that $(t, x_t) \notin P$ for $t_P \le t < \beta$.*

Proof. If the result is false then there is a sequence $\{t_n\} \to \beta^-$ and a compact subset P of Q with $(t_n, x_{t_n}) \in P$. Since P is compact, there is a subsequence, say $\{t_n\}$ again, and ψ with $(\beta, \psi) \in P$ such that $(t_n, x_{t_n}) \to (\beta, \psi)$ as $n \to \infty$. If $-\alpha < -\varepsilon < 0$, then

$$\lim_{n \to \infty} \left[\sup_{s \in [-\alpha, -\varepsilon]} |x_{t_n}(s) - \psi(s)| \right] = 0.$$

But $x_t(s) = x(t + s)$ for $-\alpha \le s \le 0$, and $\alpha > 0$, so this implies that $x(\beta + s) = \psi(s)$ for $-\alpha \le s < 0$. Hence, $\lim_{t \to \beta^-} x(t) = \psi(0)$ and $x(t)$ is continuous on $[t_0 - \alpha, \beta]$ by defining $x(\beta) = \psi(0)$. Since $(\beta, x_\beta) \in P$ we can apply the existence theorem and extend the solution past β. This completes the proof.

Now, we are wanting to say that bounded solutions are continuable to $t = \infty$. And this is true if F is completely continuous (that is, if $F \in C$ and takes closed bounded subsets of $[0, \infty) \times C$ into bounded subsets of R^n).

Proposition 3.5.2. *Let Q be an open subset of $[0, \infty) \times C$, F continuous on Q, and let F take closed bounded subsets of Q into bounded subsets of R^n. If x is a solution of (3.5.2) on an interval $[t_0 - \alpha, \beta)$, $0 \leq t_0 < \beta < \infty$, and if x cannot be continued past β, then for any closed bounded set $P \subset Q$, there is a point t_P such that $(t, x_t) \notin P$ for $t_P \leq t < \beta$.*

Proof. If the result is false, then there is a closed and bounded subset P of Q and a sequence $\{t_n\} \to \beta^-$ with $(t_n, x_{t_n}) \in P$ for all n. Thus, x_{t_n} is bounded for every n. Since $\alpha > 0$ and $t_n \to \beta^-$, it follows that $x(t)$ is bounded for $t_0 - \alpha \leq t < \beta$. Since F takes bounded sets into bounded sets, there is an $M > 0$ with $|F(s, \phi_s)| \leq M$ for (s, ϕ_s) in the closure of $\{(t, x_t) \mid t_0 - \alpha \leq t < \beta\}$. Integrating (3.5.2) from t to $t + s$ with $s \geq 0$ and $t + s < \beta$ yields

$$|x(t + s) - x(t)| = \left| \int_t^{t+s} F(u, x_u) \, du \right| \leq Ms.$$

This means that x is uniformly continuous on $[t_0 - \alpha, \beta)$. Hence, (t, x_t) is in a compact subset of Q. By the previous result, we have a contradiction.

Remark 3.5.3. There is no good reason to restrict $F(t, x_t)$ to $[0, \infty) \times C$; when we seek periodic solutions we will define F on $(-\infty, \infty) \times C$; when we discuss stability we will define F on $[t_0, \infty) \times C_H$. Likewise, it is clear from the proof of existence that we are discussing small subsets of $[0, \infty) \times C$ and F can be defined on $[t_0, t_0 + a] \times \widetilde{C}$ where $a > 0$ and $\widetilde{C}$ is a subset of C. On the other hand we are constrained to discuss solutions in the future rather than for $t < t_0 - \alpha$, unless a very different approach is to be used.

Remark 3.5.4. The notation x_t is very handy when discussing continuity of $F(t, x_t)$ and in constructing existence theorems. The notation is dreadful in examples and in Liapunov theory. We know of no author who has been able to use the notation consistently in such a context. Ordinarily, one feels compelled to be inconsistent and think of x_t as an element of

$$C(t) = \{\phi; [t - \alpha, t] \to R^n \mid \phi \text{ is continuous}\}.$$

And that is what we systematically do in the next chapter.

Definition 3.5.5. *A functional $F : [0, \infty) \times C \to R^n$ is said to be* locally Lipschitz in ϕ *if for each $M > 0$ there exists $K > 0$ such that $\phi, \psi \in C_M$ and $t \in [0, M]$ imply that*

$$|F(t, \phi) - F(t, \psi)| \leq K\|\phi - \psi\|.$$

Definition 3.5.6. *Solutions of* (3.5.2) *depend continuously on initial functions to the right if for each* $\phi \in C$ *and each* $\varepsilon > 0$ *and* $J > 0$, *if* $x(t, t_0, \phi)$ *is a solution of* (3.5.2) *defined on an interval* $[t_0, t_0 + J]$, *there exists a* $\delta > 0$ *such that* $[\psi \in C, \|\phi - \psi\| < \delta, t_0 \le t \le t_0 + J]$ *imply that any solution* $x(t, t_0, \psi)$ *satisfies*

$$|x(t, t_0, \phi) - x(t, t_0, \psi)| < \varepsilon.$$

Remark 3.5.5. If Def. 3.5.6 is satisfied then there is at most one solution $x(t, t_0, \phi)$ for a given $\phi \in C$.

Theorem 3.5.4. *If* F *is continuous on* $[0, \infty) \times C$ *and satisfies a local Lipschitz condition in* ϕ, *then solutions of* (3.5.2) *depend continuously on initial functions to the right.*

Proof. Let $x(t, t_0, \phi)$ be a solution of (3.5.2) on $[t_0, t_0 + J]$. Then for $\psi \in C$ and $x(t, t_0, \psi)$ defined on $[t_0, t_0 + \beta]$ with $\beta \le J$ we can find an M with $\|x_t(t_0, \phi)\| \le M$ and $\|x_t(t_0, \psi)\| \le M$ if $t_0 \le t \le t_0 + \beta$. We can also find a Lipschitz constant L for C_M and for $t_0 \le t \le t_0 + J$. Thus, for $t_0 \le t \le t_0 + \beta$ we have

$$|x(t, t_0, \psi) - x(t, t_0, \phi)| \le |\phi(0) - \psi(0)|$$
$$+ \int_{t_0}^{t} |F(s, x_s(t_0, \phi)) - F(s, x_s(t_0, \psi))| ds$$

$$\le |\phi(0) - \psi(0)| + \int_{t_0}^{t} L \|x_s(t_0, \phi) - x_s(t_0, \psi)\| ds$$

$$\le |\phi(0) - \psi(0)| + L \int_{t_0}^{t} \left(\|\phi - \psi\| + \sup_{t_0 \le u \le s} |x(u, t_0, \phi) - x(u, t_0, \psi)| \right) ds$$

and because the right-hand side is increasing,

$$\sup_{t_0 \le u \le t} |x(u, t_0, \phi) - x(u, t_0, \psi)| \le |\phi(0) - \psi(0)|$$

$$+ L\beta \|\phi - \psi\| + L \int_{t_0}^{t} \sup_{t_0 \le u \le s} |x(u, t_0, \phi) - x(u, t_0, \psi)| ds.$$

By Gronwall's inequality

$$\sup_{t_0 \le u \le t} |x(u, t_0, \phi) - x(u, t_0, \psi)| \le (L\beta + 1) \|\phi - \psi\| e^{L(t - t_0)}$$

and so

$$|x(t, t_0, \phi) - x(t, t_0, \psi)| \le (L\beta + 1) \|\phi - \psi\| e^{L(t - t_0)}.$$

In view of this inequality and the fact that $x(t, t_0, \phi)$ is defined for $t_0 \le t \le t_0 + J$, we readily argue that we may make $\|\phi - \psi\|$ so small as to keep

$x(t, t_0, \psi)$ in a compact set containing $x(t, t_0, \phi)$ so that the same Lipschitz constant L will hold for $t_0 \leq t \leq t_0 + J$ and that we may then take $\beta = J$. That will complete the proof.

Exercise 3.5.1. In the proof of Theorem 3.5.3 examine the set $S(a, b, A)$ and construct a complete metric space. Take the t-interval so small that P will be a contraction mapping if F satisfies a local Lipschitz condition. Then formulate and prove an existence and uniqueness theorem for (3.5.2).

Next, we consider the nonlinear system

$$(3.5.3) \qquad x' = h(t, x) + \int_{-\infty}^{t} q(t, s, x(s)) \, ds$$

in which $h : (-\infty, \infty) \times R^n \to R^n$ and $q : (-\infty, \infty) \times (-\infty, \infty) \times R^n \to R^n$ with both functions being continuous. At times we will ask that h and q be Lipschitz in x in the following sense. The function h satisfies Definition 3.5.3. Also, for each $M > 0$ there exists $K > 0$ such that if $-M \leq s \leq t \leq M$ and $|x_i| \leq M$ for $i = 1, 2$, then

$$|q(t, s, x_1) - q(t, s, x_2)| \leq K|x_1 - x_2|.$$

For a given $t_0 \in R$ and a continuous initial function $\phi : (-\infty, t_0) \to R^n$ we seek a continuous solution $x(t, t_0, \phi)$ satisfying (3.5.3) for $t \in [t_0, t_0 + \beta)$ for some $\beta > 0$ with $x(t, t_0, \phi) = \phi(t)$ for $t \leq t_0$.

The following is our basic assumption for such equations and we will have several results giving conditions under which it holds. We ask that for each $t_0 \in R$ there exists a nonempty convex subset $B(t_0)$ of the space of continuous functions $\phi : (-\infty, t_0] \to R^n$ such that

$$(3.5.4) \qquad \phi \in B(t_0) \quad \text{implies} \quad \int_{-\infty}^{t_0} q(t, s, \phi(s)) \, ds \overset{\text{def}}{=} Q(t, t_0, \phi)$$

$$\text{is continuous on} \quad [t_0, \infty).$$

An integration of (3.5.3) yields

$$(3.5.5) \qquad \begin{aligned} x(t) &= \phi(t_0) + \int_{t_0}^{t} h(s, x(s)) \, ds \\ &+ \int_{t_0}^{t} \int_{t_0}^{u} q(u, s, x(s)) \, ds \, du + \int_{t_0}^{t} Q(u, t_0, \phi) \, du. \end{aligned}$$

Theorem 3.5.5. *Suppose that for each $t_0 \in R$, $B(t_0)$ exists and (3.5.4) holds. If h and q are continuous and locally Lipschitz in x, then for each t_0 and each $\phi \in B(t_0)$, there is a unique solution $x(t, t_0, \phi)$ of (3.5.3) defined on an interval $[t_0, t_0 + \beta)$ for some $\beta > 0$.*

Proof. For a given t_0 let $\phi \in B(t_0)$ and let β_1 be a positive number. Now, for $|x - \phi(t_0)| \leq 1$ and $t_0 \leq s \leq t \leq t_0 + \beta_1$ there is an $M > 0$ with

$$\text{each of} \quad \beta_1, \ |q(t, s, x)|, \ |h(t, x)|, \ |Q(t, t_0, \phi)| \leq M/3.$$

Let $\beta > 0$, $\beta < \beta_1$, $\beta < 1/M$. Consider the complete metric space $(\mathcal{S}, \rho)$ of continuous functions $x : (-\infty, t_0 + \beta] \to R^n$ with $x(t) = \phi(t)$ on $(-\infty, t_0]$,

$$|x(t_1) - x(t_2)| \leq M|t_1 - t_2| \quad \text{for} \quad t_0 \leq t_i \leq t_0 + \beta$$

and with

$$\rho(x_1, x_2) = \max_{t_0 \leq t \leq t_0 + \beta} |x_1(t) - x_2(t)|.$$

In the (t, s, x) space under consideration there is a Lipschitz constant L for both h and q. Define a mapping $P : \mathcal{S} \to \mathcal{S}$ by $x \in \mathcal{S}$ implies

$$(Px)(t) = \phi(t) \quad \text{for} \quad -\infty < t \leq t_0$$

and

$$(Px)(t) = \phi(t_0) + \int_{t_0}^t h(s, x(s)) \, ds$$

$$+ \int_{t_0}^t \int_{t_0}^u q(u, s, x(s)) \, ds \, du + \int_{t_0}^t Q(u, t_0, \phi) \, du$$

for $t_0 \leq t \leq t_0 + \beta$. Now

$$|(Px)(t) - \phi(t_0)| \leq M\beta \leq 1$$

and

$$|(Px)(t_1) - (Px)(t_2)| \leq \left| \int_{t_1}^{t_2} h(s, x(s)) \, ds \right| + \left| \int_{t_1}^{t_2} \int_{t_0}^u q(u, s, x(s)) \, ds \, du \right|$$

$$+ \left| \int_{t_1}^{t_2} Q(u, t_0, \phi) \, du \right| \leq M|t_1 - t_2|.$$

To see that P is a contraction for small β, we let $x_1, x_2 \in \mathcal{S}$ and have

$$\rho(Px_1, Px_2) \leq \sup_{t_0 \leq t \leq t_0 + \beta} \left[\int_{t_0}^t \int_{t_0}^u |q(u, s, x_1(s)) - q(u, s, x_2(s))| \, ds \, du \right.$$

$$\left. + \int_{t_0}^t |h(u, x_1(u)) - h(u, x_2(u))| \, du \right]$$

$$\leq L\beta^2 \|x_1 - x_2\| + L\beta \|x_1 - x_2\|$$

$$= L\beta(\beta + 1)\|x_1 - x_2\|,$$

a contraction with unique fixed point $x(t, t_0, \phi)$ in case

$$L\beta(\beta + 1) = \theta < 1.$$

This completes the proof.

Exercise 3.5.2. In the proof of Theorem 3.5.5 examine the metric space $(\mathcal{S}, \rho)$ and notice that it is, in fact, a compact subset of the Banach space of bounded continuous functions from $(-\infty, t_0 + \beta]$ into R^n with the sup norm topology. Without using the Lipschitz condition, obtain a solution using Schauder's fixed-point theorem.

The question of continual dependence of solutions on initial conditions is both difficult and, at this point, unmotivated. We will find that we want to have continual dependence in terms of a weighted norm which allows the initial functions to be very far apart. The problem will be treated in Section 4.3.

We are interested in knowing that a solution of (3.5.3) which remains bounded can be continued for all future time.

Proposition 3.5.3. *Let* $x(t, t_0, \phi)$ *be a solution of* (3.5.3) *on an interval* $[t_0, \beta)$ *with* $\int_{-\infty}^{t_0} q(t, s, \phi(s)) \, ds$ *continuous for* $t_0 \leq t < \infty$. *If* $x(t) = x(t, t_0, \phi)$ *cannot be continued past* β, *then* $\limsup_{t \to \beta-} |x(t)| = +\infty$.

Proof. If the proposition is false, then there is an $M > 0$ with $|x(t)| \leq M$ on $[t_0, \beta)$. Thus, q being continuous for $t_0 \leq s \leq t \leq \beta$ and $|x| \leq M$ implies that

$$|q(t, s, x(s))| \leq H \quad \text{for} \quad t_0 \leq s \leq t < \beta \quad \text{and some} \quad H > 0;$$

also, $|h(t, x(t))| \leq H$ for $t_0 \leq t < \beta$. It follows that $|x'(t)| \leq R$ for $t_0 \leq t < \beta$ and some $R > 0$. Thus, $\lim_{t \to \beta-} |x(t)|$ exists and $x(t)$ can be uniquely extended to a continuous function on $[t_0, \beta]$ which satisfies the integral equation (3.5.5) on $[t_0, \beta]$. Also, for Q defined by (3.5.4) we have

$$Q(t, \beta, x) \overset{\text{def}}{=} Q(t, t_0, \phi) + \int_{t_0}^{\beta} q(t, s, x(s)) \, ds$$

continuous for $t \geq \beta$. We can, therefore, apply the existence theorem and obtain a continuation past β. This completes the proof.

Our final equation of interest consists of a system of Volterra functional differential equations

(3.5.6) $x' = F(t, x(s); \alpha \leq s \leq t)$

written as

$$(3.5.6) \qquad\qquad x' = F(t, x(\cdot))$$

which is defined and takes values in R^n whenever $t \geq t_0$ and $x : [\alpha, t] \to R^n$ is continuous and bounded. Here, $t_0 \geq \alpha$ is a fixed finite number and α is a fixed number, $\alpha \geq -\infty$. When $\alpha = -\infty$ we mean, of course, that $x : (\alpha, t] \to R^n$. The first coordinate in $F(t, x(\cdot))$ always establishes the domain of $x(\cdot)$ as $[\alpha, t]$. Since we only consider functions x bounded on $[\alpha, t]$, then $\sup_{s \in [\alpha, t]} |x(s)|$ always exists as a finite number. Since we let ϕ be unbounded for (3.5.3), that system is not a special case of (3.5.6).

For any $t \geq t_0$, by

$$(X(t), \| \cdot \|)$$

we shall mean the space of continuous functions $\phi : [\alpha, t] \to R^n$ with

$$\|\phi\| = \sup_{\alpha \leq s \leq t} |\phi(s)|$$

and $| \cdot |$ is any norm on R^n. The symbol

$$X_H(t)$$

denotes those $\phi \in X(t)$ with $\|\phi\| \leq H$. When it is clear that there will be no confusion between (3.5.6) and (3.5.2) we will denote by ϕ_t an element $\phi \in X(t)$; in particular, this will indicate that when we write $\|\phi_t\|$ then we mean $\phi_t \in X(t)$ and the supremum is taken over the interval $[\alpha, t]$.

It is supposed that $F(t, x(\cdot))$ is a continuous function of t for $t_0 \leq t < \infty$ whenever $x : [\alpha, \infty) \to R^n$ is continuous. In fact, a bit more is needed; we suppose that whenever $\{\psi_n\}$ is a sequence in $X_H(t)$ converging to some $\psi \in X_H(t)$ in the supremum norm, then $F(t, \psi_n(\cdot)) \to F(t, \psi(\cdot))$.

It is supposed that F takes bounded sets into bounded sets in the following sense; for each $T > t_0$ and each $H > 0$ there exists $M > 0$ such that $t_0 \leq t \leq T$ and $\phi \in X_H(t)$ implies $|F(t, \phi(\cdot))| \leq M$.

We say that F satisfies a local Lipschitz condition in $x(\cdot)$ if for each $T > t_0$ and each $H > 0$ there exists $K > 0$ such that

$$|F(t, x_t) - F(t, y_t)| \leq K \|x_t - y_t\|$$

when $t_0 \leq t \leq T$ and $x_t, y_t \in X_H(t)$.

The following result, adapted from Driver (1962), is the basic existence and uniqueness theorem for (3.5.6). An easy exercise follows the theorem yielding existence without uniqueness. A Lipschitz argument of the type seen several times already will give continual dependence of solutions on initial functions.

Theorem 3.5.6. *Let F be continuous and satisfy a local Lipschitz condition in x and let $\phi \in X(t_1)$, $t \geq t_0$. Then there exists $h > 0$ and a unique solution $x(t, t_1, \phi)$ of (3.5.6) on $[t_1, t_1 + h)$ with $x(t, t_1, \phi) = \phi(t)$ for $t \leq t_1$.*

Proof. We will find an $h > 0$ and $M > 0$ and consider the complete metric space (S, ρ) of continuous functions $\psi : [\alpha, t_1 + h] \to R^n$ with $\psi(t) = \phi(t)$ on $[\alpha, t_1]$ and

$$|\psi(u) - \psi(v)| \leq M|u - v|$$

for $t_1 \leq u$, $v \leq t_1 + h$. The metric ρ is defined by

$$\rho(\psi_1, \psi_2) = \sup_{t_1 \leq s \leq t_1 + h} |\psi_1(s) - \psi_2(s)|.$$

Now h and M are defined as follows. First, there is a $G > 0$ with $|\phi(t)| \leq G$ for $\alpha \leq t \leq t_1$. Next, if $\overline{\phi}(t) = \phi(t)$ for $[\alpha, t_1 + 1]$ and $\overline{\phi}(t) = \phi(t_1)$ for $t_1 \leq t \leq t_1 + 1$, then $\overline{\phi}(t)$ is continuous on $[\alpha, t_1 + 1]$, while $F(t, \overline{\phi}(\cdot))$ is a continuous function of t on $[t_1, t_1 + 1]$ so there is an M_1 with $|F(t, \overline{\phi}(\cdot))| \leq M_1$ on that interval. It is then possible, using continuity of F, to find $b > 0$ and $M \geq M_1$ with $|F(t, \psi(\cdot))| \leq M$ if $t_1 \leq t \leq t_1 + 1$ whenever $\psi \in X(t_1 + 1)$ and $\|\psi - \overline{\phi}\| \leq b$. We initially take

$$0 < h \leq \min[1, b/M],$$

but h will later be further restricted.

Now, define $P : S \to S$ by $\psi \in S$ implies

$$(Px)(t) = \begin{cases} \phi(t) & \text{for } \alpha \leq t \leq t_1, \\ \phi(t_1) + \int_{t_1}^t F(s, \psi(\cdot))\, ds & \text{for } t_1 \leq t \leq t_1 + h. \end{cases}$$

To see that $P : S \to S$, note that $P\psi$ is continuous because $F(s, \psi(\cdot))$ is continuous. Also,

$$|(P\psi)(u) - (P\psi)(v)| \leq \left| \int_v^u F(s, \psi(\cdot))\, ds \right| \leq M|u - v|.$$

Finally, for $t_1 \leq t \leq t_1 + h$ we have

$$|(P\psi)(t) - \overline{\phi}(t)| \leq \int_{t_1}^t |F(s, \psi(\cdot))|\, ds \leq M|t - t_1| \leq Mh$$

$$\leq Mb/M = b.$$

To see that P is a contraction, note that there is a Lipschitz constant K for the set S and $t_1 \leq t \leq t_1 + h$ so that

$$\rho(P\psi_1, P\psi_2) = \sup_{t_1 \leq t \leq t_1 + h} \left| \int_{t_1}^t [F(s, \psi_1(\cdot)) - F(s, \psi_2(\cdot))]\, ds \right|$$

$$\leq Kh\|\psi_1 - \psi_2\| = Kh\rho(\psi_1, \psi_2)$$

so that if $\theta = Kh < 1$, then P is a contraction with unique fixed point which is the solution. This completes the proof.

Exercise 3.5.3. Review the properties of S and P in the last proof. Use Schauder's fixed-point theorem to prove the existence of a (non-unique) solution without the Lipschitz condition.

Again, we wish to say that if $x(t)$ is a bounded solution then it can be continued for all future time.

Proposition 3.5.4. *Let $F(t, \phi(\cdot))$ be continuous for $t_0 \leq t < \infty$ and $\phi \in X(t)$. Suppose that for each $\gamma > t_0$ and each $H > 0$ there is an M such that $t_0 \leq t \leq \gamma$ and $\phi \in X_H(t)$ imply $|F(t, \phi(\cdot))| \leq M$. If $x(t) = x(t, t_1, \phi)$ is a solution of (3.5.6) on $[t_1, \beta)$ which is noncontinuable then for any $H > 0$ there is a sequence $\{t_n\} \to \beta^-$ with $|x(t_n)| > H$.*

Proof. If the result is false then there is an H such that $|x(t)| \leq H$ for $\alpha \leq t < \beta$. It then follows that $|F(t, x(\cdot))| \leq M$ for $t_1 \leq t < \beta$. Now for $t_1 \leq t \leq t + s < \beta$ we have, upon integration of (3.5.6), that

$$|x(t+s) - x(t)| = \left| \int_t^{t+s} F(u, x(\cdot)) \, du \right| \leq Ms.$$

Hence, x is Lipschitz on $[t_1, \beta)$. For $\{t_n\}$ monotone increasing to β, we see that $\{x(t_n)\}$ is a Cauchy sequence since, for $m > n$, for $t = t_n$, and for $t + s = t_m$, we then have

$$|x(t_m) - x(t_n)| \leq M|t_m - t_n| \to 0$$

as $n, m \to \infty$. Hence, $\lim_{n\to\infty} x(t_n)$ exists and we can uniquely extend x to a Lipschitz function on $[t_1, \beta]$. In addition,

$$x(\beta) = x(t_1) + \int_{t_1}^{\beta} F(s, x(\cdot)) \, ds.$$

The existence theorem will then apply to the initial value problem with

$$\phi(t) = x(t) \qquad \text{for} \qquad \alpha \leq t \leq \beta,$$

yielding an extension of $x(t)$. This completes the proof.

Chapter 4

Limit Sets, Periodicity, and Stability

4.1 Ordinary Differential Equations

4.1.1 Introduction and Motivation

We consider the system of ordinary differential equations

$$(4.1.1) \qquad x' = F(t, x)$$

in which $F : (-\infty, \infty) \times R^n \to R^n$ is continuous. Existence results using sophisticated fixed-point theorems were obtained in Chapter 3. At the end of this section we give an elementary proof that for each (t_0, x_0) there is a solution $x(t, t_0, x_0)$ on an interval $\alpha < t < \beta$ containing t_0 and

(i) if $\alpha > -\infty$ then $\lim_{t \to \alpha^+} |x(t, t_0, x_0)| = +\infty$,

(ii) if $\beta < \infty$ then $\lim_{t \to \beta^-} |x(t, t_0, x_0| = +\infty$.

Moreover, if F satisfies a local Lipschitz condition in x then $x(t, t_0, x_0)$ is unique and is continuous in x_0.

Virtually never can one find the solutions of (4.1.1) and, in the rare cases when they are found, the solutions tend to be very complicated. Thus, one is interested in the qualitative behavior of solutions of (4.1.1) and the central concept is that of an ω-limit point.

Definition 4.1.1. *Let* $x(t, t_0, x_0)$ *be a solution of* (4.1.1) *on* $[t_0, \infty)$. *A point y is an ω-limit point of this solution if there is a sequence* $\{t_n\} \uparrow +\infty$ *with* $\lim_{n \to \infty} x(t_n, t_0, x_0) = y$.

The set of ω-limit points of $x(t, t_0, x_0)$ is called the ω-*limit set*, denoted by $\Gamma^+(t_0, x_0)$. Under reasonable conditions a solution converges to its ω-limit set; this means that if we can locate the ω-limit set, then we will have a better understanding of the behavior of solutions. We note that if $x(t, t_0, x_0)$ is bounded for $t \geq t_0$, then $\{x(n, t_0, x_0)\}$ is bounded (n is a positive integer) and so there is a convergent subsequence $\{x(n_k, t_0, x_0)\}$; hence, bounded solutions have nonempty ω-limit sets.

Frequently, one can detect that part of the differential equation is unrelated to the ω-limit set so that one is able to concentrate on the important part of the equation. In order to motivate our emphasis on periodic solutions we will state without proof a number of basic results on limit sets. These results and extensive references are found in Yoshizawa (1966, pp. 52–59).

Proposition 4.1.1. *Let $x(t, t_0, x_0)$ be a solution of* (4.1.1) *which is bounded for $t \geq t_0$. Then $\Gamma^+(t_0, x_0)$ is compact and $x(t, t_0, x_0) \to \Gamma^+(t_0, x_0)$ as $t \to \infty$.*

Definition 4.1.2. *Let*

$$(4.1.2) \qquad\qquad x' = H(x)$$

with $H : R^n \to R^n$ being continuous. A set $M \subset R^n$ is a semi-invariant set if for each $x_0 \in M$, at least one solution $x(t, 0, x_0)$ remains in M for all $t \geq 0$.

Proposition 4.1.2. *Let $x(t, t_0, x_0)$ be a bounded solution of* (4.1.1) *for $t \geq t_0$ and approach a closed set $L \subset R^n$. Suppose also that $F(t, x)$ satisfies:*

(i) *$F(t, x) \to H(x)$ for $x \in L$ as $t \to \infty$ and on any compact subset of L the convergence is uniform.*

(ii) *For each $\varepsilon > 0$ and each $y \in L$, there is a $\delta(\varepsilon, y) > 0$ and a $T(\varepsilon, y) > 0$ such that if $|x - y| < \delta$ and $t \geq T$, then $|F(t, x) - F(t, y)| < \varepsilon$.*

Under these conditions $\Gamma^+(t_0, x_0)$ is a semi-invariant set for (4.1.2) *and $\Gamma^+(t_0, x_0)$ is the union of solutions of* (4.1.2). *Also, $\Gamma^+(t_0, x_0)$ is connected.*

Additional remarks of this nature can be made if $H(x)$ is replaced by $H(t, x)$ in which $H(t + T, x) = H(t, x)$ for some $T > 0$. Thus, we are interested in understanding the behavior of solutions of (4.1.1) when either

(a) $F(t + T, x) = F(t, x)$ for some $T > 0$

or

(b) $F(t, x) = H(x)$.

Our presentation proceeds as follows. Sections 4.1.2 and 4.1.3 focus on the form of the ω-limit set of a bounded solution of (4.1.2). We first look at the possibilities for the n-dimensional system. Then we specialize the discussion to a pair of first-order scalar autonomous equations. The central result for that system is the Poincaré-Bendixson theorem which says that a bounded solution is either periodic, approaches a periodic solution, or has an equilibrium point in its ω-limit set. While the result is false for higher-order systems, it isolates three areas of major importance on which this book focuses:

(a)   boundedness,

(b)   periodicity, and

(c)   stability of equilibrium points.

Once we know that (4.1.1) has a bounded solution, we are also interested in learning how other solutions relate to the bounded one. This leads to the study of equilibrium solutions. For we recall from Remark 1.1.4 in Chapter 1 that if ϕ is a solution of (4.1.1), then the transformation $y = x - \phi(t)$ maps (4.1.1) into

$$(4.1.3) \qquad\qquad y' = G(t, y)$$

with the property that $G(t, 0) \equiv 0$. That is, 0 is an equilibrium point (or *singular point* or *critical point*) for (4.1.3). Thus discussion of the behavior of solutions near ϕ reduces to a discussion of stability of $y = 0$, about which there is a large theory.

Before turning to the study of ω-limit sets we wish to give an elementary proof of existence of solutions of (4.1.1) so that Section 1.1 of Chapter 1 and Section 4.1. of Chapter 2 will be a self-contained study of the basic theory of ordinary differential equations.

Theorem. Cauchy-Euler-Peano *Let $(t_0, x_0) \in R^{n+1}$ and suppose there are positive constants a, b, and M such that if*

$$D = \big\{(t, x) \mid |t - t_0| \leq a, \ |x - x_0| \leq b\big\}$$

then $F : D \to R^n$ is continuous and $|F(t, x)| \leq M$. Then there is at least one solution $x(t) = x(t, t_0, x_0)$ of

$$(\text{IVP}) \qquad\qquad x' = F(t, x), \qquad x(t_0) = x_0,$$

and $x(t)$ is defined for $|t - t_0| \leq T$ with $T = \min[a, b/M]$.

Proof. We construct the solution on $[t_0, t_0 + T]$. Let j be a positive integer and divide $[t_0, t_0 + T]$ into j equal subintervals $t_0 < t_1 < \cdots < t_n = t_0 + T$, with $t_{k+1} - t_k = \delta_j = T/j$. Define a continuous function $x_j(t)$ on $[t_0, t_0 + T]$ as follows. Let

$$x_j(t) = x_0 + (t - t_0)F(t_0, x_0) \qquad \text{on} \qquad [t_0, t_1]$$

and define

$$x_j(t_1) = y_1.$$

Then

$$x_j(t) = y_1 + (t - t_1)F(t_1, y_1) \qquad \text{on} \qquad [t_1, t_2]$$

and

$$x_j(t_2) = y_2.$$

In this way we define a piecewise linear continuous function on $[t_0, t_0 + T]$. Because of the choices of M and T, the graph remains in D and the "slopes" are bounded by M. Thus, the sequence $\{x_j(t)\}$ is uniformly bounded and equicontinuous. By Ascoli's theorem there is a subsequence, say $\{x_j\}$ again, converging uniformly on $[t_0, t_0 + T]$ to a continuous function $x(t)$.

Convert (IVP) to the integral equation

$$x_j(t) = x(t_0) + \int_{t_0}^t \{F(s, x_j(s)) + \Delta_j(s)\}\, ds$$

where

$$\Delta_j(t) = \begin{cases} x_j'(t) - F(t, x_j(t)) & \text{if } t \neq t_i, \\ 0 & \text{if } t = t_i. \end{cases}$$

The $\Delta_j(t) \to 0$ uniformly as $j \to \infty$. Thus, if we take the limit as $j \to \infty$, by the uniform convergence we obtain

$$x(t) = x(t_0) + \int_{t_0}^t F(s, x(s))\, ds$$

and the proof is complete.

We will frequently say that if a solution remains bounded then it can be continued as a solution to $t = \infty$. Here are the details. We have the solution defined on $[t_0, t_0 + T]$ so we let $\bar{t}_0 = t_0 + T$ and $\bar{x}_0 = x(t_0 + T)$. Let $\overline{M}$ be the bound on F for $|t - \bar{t}_0| \leq a$ and $|x - \bar{x}_0| \leq b$. Use the theorem to obtain a solution $x(t, \bar{t}_0, \bar{x}_0)$ of

(IVP) $$\qquad\qquad x' = F(t, x), \qquad x(\bar{t}_0) = \bar{x}_0$$

for $|t - \bar{t}_0| \leq \min[a, b/\overline{M}]$, called a *continuation* of $x(t, t_0, x_0)$. If F is only continuous then there may be more than one continuation. Repeat

the above process.   Can we be sure that we can continue a solution to
an arbitrary $L > t_0$? Suppose that all continuations obtained in this way
satisfy $|x(t) - x_0| \le J$ for some $J > 0$. Then there is an M^* with $|F(t, x)| \le$
M^* if $|x - x_0| \le J + b$ and $t_0 \le t \le L$. Thus, each continuation is defined
on an additional interval of length $\min[a, b/M^*]$; hence, we can continue
$x(t)$ to L in a finite number of steps.

If F satisfies a local Lipschitz condition in x, then Gronwall's inequality
immediately yields uniqueness; moreover, it yields continuity of $x(t, t_0, x_0)$
in x_0 in the following sense: if $x(t, t_0, x_0)$ is defined on $[t_0, t_0 + L]$ then for
each $\varepsilon > 0$ there is a $\delta > 0$ such that

$$\left[\, |x_0 - x_1| < \delta \qquad \text{and} \qquad t_0 \le t \le t_0 + L \, \right]$$

imply that $|x(t, t_0, x_0) - x(t, t_0, x_1)| < \varepsilon$.

4.1.2   Dynamical Systems

We consider the system

(4.1.2) $$x' = H(x)$$

and assume now that H satisfies a local Lipschitz condition in x. Since
(4.1.2) is autonomous, the unique solution through x_0 may be expressed
as $x(t, x_0)$ or $\phi(t, x_0)$. The locus of points in R^n of $x(t, x_0)$ is called the
orbit of the solution, while the directed path is the *trajectory*. By making
a simple change of variable we can replace (4.1.2) by a system whose orbits
coincide with those of (4.1.2) and which can be extended on $(-\infty, \infty)$.

Definition 4.1.3. *Let G be an open set in R^n and $H, g : G \to R^n$. Then
H and g are said to be orbitally equivalent if the systems $x' = H(x)$ and
$y' = g(y)$ have the same solution curves (including singular points) in G.*

Theorem 4.1.1. *Let $H : R^n \to R^n$ and $h : R^n \to (0, \infty)$ with H and h
continuous. Then H and Hh are orbitally equivalent.*

Proof. Let ϕ be a solution of (4.1.2) on an interval (a, b) with $t_0 \in (a, b)$.
Define $r(t) = \int_{t_0}^{t} [ds/h(\phi(s))]$. As $h(x) > 0$, r is strictly increasing and so
there is an interval (c, d) onto which r maps (a, b) in a one-to-one fashion.
Define $\psi(t)$ on (c, d) by $\psi(s) = \phi(t)$ with $s = r(t)$ and $t = r^{-1}(s)$. Clearly,
ψ and ϕ define the same curve in R^n. Now

$$H(\psi(s)) = H(\phi(t)) = \frac{d}{dt}\phi(t) = \frac{d}{dt}\psi(r(t)) = \frac{d}{dt}\psi(s)$$

$$= \frac{d\psi(s)}{ds}\frac{ds}{dt} = \frac{d\psi(s)}{ds}\frac{1}{h(\phi(t))} = \frac{d\psi(s)}{ds}\frac{1}{h(\psi(s))} \, .$$

Hence, $d\psi(s)/ds = H(\psi(s))h(\psi(s))$ so that $\psi(s)$ is a solution of

$$(4.1.4) \qquad\qquad y' = H(y)h(y).$$

If we replace h by $1/h$ in the definition of r, then for each solution of (4.1.4) we obtain a solution of (4.1.2). That will complete the proof.

Theorem 4.1.2. Vinograd *If H is continuous on $R^n \to R^n$, then there is a function $g : R^n \to R^n$ with H and g orbitally equivalent and each solution of $y' = g(y)$ can be defined on $(-\infty, \infty)$.*

Proof. Define $g(y) = H(y)/[1 + |H(y)|]$ where $|\cdot|$ is the Euclidean length. Then $|g(y)| < 1$ and so any solution $\psi(t)$ of $y' = g(y)$ satisfies $\psi(t) = \psi(0) + \int_0^t g(\psi(s))\, ds$ yielding $|\psi(t)| \leq |\psi(0)| + |t|$. Thus, ψ cannot tend to infinity in finite time and so solutions can be defined on $(-\infty, \infty)$. This completes the proof.

It has proved to be very fruitful to ignore the specific properties of (4.1.2) and study instead just what the possible behavior of solutions can be when it is assumed that solutions behave in a regular fashion relative to continuity.

Definition 4.1.4. *Equation (4.1.2) defines a dynamical system if for each P in R^n*

(a) *the solution $\phi(t, P)$ satisfies $\phi(0, P) = P$,*

(b) *$\phi(t, P)$ is defined for $-\infty < t < \infty$,*

(c) *$\phi(t, P)$ is continuous on $(-\infty, \infty) \times R^n$, and*

(d) *$\phi(t_1 + t_2, P) = \phi(t_2, \phi(t_1, P))$.*

Note that (a) holds by definition of a solution through P; (b) can be taken care of by going to an orbitally equivalent system; (c) is obtained from continual dependence of solutions on initial conditions under proper assumptions on H; and (d) is a statement of uniqueness of solutions.

Covering Assumption. *Throughout this section it will be assumed that (4.1.2) defines a dynamical system.*

Definition 4.1.5. *A point Q is an ω-limit point of $\phi(t, P)$ if there is a sequence $\{t_n\} \uparrow \infty$ with $\lim \phi(t_n, P) = Q$. A point Q is an α-limit point of $\phi(t, P)$ if there is a sequence $\{t_n\} \uparrow \infty$ with $\lim \phi(-t_n, P) = Q$.*

If, for example, a solution $\phi(t, P) \to 0$ as $t \to \infty$, then 0 is the ω-limit point of $\phi(t, P)$.

For this autonomous equation we denote by $A(P)$ and $\Omega(P)$ the α- and ω-limit points, respectively, of $\phi(t, P)$. We reserve the notation $\Gamma^+(t_0, x_0)$ for nonautonomous systems.

Theorem 4.1.3. *If Y is an ω-limit point of $\phi(t, P)$, so is each point of $\phi(t, Y)$.*

Proof. For some sequence $\{t_n\}$ we have $\lim \phi(t_n, P) = Y$ as $t_n \to \infty$. Then $\phi(t_n + \bar{t}, P) = \phi(\bar{t}, \phi(t_n, P))$ by (d). But by (c) this yields $\lim \phi(\bar{t}, \phi(t_n, P)) = \phi(\bar{t}, Y)$, thereby completing the proof as $\bar{t}$ is arbitrary.

Definition 4.1.6. *A set S is invariant if $P \in S$ implies $\phi(t, P) \in S$ for $-\infty < t < \infty$. Also, S is positively invariant if $P \in S$ implies $\phi(t, P) \in S$ for $0 \leq t < \infty$.*

If, for example, P is a singular point, then P is invariant.

Exercise 4.1.1. Prove that $\Omega(P)$ is closed.

In Chapter 1 we mentioned the concept of an entire system being Lagrange stable. It is also fruitful to speak of a single solution being Lagrange stable.

Definition 4.1.7. *$\phi(t, P)$ is L^+-stable (called Lagrange stable) if $\phi(t, P)$ is bounded for $t \geq 0$.*

Theorem 4.1.4. *If $\phi(t, P)$ is L^+-stable, then $\Omega(P)$ is connected.*

Proof. By Exercise 4.1.1, $\Omega(P)$ is closed. By hypothesis, $\phi(t, P)$ is bounded for $t > 0$ and so $\Omega(P)$ is compact. If $\Omega(P)$ is not connected then one may argue that $\Omega(P) = M \cup N$ where M and N are nonempty disjoint compact sets. Let the distance from M to N be $d > 0$ and let R and H be $d/3$ neighborhoods of M and N, respectively. Now there are sequences $\{t'_n\}$ and $\{t''_n\} \uparrow +\infty$ with $t'_n < t''_n$, $\phi(t'_n, P) \in R$, and $\phi(t''_n, P) \in H$. As ϕ is connected, there is a sequence $\{t^*_n\}$ with $t'_n < t^*_n < t''_n$ and $\phi(t^*_n, P) \in (R \cup H)^C$. As $\{\phi(t^*_n, P)\}$ is bounded, a subsequence converges. Thus, its limit is in $\Omega(P)$ and lies outside $R \cup H$, a contradiction.

Corollary. *For (4.1.2) a second-order system $(n = 2)$, if $\Omega(P)$ contains a nontrivial periodic orbit, then it contains nothing else.*

In summary, if $\phi(t, P)$ is L^+-stable, then $\Omega(P)$ is compact, connected, invariant, and the union of orbits. In fact, $\phi(t, P)$ converges to $\Omega(P)$.

If $\phi(t, P)$ is not bounded, then $\Omega(P)$ need not be connected. In Fig. 4.1, $\Omega(P)$ consists of the two straight lines, while $\phi(t, P)$ spirals unboundedly.

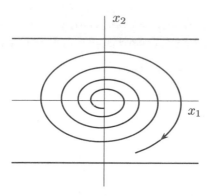

Fig. 4.1

Definition 4.1.8. *If* $\Omega(P)$ *is empty, then* $\phi(t, P)$ *is* departing *in the positive direction. If* $\Omega(P) \neq \emptyset$, *but contains no points on the orbit of* $\phi(t, P)$, *then it is said to be* asymptotic *in the positive direction.*

Definition 4.1.9. $\phi(t, P)$ *is* stable in the sense of Poisson *in the positive direction if it has* ω-*limit points on* $\phi(t, P)$.

If there exists $T > 0$ such that $\phi(t+T, P) = \phi(t, P)$ for all t then $\phi(t, P)$ is periodic and each point of $\phi(t, P)$ is an ω-limit point of $\phi(t, P)$. If P is a singular point or if $\phi(t, P)$ is a periodic solution, then $\phi(t, P)$ is Poisson stable. We will see that for two-dimensional systems these are the only possibilities.

Theorem 4.1.5. *If for every* $\varepsilon > 0$ *there exists a* Q *with* $\phi(t, Q)$ *in* $S(P, \varepsilon)$ *for all* $t \geq 0$ *or all* $t \leq 0$, *then* P *is a singular point.*

Proof. If P is not a singular point then there exists $\bar{t} > 0$ such that $\overline{P} = \phi(\bar{t}, P) \neq P$ (i.e., if P is not a singular point, then $\phi(t, P)$ moves). Also, $P^* = \phi(-\bar{t}, P) \neq P$, because if $\phi(-\bar{t}, P) = P$, then ϕ would be periodic and $\overline{P} = P$. But $\phi(t, P)$ is continuous in P and t by (c) (in the definition of a dynamical system) and so $\phi(t, P)$ is a continuous function of P uniformly for $-\bar{t} \leq t \leq \bar{t}$ and P in some compact set.

Let $d = \min[\rho(P, \overline{P}), \rho(P, P^*)]$ where $\rho(P_1, P_2)$ is the Euclidean distance from P_1 to P_2. By the uniform continuity there exists $\delta = \delta(d) > 0$ such that $\rho(\phi(t, P), \phi(t, Q)) < d/3$ if $\rho(P, Q) < \delta$ for $-\bar{t} \leq t \leq \bar{t}$. Take $\delta < d/3$. Then $\phi(t, Q)$ for any Q in $S(P, \delta)$ does not remain in $S(P, \delta)$ for $|t| = |\bar{t}|$. This completes the proof.

4.1.3   Two-Dimensional Systems, The Poincaré-Bendixson Theorem

It is convenient to express our system as

$$(4.1.5) \qquad \begin{aligned} x' &= P(x, y), \\ y' &= Q(x, y) \end{aligned}$$

where x and y are scalars, P and Q are at least continuous, and (4.1.5) defines a dynamical system.

Lemma 4.1.1. *If X is a point in R^2 which is not a singular point then there exists $\varepsilon > 0$ such that the ε-neighborhood of X, say $S(X, \varepsilon)$, contains no singular points of (4.1.5) either in its interior or on its boundary and the angle between the vectors $[P(X), Q(X)]$ and $[P(Z), Q(Z)]$ is smaller than $\pi/4$ for any Z in $S(X, \varepsilon)$.*

Proof. If θ is the angle, then $\cos \theta$ is the inner product of the vectors divided by their lengths. As P and Q are continuous and X is not a singular point, we may make the quotient as near 1 as desired by taking Z sufficiently near X.

A neighborhood $S(X, \varepsilon)$ satisfying Lemma 4.1.1 is called a *small neighborhood* of X.

If $\phi(t, Y)$ is the solution of (4.1.5) with $\phi(0, Y) = Y$, then $\phi^+(Y)$ denotes $\phi(t, Y)$ for $t \geq 0$, while $\phi^-(Y)$ denotes $\phi(t, Y)$ for $t \leq 0$.

Lemma 4.1.2. *Let $S(x, \varepsilon)$ be a small neighborhood of X and let N and N' be the points where the normal to $\phi(t, X)$ at X intersects the boundary of $S(X, \varepsilon)$. There exists δ with $0 < \delta < \varepsilon$ such that for each Y in $S(X, \delta)$, either $\phi^+(Y)$ or $\phi^-(Y)$ intersects the line segment NN' before leaving $S(X, \varepsilon)$.*

Proof. Choose $\delta < \varepsilon/2$. Let Y be a point in $S(X, \delta)$ and draw a vector K through Y parallel to the tangent of $\phi(t, X)$ at X. Construct vertical right angles through Y with K as their bisector. By Lemma 4.1.1, $\phi(t, Y)$ remains within these vertical angles so long as $\phi(t, Y)$ remains in $S(X, \varepsilon)$. See Fig. 4.2 [taken from Nemytskii and Stepanov (1960)]. Let the vertical angles intersect NN' at A and B. Observe that these intersections occur inside $S(X, \varepsilon)$ regardless of the location of Y in $S(X, \delta)$ as $\delta < \varepsilon/2$. That is, triangle AYB is a 45, 90, 45 triangle with altitude through Y of length smaller than $\varepsilon/2$. Therefore $\phi^+(Y)$ or $\phi^-(Y)$ intersects NN' before leaving $S(X, \varepsilon)$. Note also that we can choose ε sufficiently small that $\phi(t, X)$ does in fact leave $S(X, \varepsilon)$ as X is not a singular point. This completes the proof.

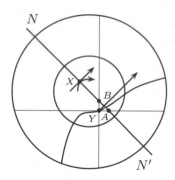

Fig. 4.2

Theorem 4.1.6. *If $\phi(t, X)$ possesses either positive or negative stability in the sense of Poisson then $\phi(t, X)$ is either a singular point or a periodic solution.*

Proof. Suppose $\phi(t, X)$ is not a singular point and is stable in the sense of Poisson in the positive direction. Let Y be an ω-limit point on $\phi(t, X)$ with $S(Y, \varepsilon)$ a small neighborhood of Y. Clearly Y is not a singular point and so the small neighborhood exists. Draw the normal NN' to $\phi(t, Y)$ through Y. Now $\phi(t, Y)$ leaves $S(Y, \varepsilon)$ and reenters again for $t > 0$ cutting NN' at a point Z; that is, Y is on $\phi(t, X)$ and Y is an ω-limit point of $\phi(t, X)$ and so $\phi(t, X)$ gets close to Y infinitely often, but $\phi(t, Y)$ has the same orbit as $\phi(t, X)$. If $Y = Z$, the proof is complete. If $Y \neq Z$, then there are two possibilities. Either Z is between Y and N, or Z is between Y and N'. See Fig. 4.3. Both cases are impossible since the direction of the field in $S(Y, \varepsilon)$ differs from the direction of $[P(Y), Q(Y)]$ by at most $\pi/4$, so $\phi(t, Y)$ cannot get closer to Y on NN' than Z. And this contradicts the fact that Y is an ω-limit point of $\phi(t, X)$. Note that $\phi(t, Y)$ cannot cross itself by uniqueness.

$S(Y, \epsilon)$

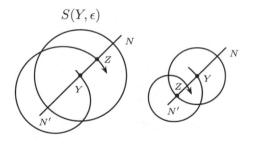

Fig. 4.3

Theorem 4.1.7. Poincaré-Bendixson *If $\phi(t, X)$ is contained in a compact set R for all $t \geq 0$ then either*

(i) $\phi(t, X)$ *is periodic,*

(ii) *there is a singular point in $\Omega(X)$, or*

(iii) $\phi(t, X)$ *approaches a periodic solution in a spiral fashion.*

Proof. Assume $\phi(t, X)$ is not periodic and that there is no singular point in $\Omega(X)$. We will show that (iii) must hold. As $\{\phi(n, X)\}$ is bounded, it has a convergent subsequence with limit A. Then $A \in \Omega(X)$ and A is not a singular point. Also, $\phi(t, A) \in \Omega(X)$ for each t and, as $\phi(t, X)$ is bounded, $\phi(t, A)$ is bounded. Hence, there is a point $B \in \Omega(A)$. As $\Omega(X)$ is closed, it follows that $B \in \Omega(X)$.

We prove next that B is actually on the orbit of $\phi(t, A)$ and hence, $\phi(t, A)$ is periodic by the previous theorem. Let $S(B, \varepsilon)$ be a small neighborhood of B with $S(B, \delta)$ as in Lemma 4.1.2. There exists $t' > 0$ such that $\phi(t', A)$ is in $S(B, \delta)$ as $B \in \Omega(A)$. Let $\phi(t, A)$ intersect NN' [the normal to $\phi(t, B)$ at B] at P_1 in $S(B, \delta)$ with, say, P_1 between B and N'. See Fig. 4.4. There are six possibilities for the next intersection P_2 of $\phi(t, A)$ with NN' in $S(B, \delta)$. See Fig. 4.5. Now $B \in \Omega(A)$ so II, III, V, and VI are impossible as $\phi(t, A)$ could not approach closer to B than P_1 or P_2 on NN'. But $\phi(t, X)$ also approaches B so I and IV are impossible as $\phi(t, X)$ cannot approach P_1. Hence, $P_1 = P_2 = B$ and $\phi(t, A)$ is periodic.

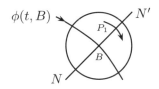

Fig. 4.4

Let $\phi(t, X)$ be inside $\phi(t, A)$. Draw NN' through A. Now $\phi(t, X)$ crosses NN' in $S(A, \varepsilon)$ at a point P as $A \in \Omega(X)$. See Fig. 4.6. But $\phi(t, X)$ must cross NN' at $t = t_n$ where t_n tends to infinity with n. Each such crossing has the property that $\phi(t_{n+1}, X)$ is between $\phi(t_n, X)$ and A. But this is also true on each normal through each point of $\phi(t, A)$. Hence, $\phi(t, X)$ approaches $\phi(t, A)$ spirally. This completes the proof.

This is just one form of the Poincaré-Bendixson theorem. If P and Q are analytic and if there are a finite number of singular points connected by orbits of $\Omega(X)$, then $\phi(t, X)$ may approach these orbits in a spiral manner. See Fig. 4.7.

The simplest form of the theorem states that if there is an annular ring which is positively invariant and contains no singular points, then there is a periodic orbit inside the ring. See Fig. 4.8.

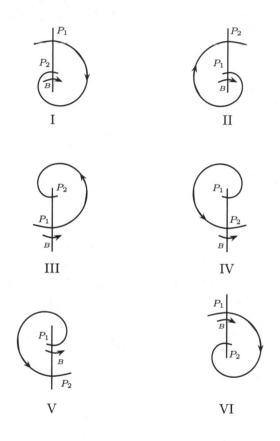

Fig. 4.5

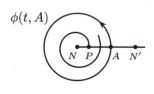

Fig. 4.6

Fig. 4.7

Fig. 4.8

Example 4.1.1. Consider the scalar equation

(4.1.6) $$x'' + f(x, x')x' + g(x) = 0$$

with f and g locally Lipschitz, $f(x, y) > 0$ if $y \neq 0$ and $xg(x) > 0$. This equation is equivalent to the system

$$x' = y,$$
$$y' = -f(x, y)y - g(x).$$

The function V defined by

$$V(x, y) = \frac{y^2}{2} + \int_0^x g(s)\, ds$$

is a positive definite autonomous Liapunov function for the system as $V' = -f(x, y)y^2 \leq 0$. Hence, 0 is Liapunov stable. Thus, if a solution starts near 0, it is bounded for $t \geq 0$. We then deduce the possible behavior of such a solution from the Poincaré-Bendixson theorem. In particular, we see that there could be no periodic solution except 0; for if so, then $p(t) = f(x(t), y(t))y^2(t)$ is a periodic function of period T for some $T > 0$, and as the solution cannot satisfy $y(t) \equiv 0$, we have $\int_0^T p(t)\, dt = M > 0$ for some M. Then $\int_0^{nT} p(t)\, dt = nM$ and so $V \to -\infty$, an impossibility.

Example 4.1.2. Consider the system

(4.1.7)
$$x' = y + x(1 - x^2 - y^2),$$
$$y' = -x + y(1 - x^2 - y^2).$$

The function $V(x, y) = x^2 + y^2$ satisfies $V' < 0$ on $x^2 + y^2 = 2$ and $V' > 0$ on $x^2 + y^2 = \frac{1}{2}$. Thus, V is decreasing on $x^2 + y^2 = 2$ along any solution intersecting that set, while V increases on $x^2 + y^2 = \frac{1}{2}$. Hence, that annulus is positively invariant. Since it is free of singular points it therefore contains a periodic orbit.

Example 4.1.3. Consider the equation

(4.1.8) $$x'' + f(x)x' + g(x) = 0$$

with $xg(x) > 0$ if $x \neq 0$, $f(0) < 0$, $f(x) > 0$ if $|x| \geq 1$, f and g continuous, $F(x) = \int_0^x f(x)\, ds$, $G(x) = \int_0^x g(s)\, ds$, $|F(x)| \to \infty$ as $|x| \to \infty$, and $G(x) \leq M$ for some $M > 0$ and all x. Write (4.1.8) as the "Liénard system"

(4.1.9)
$$x' = y - F(x),$$
$$y' = -g(x).$$

We construct a simple closed curve having the property that any solution intersecting it moves inside.

Let $V(x, y) = (y^2/2) + G(x)$ so that $V' = -g(x)F(x)$. Notice that, as $|F(x)| \to \infty$ as $|x| \to \infty$, there exists $c > 0$ with $g(x)F(x) > 0$ if $|x| \geq c$. Thus, V will not increase along solutions when $|x| \geq c$.

Let $|F(x)| \leq d$ if $|x| \leq c$ and consider the curve in quadrant II defined by $V(x, y) = (y^2/2) + G(x) = (d^2/2) + G(-c) + M$. Solutions intersecting that curve for $x \leq -c$ move downward across it. A point $(-c, y_1)$ is on that curve. Also

$$\frac{dy}{dx} = \frac{-g(x)}{y - F(x)} \leq P$$

for some $P > 0$ when $-c \leq x \leq 0$ and $y \geq y_1$. Thus, construct a line segment from $(-c, y_1)$ with slope P to intersect the y-axis at a point $(0, y_2)$. As $x' > 0$ on that line and as $dy/dx \leq P$, each solution intersecting the line segment crosses it from left to right. Continue with a horizontal line segment from $(0, y_2)$ to the curve $y = F(x)$ at a point (x_3, y_3). As $y' = -g(x) < 0$ along that line segment, solutions cross it downward. Continue with a vertical line downward into quadrant IV intersecting the curve $V(x, y) = (d^2/2) + G(c) + M$ at a point (x_3, y_4). See Fig. 4.9.

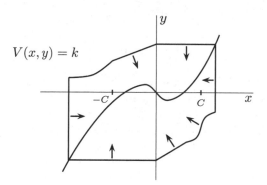

$V(x, y) = k$

Fig. 4.9

The remainder of the details of constructing the simple closed curve are just like the preceding ones and are left as an exercise.

Next notice that $V'(x, y) = -F(x)g(x)$ is positive in a neighborhood of $(0, 0)$. Thus, V increases near $(0, 0)$ along any solution. We conclude that if c is a sufficiently small constant, then $V(x, y) = c$ is a simple closed curve around $(0, 0)$ with the property that any solution which intersects it moves outside. Thus we have an annular ring which is positively invariant, free of singular points, and, therefore, the owner of a periodic orbit.

The above construction is typical of that carried out for second-order equations. Many intricate constructions may be found in Sansone and Conti (1964).

A periodic orbit is usually referred to as a *closed path*. A fundamental property of closed paths is that a singular point must lie inside such a path.

Theorem 4.1.8. *If C_0 is a closed path, then there is a singular point inside.*

Proof. Suppose C_0 is a closed path with no singular point inside. Let X be inside C_0 so that $\phi(t, X)$ is bounded and contains no singular points in its α- or ω-limit sets. Thus, $A(X)$ and $\Omega(X)$ are closed paths and, unless $\phi(t, X)$ is itself periodic, C_0 cannot serve as both $A(X)$ and $\Omega(X)$. To see this, let $C_0 = \Omega(X)$, let $Y \in C_0$, and let $S(Y, \varepsilon)$ be a small neighborhood. Let $\phi(t, X)$ intersect NN' at P_1 and P_2 in $S(Y, \varepsilon)$. See Fig. 4.10.

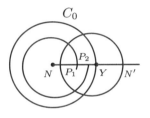

Fig. 4.10

Then $\phi^-(X)$ is bounded away from C_0 by the line segment $P_1 P_2$ together with the arc of $\phi(t, X)$ away from C_0 by the line segment $P_1 P_2$ together with the arc of $\phi(t, X)$ between P_1 and P_2. Hence, in any case, there is a closed path inside C_0, say C_1. Let B_1 be C_1 together with its interior and, generally, if C_n is any closed path inside C_0, let B_n be similarly defined. Using the above argument, we obtain a chain $B_1 \supset B_2 \supset \cdots \supset B_n$. Consider now the collection S of all possible closed paths C_α inside C_0 with corresponding B_α. A chain Q in S is maximal if there is no other chain in S which properly contains Q. By the maximality principle, there is a maximal chain in S. As the sets in this chain are nested, closed, and bounded, there is a point P common to all of them. Clearly, P is not on the boundary of any of them, for then P is on a closed path and there would be another closed path inside, a contradiction to P being a common point. Thus, $\Omega(P)$ and $A(P)$ are closed paths, one of which must be inside each of the elements of the maximal chain. Again this contradicts P being a common point and the proof is complete.

Thus, we see that no solution may forever remain in a compact convex region free of singular points.

The following result offers a very simple method for determining the absence of closed paths.

Theorem 4.1.9. Bendixson's criterion   *Let P and Q have continuous first partial derivatives in a simply connected domain G and suppose that $\partial P/\partial x + \partial Q/\partial y$ does not change sign in G and does not vanish identically in any open subset of G. Then there are no closed paths in G.*

Proof. Suppose there is a closed path L around a region R with $\phi(t) = (x(t), y(t))$ being the periodic solution of period $T > 0$. Then the line integral

$$\oint (P(x,y)\,dy - Q(x,y)\,dx)$$

$$= \int_0^T (P(\phi(t))Q(\phi(t)) - Q(\phi(t))P(\phi(t)))\,dt = 0.$$

By Green's theorem for the plane

$$\iint_R \left(\frac{\partial P}{\partial x} + \frac{\partial Q}{\partial y}\right)\,dx\,dy = \int_L (P\,dy - Q\,dx) = 0.$$

This last equation is impossible as $\partial P/\partial x + \partial Q/\partial y$ does not change sign or vanish on R. This completes the proof.

Notice that for a linear system with constant coefficients, say $z' = Az$, $\text{tr}\,A = \partial P/\partial x + \partial Q/\partial y$ which is the sum of the characteristic roots.

In the equation $x'' + f(x)x' + g(x) = 0$ with $f(x) > 0$, we obtain the

(4.1.10)
$$\begin{aligned} x' &= y, \\ y' &= -f(x)y - g(x) \end{aligned}$$

and find $\partial P/\partial x + \partial Q/\partial y = -f(x) < 0$. We therefore conclude that there are no closed paths.

Bendixson's criterion has been extended in two interesting ways. First, suppose there is an annular region G in which $\partial P/\partial x + \partial Q/\partial y$ does not change sign or vanish identically in any open subset. Then there is at most one closed path lying entirely in G. The result is proved by supposing the existence of two such paths, say C_1 and C_2, with region R between them; form a curve L by cutting from C_1 to C_2 and use Green's theorem for a contradiction as before.

The second extension is called the Bendixson-duLac theorem. Suppose there is a smooth function $B(x,y)$ such that

$$[\partial(P(x,y)B(x,y))/\partial x] + [\partial(Q(x,y)B(x,y))/\partial y]$$

does not change sign or vanish identically in any open subset of a simply connected domain G. Then there are no closed paths lying entirely in G. The result is proved by considering $\int_L B(P\,dy - Q\,dx)$ and applying Green's theorem.

Exercise 4.1.2. State and prove the last two results outlined above.

Exercise 4.1.3. Consider the system

$$x' = Ax - xy - hx^2 + hy^2,$$
$$y' = xy - y - hy^2 + hRA$$

where A, h, and R are positive constants.

(a) Find a function $B(x, y)$ enabling one to show that there are no closed paths in the first quadrant.

(b) For $B = 1$ obtain a condition ensuring that there are no closed paths in the first quadrant.

(c) Use the fact that $x' + y'$ is negative for $x^2 + y^2$ large in the first quadrant to show that all solutions in the first quadrant are bounded.

Exercise 4.1.4. The system

$$x' = x - 2xy,$$
$$y' = -y + xy$$

arises in population studies.

(a) Use uniqueness to show that no solution starting in the open quadrant I can leave quadrant I.

(b) Locate all equilibrium points.

(c) Make the transformation $u = \ln x$ and $v = \ln 2y$ to simplify the system and show that all paths in quadrant I are closed.

4.1.4   Periodic Solutions

We now consider the n-dimensional system

(4.1.11) $x' = F(t, x)$

in which $F : (-\infty, \infty) \times R^n \to R^n$ is continuous and satisfies a local Lipschitz condition in x and there is $T > 0$ with $F(t + T, x) = F(t, x)$. One may view (4.1.11) as the limit of another system, as in Proposition 4.1.2; but, in any event, we are interested in understanding the ω-limit sets of bounded solutions. The first really concrete result was the Poincaré-Bendixson theorem for two dimensional autonomous sytems. That theorem is ordinarily applied by constructing a Jordan curve bounding all solutions starting inside or on it. But if one is going to construct such a curve then a system which is much more general may be considered instead. Massera (1950) and Levinson (1944) had better ideas. In preparation for their results we give a lemma concerning the search for periodic solutions.

Lemma 4.1.3. *Let* $F(t + T, x) \equiv F(t, x)$ *for some* $T > 0$.

(i) *If* $x(t)$ *is a solution of* (4.1.11) *then* $x(t + T)$ *is also a solution of* (4.1.11).

(ii) *Equation* (4.1.11) *has a* T*-periodic solution if and only if there is a* (t_0, x_0) *with* $x(T + t_0, t_0, x_0) = x_0$.

Note. If $F(t + T, x) \equiv F(t, x)$, then for each integer m, $F(t + mT, x) \equiv F(t, x)$ and Lemma 4.1.3 can be stated in terms of mT-periodic solutions.

Proof. To prove (i), we let $q(t) = x(t + T)$ so that $q'(t) = x'(t + T) = F(t + T, x(t + T)) = F(t, q(t))$. To prove (ii), note that if $x(t, t_0, x_0)$ is T-periodic, then $x(t_0 + T, t_0, x_0) = x_0$. On the other hand, if $x(T + t_0, t_0, x_0) = x_0$, then $q(t) = x(t + T, t_0, x_0)$ is also a solution and since $q(t_0) = x_0$, by uniqueness $x(t + T, t_0, x_0) = q(t) = x(t, t_0, x_0)$, and the proof is complete.

Theorem 4.1.10. Massera *Let* (4.1.11) *be a scalar equation. If it has a solution bounded in the future, then it has a* T*-periodic solution.*

Proof. Let $x(t)$ be the solution defined on $[0, \infty)$ with $|x(t)| \leq M$. Then the sequence of functions $\{x_n(t)\}$ defined by $x_n(t) = x(t + nT)$ on $[0, \infty)$ with $n = 1, 2, \ldots$ are also solutions of (4.1.11) satisfying $|x_n(t)| \leq M$. If for some n we have $x_n(0) = x_{n+1}(0)$, then by uniqueness we have $x_n(t) = x_{n+1}(t)$ or $x(t + nT) = x(t + (n + 1)T)$ so that $x(t)$ is T-periodic.

To be definite, we suppose $x(0) < x_1(0)$. Then, by uniqueness, $x(t) < x_1(t)$ for all $t \geq 0$. Hence, for $t = nT$ we get $x_n(0) < x_{n+1}(0)$ and so $x_n(t) < x_{n+1}(t)$ for all $t \geq 0$. Thus, $\{x_n(t)\}$ is an increasing bounded sequence which converges to some $\tilde{x}(t)$ as $n \to \infty$. Also $|F(t, x)| \leq J$ for $0 \leq t < \infty$ and $|x| \leq M$, so $|x'_n(t)| \leq J$ and we have $|x_n(t) - x_n(s)| \leq J|t - s|$ for $t, s \geq 0$ and all n. On any compact t-interval the sequence $\{x_n(t)\}$ has a subsequence converging uniformly by Ascoli's theorem. Since the sequence is monotone, it is itself uniformly convergent on any compact interval. Thus

$$x_n(t) = x_n(0) + \int_0^t F(s, x_n(s)) \, ds$$

implies that the limit function $\tilde{x}(t)$ is a solution. But $\tilde{x}(T) = \lim x_n(T) = \lim x_{n+1}(0) = \tilde{x}(0)$, yielding the result.

In preparation for the next result we state a fixed point theorem of Brouwer (1912) which is not to be confused with the result known as Brouwer's fixed-point theorem discussed in Chapter 3.

Theorem. Brouwer   *Let G be a simply connected plane open domain and P a homeomorphism of G into itself which is sense preserving. If there is an x_0 in G and a subsequence of $\{x_0, Px_0, P^2x_0, \ldots\}$ converging to a point in G, then P has a fixed point in G.*

Theorem 4.1.11. Massera   *Let $n = 2$ in (4.1.11) and suppose that all solutions of (4.1.11) can be continued to $t = \infty$. If (4.1.11) has a solution bounded in the future, then it has a T-periodic solution.*

Proof. Let $P : R^2 \to R^2$ be defined by $Px_0 = x(T, 0, x_0)$. Then P is a homeomorphism of the plane into itself because of uniqueness (backward and forward in time) and the fact that all solutions can be continued. Moreover, if C is a simple closed curve and if we look at the set $C(t) = \{x(t, 0, x_0) | x_0 \in C, \ t \geq 0\}$ we see that $C(t)$ is a simple closed curve for each t and so P is sense preserving. If $x_0 = x(0)$, where $x(t)$ is the bounded solution, then $\{x_0, x(T), x(2T), \ldots\}$ is a bounded sequence with a convergent subsequence. By Brouwer's theorem, P itself has a fixed point. This completes the proof. (Here, $C(t)$ is in the $t = $ constant plane.)

Problem. Massera has an example showing that the continuation hypothesis can not be eliminated. Can one use Vinograd's theorem to partially circumvent this difficulty?

We now illustrate how one may show that solutions exist in the future.

Example 4.1.4.   Consider the scalar equation

$$x'' + f(x, x')x' + g(x) = e(t)$$

in which all functions are continuous, $f(x, y) \geq 0$, and $xg(x) > 0$ if $x \neq 0$. Write this as the system

$$x' = y,$$
$$y' = -f(x, y)y - g(x) + e(t).$$

Define a function

$$V(x, y) = y^2 + 2 \int_0^x g(s)\, ds$$

and notice that $V(x, y) \geq 0$ and if $(x(t), y(t))$ is a solution then

$$dV(x(t), y(t))/dt \leq -2f(x, y)y^2 + 2|y|\,|e(t)|$$
$$\leq 2|e(t)|V^{1/2}.$$

Hence, for $V(x(t), y(t)) = V(t)$ we have

$$V'(t)V^{-1/2}(t) \le 2|e(t)|$$

so that

$$V^{1/2}(t) \le V^{1/2}(t_0) + \int_{t_0}^{t} |c(s)|\, ds.$$

This means that for any fixed $t \ge t_0$ then $V(t)$ is bounded. Hence, $y(t) = x'(t)$ is bounded, so $x(t)$ is bounded. We then can say that for each $t_1 > t_0$ there is an M with $|x(t)| + |y(t)| \le M$ on $[t_0, t_1]$. And this means that solutions are defined for all future time.

Levinson's idea was very fruitful. It led to the following two concepts and inspired many beautiful fixed-point theorems.

Definition 4.1.10. *Solutions of* (4.1.11) *are* uniform bounded *if for each* $B_1 > 0$ *there exists* $B_2 > 0$ *such that* $[t_0 \in R, |x_0| \le B_1, t \ge t_0]$ *imply* $|x(t, t_0, x_0)| < B_2$.

Definition 4.1.11. *Solutions of* (4.1.11) *are* uniform ultimate bounded *for bound* B *if for each* $B_3 > 0$ *there exists* $S > 0$ *such that* $[t_0 \in R, |x_0| \le B_3, t \ge t_0 + S]$ *imply* $|x(t, t_0, x_0)| < B$.

Note. In the following discussion of periodic solutions it always suffices to take $t_0 = 0$, which we do. The term "uniform" then is usually replaced by "equi."

Theorem 4.1.12. *If solutions of* (4.1.11) *are uniform ultimate bounded for bound* B, *then there is a positive integer* m *such that* (4.1.11) *has an* mT-periodic solution.

Proof. Let $K = \{x \in R^n \,\|\, |x| \le B\}$. Define a mapping $P : K \to K$ by $x_0 \in K$ implies $Px_0 = x(mT, 0, x_0)$ where m is chosen so large that $|x(mT, 0, x_0)| \le B$. Now P is continuous in x_0 by the Lipschitz condition and so $Px_0 = x_0$ for some $x_0 \in K$. Thus, by Lemma 4.1.3, $x(t, 0, x_0)$ is mT-periodic.

Example 4.1.5. Let

$$x' = A(t)x + f(t, x)$$

where A is an $n \times n$ matrix, $A(t + T) = A(t)$, $f(t + T, x) = f(t, x)$, A and f are continuous, f is locally Lipschitz in x, $|f(t, x)| \le M$ for some $M > 0$, and all solutions of $q' = A(t)q$ tend to zero as $t \to \infty$. Then there is an mT-periodic solution.

Proof. By Floquet theory there is a T-periodic matrix P and a constant matrix J with $P(t)e^{Jt}$ being the principal matrix solution of $q' = A(t)q$. Since solutions tend to zero, $|e^{Jt}| \leq \alpha e^{-\beta t}$ for some α and β positive. If $x(t) = x(t, 0, x_0)$, then identify $f(t, x(t)) = g(t)$, a nonhomogeneous term, so that by the variation of parameters formula we have

$$x(t) = P(t)e^{Jt}x_0 + \int_0^t P(t)e^{J(t-s)}P^{-1}(s)g(s)\,ds$$

or

$$|x(t)| \leq |P(t)|\alpha e^{-\beta t}|x_0| + H\int_0^t e^{-\beta(t-s)}\,ds$$

where $H = (\sup|P(t)|)(\sup|P^{-1}(t)|)\alpha M$. Thus, $|x(t)| \leq |P(t)|\alpha e^{-\beta t}|x_0| + (H/\beta)$. If we let $B = (H/\beta)+1$ then solutions are uniform ultimate bounded for bound B and there is an mT-periodic solution by Theorem 4.1.12.

A result of Browder enables one to sharpen the conclusion of Theorem 4.1.12. Our presentation, however, uses Horn's theorem because it applies readily to delay equations.

Theorem 4.1.13. *If solutions of* (4.1.11) *are uniform bounded and uniform ultimate bounded for bound B, then* (4.1.11) *has a T-periodic solution.*

Proof. For the $B > 0$ of Def. 4.1.11, let $B_1 = B$ in Def. 4.1.10 and find $B_2 > 0$ such that $[|x_0| \leq B$ and $t \geq 0]$ imply $[x(t, 0, x_0)| < B_2$; then find $K_1 > 0$ such that $[|x_0| \leq B$ and $t \geq K_1]$ imply $[x(t, 0, x_0)| < B$. For the $B_2 > 0$ find $B_3 > 0$ and $K_2 > 0$ such that $[|x_0| \leq B_2$ and $t \geq 0]$ imply $[x(t, 0, x_0)| < B_3$ (by Def. 4.1.10), while $[|x_0| < B_2$ and $t \geq K_2]$ imply $|x(t, 0, x_0)| < B$ (by Def. 4.1.11). Define

$$S_0 = \{x \in R^n \mid |x| \leq B\},$$
$$\widetilde{S}_1 = \{x \in R^n \mid |x| < B_2\},$$
$$S_2 = \{x \in R^n \mid |x| \leq B_3\},$$

and

$$S_1 = \widetilde{S}_1 \cap S_2.$$

The last step is a formality in this finite-dimensional case as $S_1 = \widetilde{S}_1$.

Define $P : S_2 \to R^n$ by $x_0 \in S_2$ implies $Px_0 = x(T, 0, x_0)$. Then P is continuous because $x(T, 0, x_0)$ is continuous in x_0. One readily verifies that

$$P^j x_0 = x(jT, 0, x_0)$$

for any integer $j > 0$ and any x_0.

By choice of B, B_2, and B_3 if we pick an integer m with $mT > K_2$ then we have

$$P^j S_0 \subset S_1 \quad \text{for all} \quad j$$

and

$$P^j S_1 \subset S_0 \quad \text{for all} \quad j \geq m.$$

Certainly, $P^j S_1 \subset S_2$ for all j. By Horn's fixed-point theorem P has a fixed point and the proof is complete.

Principle 4.1.11. *If we analyze the implications of Theorem 4.1.13 we see that*

(a) *Solutions are continuous in x_0.*

(b) *Solutions are uniform bounded and uniform ultimate bounded for bound B.*

(c) *For each (t_0, x_0) there is a unique solution $x(t, t_0, x_0)$.*

(d) *If $x(t)$ is a solution so is $x(t + T)$.*

(e) *If $x(t)$ is a bounded solution, then $x'(t)$ is bounded.*

If these conditions are properly interpreted for

$$x' = h(t, x) + \int_{-\infty}^{t} q(t, s, x(s)) \, ds \, ,$$

then this equation also has a T-periodic solution and the proof will be virtually indistinguishable from the one just given.

One of the principal tools in showing uniform boundedness and uniform ultimate boundedness is Liapunov's direct method. The following is a brief introduction to the central idea. Comprehensive treatment for ordinary differential equations is found in Yoshizawa (1966) and for Volterra equations may be found in Burton (1983a). Extensive discussions of periodic solutions are found in Cronin (1964), Hale (1963), Sansone and Conti (1964), and Yoshizawa (1966, 1975).

4.1.5   Stability, Boundedness, Limit Sets

We consider the system

$$(4.1.12) \qquad\qquad x' = F(t, x), \qquad F(t, 0) = 0,$$

in which $F : (0, \infty) \times D \to R^n$ is continuous and D is an open subset of R^n with 0 in D. Thus, $x(t) = 0$ is a solution of (4.1.12) and we center our attention on the behavior of solutions which start near zero. The basic stability definitions were given in Chapter 1 and properties were developed primarily for linear systems. We will see that several properties are quite different for nonlinear systems. From Chapter 3 we know that for each $(t_0, x_0) \in [0, \infty) \times D$ there is at least one solution $x(t, t_0, x_0)$ on an interval $[t_0, \alpha)$ and, if $\alpha \neq \infty$, then $x(t, t_0, x_0) \to \partial D$ as $t \to \alpha^-$.

Definition 4.1.12. *The solution $x(t) = 0$ of* (4.1.12) *is*

(a) *stable if, for each $\varepsilon > 0$, and $t_0 \geq 0$, there is a $\delta > 0$ such that $[|x_0| < \delta$ and $t \geq t_0]$ imply that $|x(t, t_0, x_0)| < \varepsilon$,*

(b) uniformly stable *if it is stable and δ is independent of $t_0 \geq 0$,*

(c) asymptotically stable *if it is stable and if, for each $t_0 \geq 0$, there is an $\eta > 0$ such that $|x_0| < \eta$ implies that $|x(t, t_0, x_0)| \to 0$ as $t \to \infty$ (if, in addition, all solutions tend to zero, then $x = 0$ is* asymptotically stable in the large *or is* globally asymptotically stable*),*

(d) uniformly asymptotically stable *if it is uniformly stable and if there is an $\eta > 0$ such that, for each $\gamma > 0$, there is a $T > 0$ such that $[|x_0| < \eta, t_0 \geq 0$, and $t \geq t_0 + T]$ imply that $|x(t, t_0, x_0)| < \gamma$. (If η may be made arbitrarily large, then $x = 0$ is* uniformly asymptotically stable in the large.*)*

Example 4.1.6. Let $a : [0, \infty) \to R$ be continuous and let

$$x' = a(t)x$$

where

(a) $a(n) = 0$ for each nonnegative integer n,

(b) $a(t) > 0$ for $2n < t < 2n + 1$,

(c) $a(t) < 0$ for $2n + 1 < t < 2n + 2$,

(d) $\int_{2n}^{2n+1} a(t)\, dt + \int_{2n+2}^{2n+3} a(t)\, dt = \int_{2n+1}^{2n+2} -a(t)\, dt$,

and

(e) $\int_{2n}^{2n+1} a(t)\, dt \to \infty$ as $n \to \infty$.

Then $x = 0$ is stable, asymptotically stable, and globally asymptotically stable, but it is not uniformly stable.

A sample graph for $a(t)$ is given in Fig. 4.11.

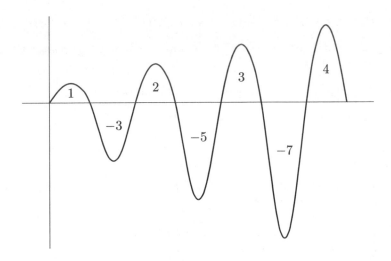

Fig. 4.11

Example 4.1.7. The zero solution of the scalar equation

$$x' = -x + x^3$$

is uniformly asymptotically stable; if $x_0 > 1$ is given, then there exists $T > 0$ such that $\lim_{t \to T^-} x(t, 0, x_0) = \infty$. See Fig. 4.12.

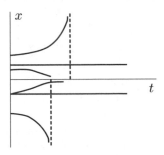

Fig. 4.12

When F is sufficiently smooth then all stability properties in Def. 4.1.12, except (c), have been characterized by Liapunov functions. The next definition will be reconciled with that given in Chapter 1 after the proof of Theorem 4.1.14.

Definition 4.1.13. *A continuous function* $W : [0, \infty) \to [0, \infty)$ *with* $W(0) = 0$, $W(s) > 0$ *if* $s > 0$, *and* W *strictly increasing is a wedge.* (*We denote wedges by* W *or* W_i, *where* i *is an integer.*)

Definition 4.1.14. *A function* $U : [0, \infty) \times D \to [0, \infty)$ *is called*

(a) positive definite *if* $U(t, 0) = 0$ *and if there is a wedge* W_1 *with* $U(t, x) \geq W_1(|x|)$,

(b) decrescent *if there is a wedge* W_2 *with* $U(t, x) \leq W_2(|x|)$,

(c) negative definite *if* $-U(t, x)$ *is positive definite,*

(d) radially unbounded *if* $D = R^n$ *and there is a wedge* $W_3(|x|) \leq U(t, x)$ *and* $W_3(r) \to \infty$ *as* $r \to \infty$*, and*

(e) mildly unbounded *if* $D = R^n$ *and if, for each* $T > 0$*,* $U(t, x) \to \infty$ *as* $|x| \to \infty$ *uniformly for* $0 \leq t \leq T$.

Definition 4.1.15. *A continuous function* $V : [0, \infty) \times D \to [0, \infty)$ *that is locally Lipschitz in* x *and satisfies*

$$(4.1.13) \quad V'_{(4.1.12)}(t, x) = \limsup_{h \to 0^+} \left[V(t + h, x + hF(t, x)) - V(t, x) \right]/h \leq 0$$

on $[0, \infty) \times D$ *is called a* Liapunov function *for* (4.1.12).

If V has continuous first partial derivatives then (4.1.13) becomes

$$V'(t, x) = \operatorname{grad} V(t, x) \cdot F(t, x) + (\partial V/\partial t) \leq 0.$$

We show in Section 4.2 that

$$V'(t, x(t)) = \limsup_{h \to 0^+} [V(t + h, x(t + h)) - V(t, x(t))]/h = V'_{(4.1.12)}(t, x).$$

Certainly, when V has continuous first partial derivatives, then the chain rule gives us this relation. Moreover, $V(t, x(t))$ is nonincreasing if and only if $V'_{(4.1.12)}(t, x) \leq 0$.

The next theorem is a summary of standard results yielding stability properties by means of Liapunov functions. One may note that (d) and (e) do not require $F(t, 0) = 0$.

Theorem 4.1.14. *Suppose there is a Liapunov function* V *for* (4.1.12).

(a) *If* V *is positive definite, then* $x = 0$ *is stable.*

(b) *If* V *is positive definite and decrescent, then* $x = 0$ *is uniformly stable.*

(c) *If* V *is positive definite and decrescent, while* $V'_{(4.1.12)}(t, x)$ *is negative definite, then* $x = 0$ *is uniformly asymptotically stable. Moreover, if* $D = R^n$ *and if* V *is radially unbounded, then* $x = 0$ *is uniformly asympotically stable in the large.*

(d) *If $D = R^n$ and if V is radially unbounded, then all solutions of (4.1.12) are bounded.*

(e) *If $D = R^n$ and if V is mildly unbounded, then each solution can be continued for all future time.*

Proof. (a) We have $V'_{(4.1.12)}(t, x) \leq 0$, V continuous, $V(l, 0) = 0$, and $W_1(|x|) \leq V(t, x)$. Let $\varepsilon > 0$ and $t_0 \geq 0$ be given. We must find $\delta > 0$ such that $[|x_0| < \delta$ and $t \geq t_0]$ imply $|x(t, t_0, x_0)| < \varepsilon$. (Throughout these proofs we assume ε so small that $|x| < \varepsilon$ implies $x \in D$.) As $V(t_0, x)$ is continuous and $V(t_0, 0) = 0$, there is a $\delta > 0$ such that $|x| < \delta$ implies $V(t_0, x) < W_1(\varepsilon)$. Thus, if $t \geq t_0$, then $V' \leq 0$ implies that for $|x_0| < \delta$ and $x(t) = x(t, t_0, x_0)$ we have

$$W_1(|x(t)|) \leq V(t, x(t)) \leq V(t_0, x_0) < W_1(\varepsilon),$$

or $|x(t)| < \varepsilon$ as required.

(b) To prove uniform stability, for a given $\varepsilon > 0$ we select $\delta > 0$ such that $W_2(\delta) < W_1(\varepsilon)$, where $W_1(|x|) \leq V(t, x) \leq W_2(|x|)$. Now, if $t_0 \geq 0$ and $|x_0| < \delta$, then for $x(t) = x(t, t_0, x_0)$ and $t \geq t_0$ we have

$$W_1(|x(t)|) \leq V(t, x(t)) \leq V(t_0, x_0) \leq W_2(|x_0|) < W_2(\delta) < W_1(\varepsilon)$$

or $|x(t)| < \varepsilon$ as required.

(c) The conditions for uniform stability are satisfied. Thus, for $\varepsilon = 1$ find the δ of uniform stability and call it η in the definition of U.A.S. Now, let $\gamma > 0$ be given. We must find $T > 0$ such that $[|x_0| < \eta, t_0 \geq 0$, and $t \geq t_0 + T]$ imply $|x(t, t_0, x_0)| < \gamma$. Set $x(t) = x(t, t_0, x_0)$. Pick $\mu > 0$ with $W_2(\mu) < W_1(\gamma)$ so that if there is a $t_1 \geq t_0$ with $|x(t_1)| < \mu$ then for $t \geq t_1$ we have

$$W_1(|x(t)|) \leq V(t, x(t)) \leq V(t_1, x(t_1))$$
$$\leq W_2(|x(t_1)|) < W_2(\mu) < W_1(\gamma)$$

or $|x(t)| < \gamma$. Now, $V'(t, x) \leq -W_3(|x|)$ so that as long as $|x(t)| \geq \mu$, then $V'(t, x(t)) \leq -W_3(\mu)$; thus

$$V(t, x(t)) \leq V(t_0, x_0) - \int_{t_0}^{t} W_3(|x(s)|)\, ds$$
$$\leq W_2(|x_0|) - W_3(\mu)(t - t_0)$$
$$\leq W_2(\eta) - W_3(\mu)(t - t_0)$$

which vanishes at

$$t = t_0 + W_2(\eta)/W_3(\mu) \stackrel{\text{def}}{=} t_0 + T.$$

Hence, if $t > t_0 + T$ then $|x(t)| > \mu$ fails and we have $|x(t)| < \gamma$ for all $t \geq t_0 + T$. This proves U.A.S. The proof for U.A.S. in the large is accomplished in the same way.

(d) As V is radially unbounded, we have $V(t, x) \geq W_1(|x|) \to \infty$ as $|x| \to \infty$. Thus, given $t_0 \geq 0$ and x_0, there is an $r > 0$ with $W_1(r) > V(t_0, x_0)$. Hence, if $t \geq t_0$ and $x(t) = x(t, t_0, x_0)$, then

$$W_1(|x(t)|) \leq V(t, x(t)) \leq V(t_0, x_0) < W_1(r)$$

or $|x(t)| < r$.

(e) To prove continuation of solutions it will suffice to show that if $x(t)$ is a solution on any interval $[t_0, T)$, then there is an M with $|x(t)| < M$ on $[t_0, T)$. Now $V(t, x) \to \infty$ as $|x| \to \infty$ uniformly for $0 \leq t \leq T$. Thus, there is an $M > 0$ with $V(t, x) > V(t_0, x_0)$ if $0 \leq t \leq T$ and $|x| > M$. Hence, for $0 \leq t < T$ we have $V(t, x(t)) \leq V(t_0, x_0)$ so that $|x(t)| < M$.

The proof of Theorem 4.1.14 is complete.

We have chosen our wedges to simplify proofs. But this choice makes examples harder. One can define a wedge as a continuous function $W : D \to [0, \infty)$ with $W(0) = 0$ and $W(x) > 0$ if $x \neq 0$. That choice makes examples easier, but proofs harder. The following device is helpful in constructing a wedge $W_1(|x|)$ from a function $W(x)$.

Suppose $W : D \to [0, \infty)$, $D = \{x \in R^n : |x| \leq 1\}$, $W(0) = 0$, and $W(x) > 0$ if $x \neq 0$. We suppose there is a function $V(t, x) \geq W(x)$ and we wish to construct a wedge $W_1(|x|) \leq V(t, x)$. First, define $\alpha(r) = \min_{r \leq |x| \leq 1} W(x)$ so that $\alpha : [0, 1] \to [0, \infty)$ and α is nondecreasing. Next, define $W_1(r) = \int_0^r \alpha(s)\, ds$ and note that $W_1(0) = 0$, $W_1'(r) = \alpha(r) > 0$ if $r > 0$, and $W_1(r) \leq r\alpha(r) \leq \alpha(r)$. Thus if $|x_1| \leq 1$, then

$$V(t, x_1) \geq W(x_1) \geq \min_{|x_1| \leq |x| \leq 1} W(x) = \alpha(|x_1|)$$

$$\geq W_1(|x_1|).$$

Example 4.1.8. Write the scalar equation $x'' + h(x')g(x) = 0$ as the system

(4.1.14)
$$\begin{aligned}x' &= y, \\ y' &= -h(y)g(x)\end{aligned}$$

and assume that h and g are continuous with

(a) $h(y) > 0$ and $xg(x) > 0$ for $x \neq 0$.

We can eliminate the parameter t by writing

$$dy/dx = -h(y)g(x)/y$$

with solution

$$V(x,y) = \int_0^y [s/h(s)]\,ds + \int_0^x g(s)\,ds = \text{constant.}$$

Then the derivative of V along a solution of (4.1.14) is

$$V'(x,y) \equiv 0.$$

Because of (a), V is positive definite and decrescent so the zero solution of (4.1.14) is uniformly stable by Theorem 4.1.14(b). However, in this case, since $V' \equiv 0$ solutions move along the curves $V(x,y) = \text{constant}$ and so the solutions will be bounded if and only if V is radially unbounded; and V is radially unbounded provided that

$$\int_0^{\pm\infty} [s/h(s)]\,ds = \int_0^{\pm\infty} g(s)\,ds = +\infty.$$

Example 4.1.9. Write the scalar equation

$$x'' + (1 + e^{-t})(1 - e^{-t})h(x')g(x) = 0$$

as the system

$$x' = y,$$
$$y' = -(1 + e^{-t})(1 - e^{-t})h(y)g(x)$$

and take

$$V(t,x,y) = [1/(1 - e^{-t})]\int_0^y [s/h(s)]\,ds + (1 + e^{-t})\int_0^x g(s)\,ds.$$

Note that $V'(t,x,y) \leq 0$, that V is positive definite, and that V is decrescent when (a) of Example 4.1.8 holds and $t \geq 1$.

Exercise 4.1.5. Consider the system (with $ak > dc$)

$$x' = x[a - dx - by],$$
$$y' = y[-c + kx]$$

with equilibrium point $(\bar{x}, \bar{y})$ in quadrant I when a, b, c, d, and k are positive. Make the transformation $u = \ln(x/\bar{x})$ and $v = \ln(y/\bar{y})$ to obtain a system in (u,v). Show that $V(u,v) = k\bar{x}(e^u - u) + b\bar{y}(e^v - v) - k\bar{x} - b\bar{y}$ is positive definite with $V'(u,v) \leq 0$. Determine stability properties of $(\bar{x}, \bar{y})$.

Example 4.1.10. Write the scalar equation $x'' + f(x)x' + g(x) = 0$ as

(4.1.15)
$$x' = y,$$
$$y' = -f(x)y - g(x)$$

and assume that f and g are continuous with

(b) $f(x) > 0$ and $xg(x) > 0$ for $x \neq 0$.

Define
$$V(x, y) = y^2 + 2G(x),$$

for $G(x) = \int_0^x g(s)\, ds$, so that
$$V'(x, y) = -2f(x)y^2 \leq 0.$$

Next, write the same scalar equation as the system

(4.1.16)
$$\begin{aligned} x' &= z - F(x), \\ z' &= -g(x), \end{aligned}$$

for $F(x) = \int_0^x f(s)\, ds$, and define $W(x, z) = z^2 + 2G(x)$ so that
$$W'(x, z) = -2F(x)g(x) \leq 0.$$

Now, combine the results and write
$$U(x, y) = y^2 + [y + F(x)]^2 + 4G(x)$$

with the derivative of U along solutions of (4.1.15) satisfying
$$U'(x, y) = -2[f(x)y^2 + F(x)g(x)]$$

which is negative definite.   A careful analysis will allow us to say that the zero solution of (4.1.15) is uniformly asymptotically stable in the large provided that
$$\int_0^{\pm\infty} [f(x) + |g(x)|]\, dx = \pm\infty,$$

using Theorem 4.1.14(c). It turns out that this same condition is necessary and sufficient for all solutions to be bounded (cf. Burton, 1965).

Example 4.1.11. Conti-Wintner Consider the system (4.1.12) and suppose there are continuous functions
$$\lambda : [0, \infty) \to [0, \infty) \quad \text{and} \quad \omega : [0, \infty) \to [1, \infty)$$

with $\int_0^\infty [ds/\omega(s)] = +\infty$. If $|F(t, x)| \leq \lambda(t)\omega(|x|)$, then any solution of (4.1.12) can be continued to $t = \infty$.

Proof. Let $x(t)$ be a solution of (4.1.12) on $[t_0, \alpha)$ and define
$$V(t, x) = \left\{ \int_0^{|x|} [ds/\omega(s)] + 1 \right\} \exp\left[-\int_0^t \lambda(s)\, ds. \right].$$

Then

$$V'(t, x) \leq -\lambda(t)V(t, x) + \left[|F(t, x)|/\omega(|x|)\right] \exp\left[-\int_0^t \lambda(s)\, ds\right]$$

$$\leq -\lambda(t)V(t, x) + \lambda(t) \exp\left[-\int_0^t \lambda(s)\, ds\right] \leq 0$$

and so Theorem 4.1.14(e) is satisfied.

Example 4.1.12. Let A be an $n \times n$ constant matrix all of whose characteristic roots have negative real parts and let $G : [0, \infty) \times D \to R^n$ be continuous with

$$\lim_{|x| \to 0} |G(t, x)|/|x| = 0 \quad \text{uniformly in } t.$$

Then the zero solution of

$$(4.1.17) \qquad\qquad x' = Ax + G(t, x)$$

is U.A.S.

Proof. There is a positive definite symmetric matrix B with $A^T B + BA = -I$. Define $V(x) = x^T Bx$ and obtain

$$\begin{aligned} V'_{(4.1.17)}(x) &= -x^T x + 2x^T BG(t, x) \\ &\leq x^T x[-1 + 2|B||G(t, x)|/|x|] \\ &\leq -x^T x/2 \end{aligned}$$

for $|x|$ small enough. The result now follows from Theorem 4.1.14(c). Indeed one easily sees that

$$(4.1.18) \qquad\qquad V'(x) \leq -\alpha V(x)$$

for some $\alpha > 0$ and $|x|$ small. Thus, we have

$$V(x(t)) \leq V(x(t_0))e^{-\alpha(t-t_0)}$$

and there is a $\beta > 0$ with

$$\beta x^T x \leq V(x) \leq V(x(t_0))e^{-\alpha(t-t_0)}$$

so that

$$(4.1.19) \qquad |x(t)| \leq K(x(t_0))e^{-(\alpha/2)(t-t_0)}, \qquad K > 0,$$

a property called exponential asymptotic stability.

In preparation for the next two results the reader should review Defs. 4.1.10 and 4.1.11 concerning uniform boundedness and Theorem 4.1.13 concerning the existence of periodic solutions.

Consider the system $(4.1.12)$ without the condition $F(t,0) = 0$ and denote it by

$(4.1.12)'$ $$x' = F(t, x).$$

Theorem 4.1.15. *Let $D = R^n$ and let $H = \{x \in R^n : |x| \geq M, \ M > 0\}$. Suppose that $V : [0, \infty) \times H \to [0, \infty)$ is continuous, locally Lipschitz in x, radially unbounded, and $V'_{(4.1.12)'}(t, x) \leq 0$ if $|x| \geq M$. If there is a constant $P > 0$ with $V(t, x) \leq P$ for $|x| = M$, then all solutions of $(4.1.12)'$ are bounded.*

Proof. As in the proof of Theorem 4.1.14(d), if a solution $x(t)$ satisfies $|x(t)| \geq M$ for all t, then it is bounded. Suppose $x(t)$ is a solution with $|x(t_1)| = M$ and $|x(t)| \geq M$ on an interval $[t_1, T]$. Then

$$W_1(|x(t)|) \leq V(t, x(t))$$
$$\leq V(t_1, x(t_1)) \leq P$$

so that $|x(t)| \leq W_1^{-1}(P)$ on $[t_1, T]$. As we may repeat this argument on any such interval $[t_1, T]$, it follows that $W_1^{-1}(P)$ is a future bound for any solution entering H^c. This completes the proof.

Theorem 4.1.16. *Let $D = R^n$ and $H = \{x \in R^n \big| |x| \geq M, \ M > 0\}$. Suppose that $V : [0, \infty) \times H \to [0, \infty)$ is continuous, locally Lipschitz in x, and on $[0, \infty) \times H$ satisfies*

$$W_1(|x|) \leq V(t, x) \leq W_2(|x|) \qquad W_1(r) \to \infty \ as \ r \to \infty$$

and

$$V'_{(4.1.12)'}(t, x) \leq -W_3(|x|).$$

Then solutions of $(4.1.12)'$ are uniform bounded and uniform ultimate bounded for bound B with $W_1(B) = W_2(M + 1)$.

Proof. Let $B_1 > 0$ be given with $B_1 \geq M$. Find $B_2 > 0$ with $W_1(B_2) = W_2(B_1)$. If $t_0 \geq 0$ and $|x_0| \leq B_1$, then for each interval $[t_1, t_2]$ with $t_0 \leq t_1$, $|x(t_1, t_0, x_0)| = B_1$ and $|x(t, t_0, x_0)| \geq B_1$ on $[t_1, t_2]$ we have $x(t) = x(t, t_0, x_0)$ and

$$W_1(|x(t)|) \leq V(t, x(t)) \leq V(t_1, x(t_1))$$
$$\leq W_2(|x(t_1)|) = W_2(B_1) = W_1(B_2)$$

so $|x(t)| \leq B_2$. This is the required uniform boundedness.

Next, determine $B > 0$ with $W_1(B) = W_2(M+1)$ and let $B_3 > 0$ be given. We must find $K > 0$ such that $[t_0 \geq 0, |x_0| \leq B_3, t \geq t_0 + K]$ imply that $|x(t, t_0, x_0)| < B$. If $B_3 \leq M+1$, then $K = 0$ suffices. Thus, we suppose $B_3 > M+1$ and, using uniform boundedness, find B_4 such that $[t_0 \geq 0, |x_0| \leq B_3, t \geq t_0]$ imply that $|x(t, t_0, x_0)| \leq B_4$. We will find K such that $[t_0 > 0, |x_0| \leq B_3, t \geq t_0 + K]$ imply that $|x(t, t_0, x_0)| \leq M+1$. Now there is an $\alpha > 0$ such that $V'(t, x) \leq -\alpha$ if $M+1 \leq |x| \leq B_4$; hence, so long as $M+1 \leq |x(t)| \leq B_4$ we have

$$W_1(|x(t)|) \leq V(t, x(t)) \leq V(t_0, x_0) - \alpha(t - t_0)$$
$$\leq W_2(B_3) - \alpha(t - t_0).$$

Thus, when the right side equals $W_1(M+1)$, then we have $|x(t)| \leq M+1$. Set $W_2(B_3) - \alpha(t - t_0) = W_1(M+1)$ and obtain

$$K = t - t_0 = [W_2(B_3) - W_1(M+1)]/\alpha$$

as the desired constant. This completes the proof.

Problem. It is our view that the central problem in boundedness theory of integrodifferential equations is to extend these last two theorems to Liapunov functionals with unbounded delays in a variety of really useful ways.

Example 4.1.13. Consider the scalar equation $x'' + f(x)x' + g(x) = p(t)$ with f and g as in Example 4.1.10 and with $p(t)$ a bounded and continuous function. Write the equation as the system

(4.1.20)
$$\begin{aligned} x' &= y, \\ y' &= -f(x)y - g(x) + p(t) \end{aligned}$$

and define

$$U(x, y) = y^2 + [y + F(x)]^2 + 4G(x)$$

so that

(4.1.21)
$$\begin{aligned} U'_{(4.1.20)}(x, y) \leq &- 2[f(x)y^2 + F(x)g(x)] \\ &+ 2yp(t) + 2|y + F(x)||p(t)|. \end{aligned}$$

Note that for $f(x) = c_1 > 0$ and $g(x) = c_2 x$ with $c_2 > 0$, then

$$U'(x, y) \leq -2c_1[y^2 + c_2 x^2] + 2|y + c_1||p(t)| + 2|y||p(t)|$$

so that $U'(x, y)$ is negative for large $x^2 + y^2$. In this case the conditions of Theorem 4.1.16 hold.

Exercise 4.1.6. Recall that $U(x, y)$ is radially unbounded provided that

$$\int_0^{\pm\infty} [f(x) + |g(x)|] \, dx = \pm\infty.$$

Carefully consider (4.1.21) and give the best additional conditions on f and g to ensure that the conditions of Theorem 4.1.16 hold.

Notice that no part of Theorem 4.1.14 deals with asymptotic stability which is not uniform. The next example perfectly illustrates the difficulty.

Example 4.1.14. Let $g : [0, \infty) \to (0, 1]$ be a differentiable function with $g(n) = 1$ and $\int_0^\infty g(s) \, ds < \infty$. We will construct a function $V(t, x) = a(t)x^2$ with $a(t) > 0$ and the derivative of V along any solution of

$$(4.1.22) \qquad\qquad x' = [g'(t)/g(t)]x$$

satisfying

$$V'(t, x) = -x^2.$$

By this we will see that $V(t, x) \geq 0$ and V' negative definite do not imply that solutions tend to zero, because $x(t) = g(t)$ is a solution of (4.1.22). To find $a(t)$ we compute

$$V'_{(4.1.22)}(t, x) = a'(t)x^2 + 2a(t)[g'(t)/g(t)]x^2$$

and set $V'(t, x) = -x^2$. This yields

$$a'(t) = -2a(t)[g'(t)/g(t)] - 1$$

with solution

$$a(t) = \left[a(0)g^2(0) - \int_0^t g^2(s) \, ds\right] \Big/ g^2(t).$$

But $0 < g(t) \leq 1$ and $g \in L^1[0, \infty)$, so we pick $a(0)$ so large that $a(t) > 1$ on $[0, \infty)$. This means V and $-V'$ are positive definite, but V is not decrescent.

This example leads us naturally to ask just what can be concluded from $V(t, x)$ and $-V'(t, x)$ being positive definite. This is an old and important problem. The first person to offer an answer was Marachkov (cf. Antosiewicz, 1958, Theorem 7, p. 149), but the stream of related results continues to pour into the literature to this day.

Theorem 4.1.17. Marachkov *If $F(t, x)$ is bounded for $|x|$ bounded and if there is a positive definite Liapunov function for (4.1.12) with negative definite derivative, then the zero solution of (4.1.12) is asymptotic stable.*

Proof. There is a function $V : [0, \infty) \times D \to [0, \infty)$ with $W_1(|x|) \leq V(t, x)$ and $V'_{(4.1.12)}(t, x) \leq -W_2(|x|)$ for wedges W_1 and W_2. Also, there is a constant P with $|F(t, x)| \leq P$ if $|x| \leq m$ where m is chosen so that $|x| \leq m$ implies x is in D.

As V is positive definite and $V' \leq 0$, $x = 0$ is stable. To show asymptotic stability, let $t_0 \geq 0$ be given and let $W_1(m) = \alpha > 0$. As $V(t_0, x)$ is continuous and $V(t_0, 0) = 0$, there is an $\eta > 0$ such that $|x_0| < \eta$ implies $V(t_0, x_0) < \alpha$. Now for $x(t) = x(t, t_0, x_0)$, we have $V'(t, x(t)) \leq 0$ so

$$W_1(|x(t)|) \leq V(t, x(t)) \leq V(t_0, x_0) < W_1(m)$$

implying $|x(t)| < m$ if $t \geq t_0$. Notice that $V'(t, x(t)) \leq -W_2(|x(t)|)$ so that

$$0 \leq V(t, x(t)) \leq V(t_0, x_0) - \int_{t_0}^{t} W_2(|x(s)|) \, ds$$

from which we conclude that there is a sequence $\{t_n\} \to \infty$ with $|x(t_n)| \to 0$.

The following paragraph is called the *annulus argument* and it occurs frequently in the literature.

Now if $x(t) \not\to 0$, there is an $\varepsilon > 0$ and a sequence $\{s_n\}$ with $|x(s_n)| \geq \varepsilon$ and $s_n \to \infty$. But as $x(t_n) \to 0$ and as $x(t)$ is continuous, there is a pair of sequences $\{U_n\}$ and $\{J_n\}$ with $U_n < J_n < U_{n+1}$, $|x(U_n)| = \varepsilon/2$, $|x(J_n)| = \varepsilon$, and $\varepsilon/2 \leq |x(t)| \leq \varepsilon$ if $U_n \leq t \leq J_n$. Integrating (4.1.12) from U_n to J_n we have

$$x(J_n) = x(U_n) + \int_{U_n}^{J_n} F(s, x(s)) \, ds$$

so that

$$\varepsilon/2 \leq |x(J_n) - x(U_n)| \leq P(J_n - U_n)$$

or

$$J_n - U_n \geq \varepsilon/2P.$$

Also, if $t > J_n$, then

$$0 \leq V(t, x(t)) \leq V(t_0, x_0) - \int_{t_0}^{t} W_2(|x(s)|) \, ds$$

$$\leq V(t_0, x_0) - \sum_{i=1}^{n} \int_{U_i}^{J_i} W_2(|x(s)|) \, ds$$

$$\leq V(t_0, x_0) - \sum_{i=1}^{n} \int_{U_i}^{J_i} W_2(\varepsilon/2) \, ds$$

$$\leq V(t_0, x_0) - nW_2(\varepsilon/2)\varepsilon/2P \to -\infty$$

as $n \to \infty$, a contradiction.

Definition 4.1.16. *The argument given in the final paragraph of the proof of Theorem 4.1.17 is called the* annulus argument.

That argument is central to 40 years of research on conditions needed to conclude asymptotic stability in ordinary and functional differential equations. Marachkov's work was done in 1940 and it was extended in 1952 by Barbashin and Krasovskii (cf. Barbashin, 1968, p. 1099) in a very significant manner which allows V' to be zero on certain sets. Somewhat similar extensions were given independently by La Salle, Levin and Nohel, and Yoshizawa. The following is essentially Yoshizawa's formulation of one of those results.

Here, a scalar function $f : R^n \to [0, \infty)$ is *positive definite with respect to a set A*, if $f(x) = 0$ for $x \in A$ and for each $\varepsilon > 0$ and each compact set Q in R^n, there exists $\delta = \delta(Q, \varepsilon)$ such that $f(x) \geq \delta$ for $x \in Q \cap U(A, \varepsilon)^c$ where $U(A, \varepsilon)$ is the ε-neighborhood of A.

Theorem 4.1.18. Yoshizawa (1963) *Let $D = R^n$ and let $F(t, x)$ be bounded for x bounded. Suppose also that all solutions of (4.1.12) are bounded. If there is a continuous function $V : [0, \infty) \times R^n \to [0, \infty)$ which is locally Lipschitz in x, if there is a continuous function $W : R^n \to [0, \infty)$ which is positive definite with respect to a closed set Ω, and if $V'_{(4.1.12)}(t, x) \leq -W(x)$, then every solution of (4.1.12) approaches Ω as $t \to \infty$.*

Proof. Consider a solution $x(t)$ on $[t_0, \infty)$ which, being bounded, remains in some compact set Q for $t \geq t_0$. If $x(t) \not\to \Omega$, then there is an $\varepsilon > 0$ and a sequence $\{t_n\} \to \infty$ with $x(t_n) \in U(\Omega, \varepsilon)^c \cap Q$. As $F(t, x)$ is bounded for x in Q, there is a K with $|F(t, x(t))| \leq K$. Thus, there is a $T > 0$ with $x(t) \in U(\Omega, \varepsilon/2)^c \cap Q$ for $t_n \leq t \leq t_n + T$. By taking a subsequence, if necessary, we may suppose these intervals disjoint. Now, for this $\varepsilon/2$ there is a $\delta > 0$ with

$$V'(t, x) \leq -\delta \qquad \text{on} \qquad [t_n, t_n + T].$$

Thus, for $t \geq t_n + T$ we have

$$0 \leq V(t, x(t)) \leq V(t_0, x(t_0)) - \int_{t_0}^{t} W(x(s)) \, ds$$

$$\leq V(t_0, x(t_0)) - \sum_{i=1}^{n} \int_{t_i}^{t_i + T} W(x(s)) \, ds$$

$$\leq V(t_0, x(t_0)) - nT\delta \to -\infty,$$

a contradiction.

Usually, one shows boundedness by showing V to be radially unbounded. Also, we gather from the proof that the requirement that all solutions be bounded may be dropped and the conclusion changed to read that all bounded solutions approach Ω. Moreover, some authors let $V : [0, \infty) \times R^n \to (-\infty, \infty)$ and ask that V be bounded from below for x bounded, concluding again that bounded solutions approach Ω (cf. Haddock, 1974).

Example 4.1.15. Consider the scalar equation $x'' + f(x, x', t)x' + g(x) = 0$ in which f and g are continuous, $xg(x) > 0$ for $x \neq 0$, $f(x, y, t)$ bounded for $x^2 + y^2$ bounded, $f(x, y, t) \geq \overline{f}(x, y)$ a positive continuous function, and $G(x) = \int_0^x g(s)\, ds \to \infty$ as $|x| \to \infty$. Write the equation as the system

$$x' = y,$$
$$y' = -f(x, y, t)y - g(x)$$

and define

$$V(x, y) = y^2 + 2G(x)$$

so that

$$V'(x, y) = -2f(x, y, t)y^2 \leq -2\overline{f}(x, y)y^2.$$

Now V is radially unbounded so all solutions are bounded. Moreover, $2\overline{f}(x, y)y^2$ is positive definite with respect to the x-axis; hence, all solutions converge to the x-axis.

Ideally, we would like to say far more about an example such as Example 4.1.15. First, it is hard to construct a V which is positive definite and decrescent with V' negative definite. We would like to get by with a weaker V and, through ad hoc arguments, conclude that solutions are bounded and tend to zero. Then we would like to appeal to some fundamental theorem which says that if $x = 0$ is stable and if all solutions tend to zero, then $x = 0$ is uniformly asymptotically stable in the large. And, finally, we would like to say that since $x = 0$ is U.A.S. in the large, we may perturb the equation with a bounded function and conclude that solutions are uniform bounded and uniform ultimate bounded for bound B. The next example illustrates this for linear periodic systems.

Example 4.1.16. Consider the scalar equation $x'' + q(t)x' + x = 0$ in which $q(t + T) = q(t)$ for all t and some $T > 0$ with $q(t) \geq 1$ and continuous. Write this equation as the system

$$x' = y,$$
$$y' = -q(t)y - x,$$

define $V(x, y) = x^2 + y^2$, and conclude that $V'(x, y) = -2q(t)y^2 \leq 0$. There are two linearly independent solutions, one of which must tend to zero as may be deduced from Floquet theory; the other solution could be periodic or almost periodic. However, all solutions approach the x-axis so an almost periodic solution must be constant. Since the equation has no nontrivial constant solutions, we conclude that all solutions tend to zero. But Floquet theory then tells us that the solutions tend to zero exponentially. If the system is denoted by $X' = A(t)X$ with $Z(t, 0) = P(t)e^{Rt}$ being the PMS in which P is T-periodic and R is constant, then a solution of $Y' = A(t)Y + p(t)$ is expressed as

$$Y(t, 0, y_0) = Z(t, 0)Y(0) + \int_0^t P(t)e^{R(t-s)}P^{-1}(s)p(s)\, ds.$$

If $p(t)$ is bounded and continuous, then solutions are uniform bounded and uniform ultimate bounded for bound B.

While the theory is not as complete as we would like, much progress has been made in a parallel direction for (4.1.12). We first indicate the set of results we would like and how we would use them. Then we indicate what results are available and what needs to be done.

Program. Suppose that it can be determined that the zero solution of (4.1.12) is U.A.S. in the large. There is much evidence to indicate that, under adequate smoothness and growth conditions on F, there is a continuous function $V : [0, \infty) \times R^n \to [0, \infty)$ satisfying $W_1(|x|) \leq V(t, x) \leq W_2(|x|)$, $V'_{(4.1.12)}(t, x) \leq -cV(t, x)$ for $c > 0$, and $|V(t, x_1) - V(t, x_2)| \leq L|x_1 - x_2|$ for $(t, x_i) \in [0, \infty) \times R^n$ and L is a positive constant. Then consider

$$x' = F(t, x) + p(t)$$

with $p : [0, \infty) \to R^n$ being continuous and $|p(t)| \leq M$ for some $M > 0$. Then the derivative of V along the new system satisfies

$$V'(t, x) \leq -cV(t, x) + L|p(t)|$$

(for c a positive constant) and so

$$W_1(|x(t)|) \leq V(t, x(t))$$
$$\leq e^{-c(t-t_0)}V(t_0, x_0) + \int_0^t e^{-c(t-s)}L|p(s)|\, ds$$

from which uniform boundedness and uniform ultimate boundedness for bound B follows.

The following results, stated without proof, form the foundation for the program. The only missing part concerns the uniform Lipschitz constant L for V; in the results stated below, for each $\alpha > 0$ there is an $L(\alpha)$ such that $|x_i| \leq \alpha$ imply

$$|V(t, x_1) - V(t, x_2)| \leq L(\alpha)|x_1 - x_2|.$$

One needs to find conditions on F to ensure that L is independent of α, or, at least, estimate the size of $L(\alpha)$.

The first result concerns a periodic system

$$(4.1.23) \qquad x' = F(t, x), \qquad F(t, 0) = 0, \qquad F(t + T, x) = F(t, x)$$

for some $T > 0$. We suppose $F : (-\infty, \infty) \times R^n \to R^n$ is continuous. The result and proof are found in Yoshizawa (1966, pp. 30, 46).

Theorem 4.1.19. *If* $x = 0$ *is stable for* (4.1.23), *then it is uniformly stable. If* $x = 0$ *is asymptotically stable for* (4.1.23), *then it is uniformly asymptotically stable. If* $x = 0$ *is asymptotically stable in the large, then it is U.A.S. in the large.*

In the following results, $C_0(x)$ denotes the family of functions $F : [0, \infty) \times R^n \to R^n$ which satisfy a local Lipschitz condition in x: for each (t_0, x_0) and each $M > 0$ there is an $L > 0$ such that $\big[|t - t_0| \le M, \ t \ge 0, |x_i - x_0| \le M$ for $i = 1, 2\big]$ imply $|F(t, x_1) - F(t, x_2)| \le L|x_1 - x_2|$. Also, $\overline{C}_0(x)$ denotes the family satisfying a uniform local Lipschitz condition in x : for each $x_0 \in R^n$ and each $M > 0$ there is an $L > 0$ such that $\big[t \ge 0, \ |x_i - x_0| \le M$ for $i = 1, 2\big]$ imply $|F(t, x_1) - F(t, x_2)| \le L|x_1 - x_2|$.

Theorem 4.1.20. Yoshizawa (1966, p. 100) *If* $F(t, x) \in C_0(x)$ *and if the zero solution of* (4.1.12) *is U.A.S. in the large, then there exists* $V : [0, \infty) \times R^n \to [0, \infty)$ *with*

(a) $W_1(|x|) \le V(t, x) \le W_2(|x|)$, $W_1(r) \to \infty$ *as* $r \to \infty$,

(b) $V'_{(4.1.12)}(t, x) \le -cV(t, x)$, $c > 0$,

(c) $|V(t, x_1) - V(t, x_2)| \le h(\alpha)f(t)|x_1 - x_2|$

where h *and* f *are continuous functions and* $|x_i| \le \alpha$. *If* $F(t, x) \in \overline{C}_0(x)$, *then* $f(t) = 1$ *for all* t. *If* F *is periodic,* V *is periodic. If* F *is autonomous, then* V *is autonomous.*

Much thought has gone into the problem of perturbations.

Example 4.1.17. Levin and Nohel (1960) Consider the scalar equation

$$x'' + f(t, x, x')x' + g(x) = e(t)$$

with f, g, and e continuous, $xg(x) > 0$ for $x \ne 0$, $f(t, x, y) \ge k(x, y) > 0$ for $y \ne 0$ with $k(x, y)$ continuous, $G(x) = \int_0^x g(s) \, ds \to \infty$ as $|x| \to \infty$, and $E(t) = \int_0^t |e(s)| \, ds \le M$. Write the equation as

$$x' = y, \qquad y' = -f(t, x, y)y - g(x) + e(t).$$

Define
$$V(t, x, y) = e^{-2E(t)}\{2G(x) + y^2 + 2\}$$

and obtain
$$V'(t, x, y) \leq -2e^{-2E(\infty)}f(t, x, y)y^2.$$

Solutions are bounded and, if f is bounded for $x^2 + y^2$ bounded, then solutions approach the x-axis.

Example 4.1.18. Yoshizawa (1966, p. 39) Consider the previous example and use
$$V(t, x, y) = [y^2 + 2G(x)]^{1/2} - \int_0^t |e(s)| \, ds$$

to obtain
$$V'(t, x, y) = [\{-f(t, x, y)y^2 + ye(t)\}/[y^2 + 2G(x)]^{1/2}] - |e(t)|$$

which is nonpositive.

Exercise 4.1.7. Consider Example 4.1.13 once more and use
$$U(x, y) = y^2 + 2G(x) + \ln\{[y + F(x)]^2 + 2G(x) + 1\}.$$

Give conditions to ensure U' negative for large $x^2 + y^2$.

When F is independent of t (or, more generally, periodic in t), say

(4.1.24) $$x' = G(x)$$

where $G : R^n \to R^n$ is continuous and locally Lipschitz in x then more detailed information can be obtained from a Liapunov function. We remind the reader briefly of limit sets.

A point y is an ω-*limit point* of a solution $x(t)$ of (4.1.24) if there is a sequence $\{t_n\} \to \infty$ with $x(t_n) \to y$. The set of ω-limit points of a solution of (4.1.24) is called the ω-*limit set*. By the uniqueness, if y is in the ω-limit set of $x(t)$, then the orbit through y, say
$$\{z \in R^n : z = x(t, 0, y), \quad t \geq 0\}$$

is also in the ω-limit set.

A set A is positively invariant if $y \in A$ implies $x(t, 0, y) \in A$ for $t \geq 0$.

Theorem 4.1.21. *Let the conditions of Theorem* 4.1.18 *hold for* (4.1.24) *and let* $V = V(x)$. *Also, let* M *be the largest invariant set in* Ω. *Then every solution of* (4.1.24) *approaches* M *as* $t \to \infty$.

Proof. If $x(t)$ is a solution of (4.1.24), then it approaches Ω. Suppose there is a point y in the ω-limit set of $x(t)$ not in M. Certainly, $y \in \Omega$, and as $y \notin M$, there is a $t_1 > 0$ with $x(t_1, 0, y) \notin \Omega$. Also, there is a sequence $\{t_n\} \to \infty$ with $x(t_n) \to x(t_1, 0, y)$, a contradiction to $x(t) \to \Omega$ as $t \to \infty$. This completes the proof.

The result can be refined still more by noticing that $V(x(t)) \to c$ so the set M is restricted still more by satisfying $V(x) = c$ for some $c \geq 0$.

The ideas in Theorems 4.1.18 and 4.1.21 were extended by Hale (1965) to autonomous functional differential equations using Liapunov functionals and by Haddock and Terjéki (1983, 1984) using a Razumikhin technique.

The condition of Theorems 4.1.17 and 4.1.18 that $F(t, x)$ be bounded for $|x|$ bounded is very troublesome and so often runs counter to our intuition concerning what should be true. Clearly, the solutions of the scalar equation $x' = -(t^2 + 1)x$ should approach zero faster than those of $x' = -x$.

The idea in the next theorem was introduced by Burton (1969, 1970), and has been generalized extensively by Erhart (1973) Haddock (1972a,b, 1974), Hatvani (1978), Burton (1969, 1977, 1979a,b), and at this writing we are still receiving preprints of further generalizations (cf. Murakami, 1984a,b). It is very effective with Volterra and functional differential equations.

Theorem 4.1.22. *Let $|\cdot|$ denote Euclidean length. Suppose there is a continuous function $V : [0, \infty) \times R^n \to [0, \infty)$ which is locally Lipschitz in x and suppose that for each $\varepsilon > 0$ there exists $\delta > 0$ such that $|x| \geq \varepsilon$ implies $V'_{(4.1.12)}(t, x) \leq -\delta|F(t, x)|$.*

(a) *If V is decrescent or if δ is independent of ε, then all solutions of (4.1.12) are bounded.*

(b) *If δ is independent of ε and if for each $x_0 \in R^n - \{0\}$ there is an $\eta > 0$ and a continuous function $\gamma : [0, \infty) \to [0, \infty)$ with $\int_0^\infty \gamma(s)\,ds = \infty$ and $|F(t, x)| \geq \gamma(t)$ for $|x - x_0| \leq \eta$, then each solution of (4.1.12) tends to zero as $t \to \infty$.*

Proof. Suppose first that δ is independent of ε. Then for any solution $x(t)$ of (4.1.12) we have $V'(t, x(t)) \leq -\delta|x'(t)|$ and so $0 \leq V(t, x(t)) \leq V(t_0, x_0) - \delta \int_{t_0}^t |x'(s)|\,ds$ so that $|x(t)|$ is bounded and, in fact, has finite Euclidean length. Condition (b) simply implies that there is no $x_1 \neq 0$ such that $x(t) \to x_1$ as $t \to \infty$; for if so, then $V'(t, x(t)) \leq -\delta\gamma(t)$ for $t \geq t_2$, some $t_2 > t_0$, and this implies

$$0 \leq V(t, x(t)) \leq V(t_2, x(t_2)) - \delta \int_{t_2}^t \gamma(s)\,ds$$

$$\to -\infty \quad \text{as} \quad t \to \infty.$$

But $\{x(n)\}$ is a bounded sequence with convergent subsequence having limit x_2; since $x(t)$ has finite arclength, $x(t) \to x_2$. This means $x_2 = 0$.

To complete the proof we suppose δ depends on ε and $|x(t)| \geq \varepsilon$ on $[t_1, t_2]$. Then for $t_1 \leq t \leq t_2$ we have

$$0 \leq V(t, x(t)) \leq V(t_1, x(t_1)) - \delta \int_{t_1}^{t_2} |x'(s)| \, ds$$

$$\leq W(|x(t_1)|) - \delta \int_{t_1}^{t_2} |x'(s)| \, ds$$

so that $|x(t)|$ is bounded. This completes the proof.

Exercise 4.1.8. Improve Theorem 4.1.22 by asking that $V'(t, x) \leq -\delta |F(t, x)|/(1 + |x|)$.

The condition $V'(t, x) \leq -\delta |F(t, x)|$ is natural when V is autonomous. In that case

$$V'(x) = \text{grad } V(x) \cdot F(t, x)$$
$$= |\text{grad } V(x)| \, |F(t, x)| \cos \theta$$

where θ is the angle between grad V and F. If V is shrewdly chosen then it should be possible to satisfy the condition.

In Section 1.6 we saw examples of Liapunov functionals for Volterra equations satisfying $V'(t, x(\cdot)) \leq -\delta |x'|$. The condition was used effectively to drive solutions to zero.

In Section 4.2, Theorem 4.2.11, we will see that $V'(t, x_t) \leq -\mu |x'|$ plays a major role in uniform ultimate boundedness.

And the condition $V' \leq -\delta |x'|$ also occurs naturally when F is bounded for $|x|$ bounded. Suppose, for example, that there is a $V(t, x)$ with

$$V'(t, x) \leq -W(|x|) \qquad \text{for} \qquad |x| \leq 1$$

and

$$|F(t, x)| \leq M \qquad \text{for} \qquad |x| \leq 1.$$

Clearly, for each $\varepsilon > 0$ there is a $\delta > 0$ with $V'(t, x) \leq -\delta |x'|$ if $\varepsilon \leq |x| \leq 1$.

Example 4.1.19. Let

(4.1.25) $$x' = Ax$$

where A is an $n \times n$ constant real matrix all of whose characteristic roots have negative real parts. Let

$$V(x) = [x^T B x]^{1/2}$$

where $B^T = B$ is positive definite and

(4.1.26) $$A^T B + BA = -I.$$

Note that the zero solution is unique and that V has continuous first partial derivatives for $x \neq 0$. Thus

$$V'_{(4.1.25)}(x) = (x^T A^T B x + x^T B A x)/\{2[x^T B x]^{1/2}\}$$
$$= -x^T x/\{2[x^T B x]^{1/2}\}$$

and there is a $k > 0$ with

$$|x|/\{2[x^T B x]^{1/2}\} \geq k$$

so that

(4.1.27) $\qquad V'_{(4.1.25)}(x) \leq -k|x| \leq -\delta|Ax|, \qquad \delta > 0.$

Example 4.1.20. Let A be an $n \times n$ constant matrix all of whose characteristic roots have negative real parts, c and b be constant vectors, r be a positive constant, $f : (-\infty, \infty) \to (-\infty, \infty)$ be continuous with $\sigma f(\sigma) > 0$ for $\sigma \neq 0$, and consider the $(n+1)$-dimensional system

$$x' = Ax + bf(\sigma), \qquad \sigma' = c^T x - rf(\sigma).$$

This is called the problem of Lurie and concerns a control system. Lurie used the Liapunov function

$$V(x, \sigma) = x^T B x + \int_0^\sigma f(s)\, ds$$

in which $B = B^T$ and $A^T B + BA = -D$, where D is positive definite. The derivative is given by

$$V'(x, \sigma) = -x^T D x + f(\sigma)[2b^T B + c^T]x - rf^2(\sigma).$$

Lefschetz (1965) showed this to be negative definite if and only if $r > (Bb + c/2)^T D^{-1}(Bb + c/2)$.

Exercise 4.1.9. Let $p(t)$ be a bounded and continuous function and consider the system

$$x' = Ax + bf(\sigma) + p(t), \qquad \sigma' = c^T x - rf(\sigma)$$

with the same condition as in Example 4.1.20. Use the new Liapunov function

$$V_1(x, \sigma) = V(x, \sigma)^{1/2}$$

and prove that solutions are uniform bounded and uniform ultimate bounded.

This equation with a delay is treated by Somolinos (1977).

Example 4.1.21. Write the scalar equation

$$x''' + f(x')x'' + ax' + bx = 0$$

as the system

$$x' = y,$$
$$y' = z,$$
$$z' = -f(y)z - ay - bx.$$

Assume that $a > 0$, $b > 0$, and $f(y) \geq c > b/a$ for all y. Then

$$V(x, y, z) = az^2 + 2byz + 2b \int_0^y f(s)s\, ds + (bx + ay)^2$$

satisfies

$$V'(x, y, z) \leq -2a[f(y) - c]z^2.$$

Exercise 4.1.10. Use the theorems developed here to show that the zero solution of the system in Example 4.1.21 is uniformly asymptotically stable in the large. Then perturb the equation by a bounded continuous $p(t)$ and prove solutions are bounded (cf. Wang and Wang, 1983).

The monograph of Yoshizawa (1966) contains extensive material on stability and Liapunov functions, as do Lakshmikantham and Leela (1969), Krasovskii (1963), and Hahn (1963, 1967). Hale (1969) is a standard textbook on ordinary differential equations, while Hartman (1964) contains very extensive material on almost every aspect of the subject. Discussions of construction of Liapunov functions are found in Barbashin (1968) and Burton (1983a). Extensive treatment of existence of periodic solutions of second-order differential equations is given in Cronin (1964), Sansone and Conti (1964), Reissig *et al.* (1963), Graef (1972), and Burton and Townsend (1968, 1971).

We began this section with a discussion of limit sets. In Proposition 4.1.2 the system

(4.1.2) $x' = H(x)$

is called the *limiting equation* for

(4.1.1) $x' = F(t, x).$

The study of limiting equations is now a fairly large and active area of research. We refer the reader to Artstein (1978), Hatvani (1983), Kato and Yoshizawa (1981), and Murakami (1984b).

4.1.6 Instability

We consider once more the system

$$(4.1.12) \qquad x' = F(t,x), \qquad F(t,0) = 0$$

with $F : [0,\infty) \times D \to R^n$ being continuous and D an open subset of R^n with $0 \in D$.

If we negate the definition of stability we see that there is an $\varepsilon > 0$ and a $t_0 \geq 0$ such that for any $\delta > 0$ there is an x_0 with $|x_0| < \delta$ and there is a $t_1 > t_0$ such that $|x(t_1, t_0, x_0)| \geq \varepsilon$. Then the zero solution is unstable.

Theorem 4.1.23. *Suppose there is a continuous function $V : [0,\infty) \times D \to [0,\infty)$ which is locally Lipschitz in x and wedges W_i such that on $[0,\infty) \times D$ we have*

(a) $W_1(|x|) \leq V(t,x) \leq W_2(|x|)$

and

(b) $V'_{(4.1.12)}(t,x) \geq W_3(|x|)$.

Then the zero solution of (4.1.12) is unstable.

Proof. If the theorem is false then for $\varepsilon = \min[1, d(0, \partial D)]$ there is a $\delta > 0$ such that $|x_0| < \delta$ and $t \geq 0$ imply that $|x(t,0,x_0)| < \varepsilon$. Pick x_0 with $|x_0| = \delta/2$ and find γ with $W_2(\gamma) = W_1(\delta/2)$. Then for $x(t) = x(t,0,x_0)$ we have $V'(t,x(t)) \geq 0$ so that

$$W_2(|x(t)|) \geq V(t,x(t)) \geq V(0,x_0) \geq W_1(\delta/2) = W_2(\gamma)$$

from which we conclude that $\gamma \leq |x(t)|$ for $t \geq 0$. Thus,

$$V'(t,x(t)) \geq W_3(|x(t)|) \geq W_3(\gamma)$$

and so

$$W_2(|x(t)|) \geq V(t,x(t)) \geq V(0,x_0) + W_3(\gamma)t$$

which implies that $|x(t)| \to \infty$. This completes the proof.

Example 4.1.22. Let $x' = Ax$ with A an $n \times n$ real constant matrix all of whose characteristic roots have positive real parts. Then all characteristic roots of $-A$ have negative real parts and there is a positive definite matrix $B = B^T$ with $(-A)^T B + B(-A) = -I$ or with $A^T B + BA = I$. Thus,

$$V(x) = x^T B x$$

satisfies

$$V'(x) = +x^T x.$$

Note that Theorem 4.1.23 asks too much. In the proof we chose any x_0 satisfying $|x_0| = \delta/2$ and showed that $|x(t, 0, x_0)| < \varepsilon$ is violated for some $t > 0$. This may be called *complete instability*. A proper instability theorem should allow some solutions to remain near zero.

Theorem 4.1.24. *Suppose there is a continuous function $V : [0, \infty) \times D \to (-\infty, \infty)$ which is locally Lipschitz in x and a sequence $\{t_0, x_n\} \in [0, \infty) \times D$ with $V(t_0, x_n) > 0$ and $\lim_{n \to \infty} x_n = 0$. If there are wedges W_i with*

(a) $V(t, x) \leq W_2(|x|)$

and

(b) $V'_{(4.1.12)}(t, x) \geq W_3(|x|)$

then the zero solution of (4.1.12) *is unstable.*

Proof. If the result is false then there is an ε and δ as in the previous proof. For the $\delta > 0$ find (t_n, x_n) with $|x_n| < \delta$ and $V(t_n, x_n) > 0$. The proof is now completed just as in the previous theorem.

Lemma. *Let A be an $n \times n$ constant matrix none of whose characteristic roots have zero real parts. Then a positive definite matrix $C = C^T$ can be found so that $A^T B + BA = -C$ can be solved for $B = B^T$.*

A proof of this result may be found in Burton (1983a, pp. 130–132).

Example 4.1.23. Let

$$A = \begin{pmatrix} 1 & 1 \\ 0 & -1 \end{pmatrix}$$

and $C = \begin{pmatrix} 2 & 1 \\ 1 & 2 \end{pmatrix}$. Then

$$B = \begin{pmatrix} 1 & 2 \\ 2 & 1 \end{pmatrix}$$

yields $A^T B + BA = C$. If we consider $x' = Ax$ and $V(x) = x^T Bx = x_1^2 + x_2^2 + 4x_1 x_2$ we have $V'(x) \geq x_1^2 + x_2^2$. Theorem 4.1.24 applies showing $x = 0$ to be unstable.

Exercise 4.1.11. Consider the scalar equation $x'' - x^3 = 0$. Express it as a system and find a function satisfying Theorem 4.1.24.

The next result is due to Chetaev and there are nonautonomous versions, but so much care is required in such statements that, in the interest of clarity, we present an autonomous form.

Theorem 4.1.25. *Let $L \subset R^n$ be an open set with $0 \in \partial L$ and let B be a positive number. For $\overline{L}$ the closure of L, let $V : \overline{L} \to [0, \infty)$ and suppose that $V(x) = 0$ for $|x| \leq B$ and $x \in \partial L$. Suppose that $V(x) > 0$ for $x \in L$ and $|x| \leq B$ and that $V'_{(4.1.12)}(x) \geq W(x)$ for $x \in L$ and $|x| \leq B$ where W is a continuous scalar function with $W(x) > 0$ for $x \in L$ and $|x| \leq B$. Then the zero solution of $(4.1.12)$ is unstable.*

Proof. If the theorem is false then for $\varepsilon = B/2$ there is a $\delta > 0$ such that $|x_0| < \delta$ and $t \geq 0$ imply $|x(t, 0, x_0)| < \varepsilon$. Pick $x_0 \in L$ with $|x_0| = \delta/2$. Then $V(x_0) > 0$ and, so long as $x(t, 0, x_0) \in L$ we have $V'(x(t, 0, x_0)) > 0$ so that $V(x(t, 0, x_0)) \geq V(x_0) > 0$. This means that $x(t, 0, x_0)$ is bounded strictly away from ∂L and so $W(x(t, 0, x_0))$ is bounded strictly away from zero, say by μ. Thus, $V'(x(t, 0, x_0)) \geq \mu$ and an integration yields a contradiction. See Fig. 4.13.

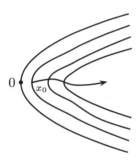

Fig. 4.13

Example 4.1.24. Consider the system

$$x' = y,$$
$$y' = x^2.$$

Let $V(x, y) = xy$ in quadrant I so that $V'(x, y) = y^2 + x^3 > 0$ in quadrant I.

4.2 Equations with Bounded Delays

4.2.1 Introduction

In this section we are concerned with equations of the form

(4.2.1) $$x' = F(t, x_t)$$

as treated in Chapter 3 under (3.5.2). We suppose that $F : (-\infty, \infty) \times C \to R^n$ where C is the set of continuous functions $\phi : [-\alpha, 0] \to R^n$, $\alpha > 0$. Here, F is continuous and takes bounded sets into bounded sets.

Recall that in Section 3.5 we let x_t denote the restriction of $x(u)$ to the interval $[t - \alpha, t]$ and then translated to $[-\alpha, 0]$. We also stated in Remark 3.5.4 that in examples and Liapunov theory it was much preferable to not translate the segment back to $[-\alpha, 0]$. Thus, we introduce the convention to be used in this chapter.

Notation   (a) For $t_1 \in R$,

$$C(t_1) = \{\phi : [t_1 - \alpha, t_1] \to R^n | \phi \quad \text{is continuous}\}.$$

In particular, $C = C(0)$.

(b) The function ϕ_t denotes an element of $C(t)$. When it is clear from the context, the subscript on ϕ will be omitted.

(c) If $\phi \in C(t)$, then $\|\phi\| = \sup_{t-\alpha \leq s \leq t} |\phi(s)|$ where $|\cdot|$ is any norm on R^n.

A typical example of (4.2.1) is

$$(4.2.1)' \qquad x' = G(t, x(t), x(t - r(t))) + \int_{t-g(t)}^{t} q(t, s, x(s)) \, ds$$

in which r and g are nonnegative and bounded by α.

Recall that to specify a solution of (4.2.1) we require a $t_0 \in R$ and a $\phi \in C(t_0)$; then we obtain a continuous function $x(t_0, \phi)$ on some interval $[t_0 - \alpha, \beta)$, $\beta > t_0$; the value of the function is denoted by $x(t, t_0, \phi)$ and it agrees with ϕ on $[t_0 - \alpha, t_0]$, while it satisfies (4.2.1) on $[t_0, \beta)$. If $F(t, \phi)$ is continuous and locally Lipschitz in ϕ, the solution is uniquely determined by ϕ and is continuous in ϕ. If solutions remain bounded, then $\beta = \infty$.

We saw in Chapter 1 that the solution space of

$$x' = A(t)x + \int_{0}^{t} C(t, s)x(s) \, ds$$

is very simple in structure and that some such equations can be reduced to ordinary differential equations. Corresponding results for (4.2.1) are exceedingly rare. To illustrate the complexity of the solution space we consider the scalar equation

$$x'(t) = ax(t) + bx(t - 1)$$

with a and b positive constants. Try for a solution of the form $x = e^{mt}$ with m constant and obtain

$$m = a + be^{-m}$$

which is called the characteristic quasipolynomial. It is known (see El'sgol'ts, 1966) that there is an infinite sequence of solutions $\{m_i\}$ with

Re $m_i \to -\infty$ as $n \to \infty$. Thus, in one step we see that the solution space is infinite dimensional and that any concept of complete instability may be impossible.

When $r(t)$ is bounded strictly away from zero and $q = 0$, then $(4.2.1)'$ can be solved by the method of steps. We illustrate the method with a simple equation.

Example 4.2.1. Consider the scalar equation

$$x' = x + x(t - 1)$$

with $\phi(t) = t$ on $[-1, 0]$. Find a solution on $[0, 3]$, say $x(t, 0, \phi)$.

Note that for $0 \leq t \leq 1$, then $-1 \leq t - 1 \leq 0$ so $x(t-1) = \phi(t-1) = t-1$ and the equation with initial condition becomes

$$x' = x + (t - 1), \qquad x(0) = 0,$$

with solution (by variation of parameters)

$$x(t, 0, \phi) = -t \qquad \text{for } 0 \leq t \leq 1.$$

Thus, the problem is solved on $[0, 1]$ and the solution $x(t) = -t$ becomes the initial function on the new initial interval $[0, 1]$. We then try to solve the equation for $1 \leq t \leq 2$ so that $0 \leq t - 1 \leq 1$ and $x(t-1) = \phi(t-1) = -(t-1)$, yielding the problem

$$x' = x - (t - 1), \qquad x(1) = -1.$$

The solution is

$$x(t) = t - 2e^{(t-1)}, \qquad 1 \leq t \leq 2.$$

Notice that $x'(t) = 1 - 2e^{(t-1)}$ and so $x'(1^+) = -1$, which agrees with $x'(1^-)$. Notice also that the derivative of the function $x(t)$ on $[-1, 2]$ fails to exist at $t = 0$. It is generally true that $x(t, t_0, \phi)$ fails to be differentiable at $t = t_0$, but smooths as t increases.

The investigator almost always feels that delays cause difficulties. But there are striking incidents of simplifications wrought by delays. A delay can cause boundedness, stability, continuation, integrability, or oscillation.

Example 4.2.2. The scalar equation

$$x' = x^2 \qquad \text{with} \qquad x(0) = 1$$

has a solution tending to infinity in finite time. But if $r(t)$ is positive for all $t \geq 0$ then

$$x' = x^2(t - r(t))$$

has all solutions continuable for all $t \geq 0$. To see this, let an initial function on $[-\alpha, 0]$ be given and let $x(t)$ be a solution defined on $[0, T)$ with $\limsup_{t \to T^-} |x(t)| = +\infty$. Then there is a $\gamma > 0$ with $r(t) \geq \gamma$ on $[0, T]$ so that $t - r(t) \leq T - \gamma$ for $t \leq T$. But $x(t)$ is continuous on $[0, T - \gamma]$ and, hence, it is bounded by some number M. Thus, $|x'(t)| \leq M^2$ for $0 \leq t \leq T$ and so $|x(t) - x(0)| \leq M^2 T$, a contradiction.

Example 4.2.3. One can prove using ideal theory (cf Kaplansky, 1957) that the scalar equation

$$x'' + tx = 0$$

cannot be solved in closed form. But if $r(t) \geq \gamma > 0$ on $[0, \infty)$, then, given an initial function, we can solve

$$x'' + tx(t - r(t)) = 0$$

by quadratures using the method of steps.

Example 4.2.4. The predator-prey system

$$x' = ax - bx^2 - cxy,$$
$$y' = -ky + dxy$$

has never been integrated. But the term dxy represents the predator's utilization of the prey which it captures. Hence, there is certainly a delay, say $T > 0$, yielding

$$x' = ax - bx^2 - cxy,$$
$$y' = -ky + dx(t - T)y(t - T).$$

Thus, given an initial function $(\phi(t), \psi(t))$ on an initial interval $[-T, 0]$, we explicitly integrate the linear equation

$$y' = -ky + d\phi(t - T)\psi(t - T) \qquad \text{for} \qquad 0 \leq t \leq T$$

and obtain $y(t) = \eta(t)$. Then the equation

$$x' = ax - bx^2 - cx\eta(t), \qquad x(0) = \phi(0)$$

is a Bernoulli equation which can be solved by quadratures.

Example 4.2.4. A feedback system with friction proportional to velocity, an external force $p(t)$, and a delayed restoring force $g(x(t - r))$, $r > 0$, may be written as

$$x'' + cx' + g(x(t - r)) = p(t).$$

If $r = 0$, it cannot be solved. But for r a positive constant, given an initial function, it becomes a linear second-order equation with constant coefficients, easily solved by the method of steps.

Much more information concerning delay equations may be found in the extensive and excellent books of Bellman and Cooke (1963), Driver (1977), El'sgol'ts (1966), Hale (1977), Krasovskii (1963), and Yoshizawa (1966). Shift properties of Laplace transforms make that method ideal for linear differential-difference equations with r constant (see Bellman and Cooke, 1963). And complex variables can be used on such equations in a particularly pleasing way.

Example 4.2.5. (El'sgol'ts, 1966) Consider again the scalar equation

$$x' = ax + bx(t - 1)$$

with characteristic quasipolynomial

$$\Phi(m) = a - m + be^{-m}.$$

Now, suppose m_1 is a root of $\Phi(m) = 0$ and suppose C is a circle in the complex plane containing m_1 and no other root of $\Phi(m) = 0$. Let $F(s)$ be an arbitrary analytic function on and inside C. Then

$$x(t) = \int_C [e^{st} F(s)/\Phi(s)] \, ds$$

is a solution. To see this, substitute it into the equation and note that we obtain

$$-\int_C [\Phi(s) e^{st} F(s)/\Phi(s)] \, ds$$

which is zero by Cauchy's integral theorem.

Moreover, if m_1 is a pole of order j then by the residue theorem

$$\int_C [e^{st} F(s)/\Phi(s)] \, ds = 2\pi i \text{Res}[e^{st} F(s)/\Phi(s)]$$

$$= [2\pi i/(j-1)!][e^{st} F(s)(s - m_1)^j/\Phi(s)]^{(j-1)}|_{s=m_1}$$

$$= P_{j-1}(t) e^{m_1 t}$$

where $P_{j-1}(t)$ is a polynomial of degree $j - 1$. Because $F(s)$ is arbitrary, $P_{j-1}(t)$ has arbitrary coefficients.

While such examples are helpful in giving us some ideas about solutions, the integrations in the method of steps and the problem of locating roots of characteristic quasipolynomials are awesome. We turn now to discussion of qualitative theory.

4.2.2 Periodic Solutions

We suppose now that there is a $T > 0$ such that $F(t, \phi)$ is T-periodic in the sense that if $x(t)$ is a solution of (4.2.1) so is $x(t + T)$. Moreover, we suppose in this section that F satisfies a local Lipschitz condition in ϕ.

Lemma 4.2.1. *Suppose that $x(t + T)$ is a solution of (4.2.1) whenever $x(t)$ is a solution. Then (4.2.1) has a T-periodic solution if and only if there is a $t_0 \in R$ and $\phi : [t_0 - \alpha, t_0] \to R^n$ with $x(t + T, t_0, \phi) = \phi(t)$ for $t_0 - \alpha \leq t \leq t_0$.*

Proof. Suppose that $x(t, t_0, \phi)$ is a T-periodic solution of (4.2.1). Then $x(t + T, t_0, \phi) \equiv x(t, t_0, \phi) = \phi(t)$ for $t_0 - \alpha \leq t \leq t_0$. Next, suppose that there is a t_0 and ϕ with

$$x(t + T, t_0, \phi) = \phi(t) \qquad \text{for} \qquad t_0 - \alpha \leq t \leq t_0.$$

Then $x(t, t_0, \phi)$ and $x(t + T, t_0, \phi)$ are both solutions with the same initial condition and so, by uniqueness, they are equal. The proof is complete.

We now extend the definitions of uniform boundedness to (4.2.1). As before, the definitions are usually stated in terms of arbitrary t_0. Here, we consistently take $t_0 = 0$. Recall now that $C(t) = \{\phi : [t - \alpha, t] \to R^n | \phi$ is continuous$\}$ and $C = C(0)$.

Definition 4.2.1. *Solutions of (4.2.1) are* uniform bounded *at $t = 0$ if for each $B_1 > 0$ there exists $B_2 > 0$ such that $[\phi \in C, \|\phi\| \leq B_1, t \geq 0]$ imply $|x(t, 0, \phi)| < B_2$.*

Definition 4.2.2. *Solutions of (4.2.1) are* uniform ultimate bounded *for bound B at $t = 0$ if for each $B_3 > 0$ there is a $K > 0$ such that $[\phi \in C, \|\phi\| \leq B_3, t \geq K]$ imply that $|x(t, 0, \phi)| < B$.*

Kato (1980a) has constructed an example of a general form $x' = H(x_t)$ in which solutions are uniform ultimate bounded but not uniform bounded. For ordinary differential equations which are autonomous or periodic in t it is known that uniform ultimate boundedness implies uniform boundedness. However, for a general ordinary differential equation Kato and Yoshizawa (1981) show that uniform ultimate boundedness does not imply uniform boundedness.

Theorem 4.2.1. *Suppose that $x(t + T)$ is a solution of (4.2.1) whenever $x(t)$ is a solution of (4.2.1). If solutions of (4.2.1) are uniform ultimate bounded for bound B at $t = 0$, then (4.2.1) has a mT-periodic solution for some positive integer m.*

Proof. The space $(C, \|\cdot\|)$ is a Banach space. For the $B > 0$ we can find mT such that $[\phi \in C, \|\phi\| \leq B$, and $t \geq mT - \alpha]$ imply that $|x(t, 0, \phi)| < B$. There exists $L > 0$ such that $t_0 \geq 0$ and $\|\phi\| \leq B$ implies $|F(t, \phi)| \leq L$. Let $S = \{\phi \in C \mid \|\phi\| \leq B, |\phi(u) - \phi(v)| \leq L|u - v|\}$. Then S is compact and convex. Define $(P\phi)(t) = x(t + (m+1)T, 0, \phi)$ for $-\alpha \leq t \leq 0$. Then $|x'| \leq L$

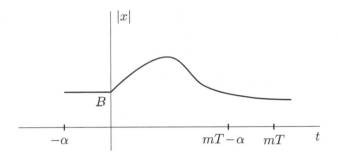

Fig. 4.14

and $P : S \to S$. See Fig. 4.14. Then P is continuous because F is Lipschitz and so P has a fixed point by Schauder's theorem. Thus, $x(t + mT, 0, \phi)$ and $x(t, 0, \phi)$ are both solutions with the same initial condition and, hence, by uniqueness they are equal.

The fixed-point theorem of Horn will now show that the period is T. This next result was proved by Yoshizawa (1966) when $T \geq \alpha$. See also the paper of Hale and Lopes (1973). The problem is discussed by Jones (1963, 1964, 1965), and by Billotti and LaSalle (1971).

Theorem 4.2.2. *Suppose that $x(t + T)$ is a solution of (4.2.1) whenever $x(t)$ is a solution of (4.2.1). If solutions of (4.2.1) are uniform bounded and uniform ultimate bounded for bound B at $t = 0$, then (4.2.1) has a T-periodic solution.*

Proof. There is a $B_2 > 0$ such that $[\phi \in C, \|\phi\| \leq B, t \geq 0]$ imply that $|x(t, 0, \phi)| < B_2$. There exists $B_3 > 0$ such that $[\phi \in C, \|\phi\| \leq B_2 + 1, t \geq 0]$ imply that $|x(t, 0, \phi)| < B_3$. Also, there is an integer m such that $[\phi \in C, \|\phi\| \leq B_2 + 1, t \geq mT - \alpha]$ imply that $|x(t, 0, \phi)| < B$. Finally, there exists $L > 0$ such that $[0 \leq t \leq mT, \|\phi\| \leq B_3]$ imply that $|x'(t, 0, \phi)| \leq L$. Let

$$S_0 = \{\phi \in C \mid \|\phi\| \leq B, |\phi(u) - \phi(v)| \leq L|u - v|\}$$
$$S_2 = \{\phi \in C \mid \|\phi\| \leq B_3, |\phi(u) - \phi(v)| \leq L|u - v|\}$$

and

$$S_1 = \{\phi \in C \mid \|\phi\| < B_2 + 1\} \cap S_2.$$

Then the S_i are convex, S_0 and S_2 are compact by Ascoli's theorem, and S_1 is open in S_2.

Define $P : S_2 \to C$ by $\phi \in S_2$ implies

$$P\phi = x(t+T, 0, \phi) \qquad \text{for} \qquad -\alpha \le t \le 0.$$

Now $x(t+T, 0, \phi)$ is a solution for $t \ge 0$ and its initial function is ϕ. Hence,

$$(*) \qquad x(t+T, 0, \phi) = x(t, 0, P\phi)$$

by the uniqueness theorem. Next,

$$P^2\phi = x(t+T, 0, P\phi) \qquad \text{for} \qquad -\alpha \le t < 0$$

and $x(t+T, 0, P\phi)$ is a solution with initial function $P\phi$. Hence

$$(**) \qquad x(t+T, 0, P\phi) = x(t, 0, P^2\phi)$$

by uniqueness. Now in $(*)$ let t be replaced by $t+T$ so that

$$x(t+2T, 0, \phi) = x(t+T, 0, P\phi) = x(t, 0, P^2\phi)$$

by $(**)$. In general,

$$x(t+kT, 0, \phi) = x(t, 0, P^k\phi).$$

By construction of S_1 and S_2 we have $P^j S_1 \subset S_2$ for $1 \le j \le m$. By choice of m we have $P^j S_1 \subset S_0$ for $j \ge m$. Also, $P^j(S_0) \subset S_1$ for all j. See Fig. 4.15. By Horn's theorem P has a fixed point $\phi \in S_0$ so by Lemma 4.2.1 we conclude that $x(t, 0, \phi)$ is T-periodic. This completes the proof.

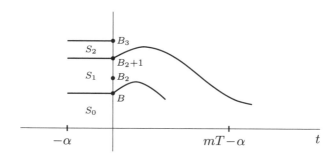

Fig. 4.15

We now consider a theory which will allow us to show the required boundedness.

4.2.3 Stability and Boundedness

We now extend the stability definitions of $x' = F(t, x)$ to the system (4.2.1). Recall that (4.2.1) was defined for $-\infty < t < \infty$ because we had in mind the search for periodic solutions. But in the study of stability it is customary to define (4.2.1) for $t \geq t_0$, for some fixed t_0. We understand that if $t_0 = -\infty$, then the notation $t \geq t_0$ means $t > t_0$.

Recall that

$$C(t) = \{\phi : [t - \alpha, t] \to R^n \mid \phi \text{ is continuous}\},$$

that ϕ_t denotes the ϕ in the particular $C(t)$, and that $\|\phi_t\| = \max_{t-\alpha \leq s \leq t} |\phi(t)|$ where $|\cdot|$ is a convenient norm on R^n. A solution is denoted by $x(t_0, \phi_{t_0})$ and its value by $x(t, t_0, \phi_{t_0})$ or, if no confusion is likely, by $x(t_0, \phi)$ and $x(t, t_0, \phi)$.

Definition 4.2.3. *Let $x(t) = 0$ be a solution of (4.2.1).*

(a) *The zero solution of (4.2.1) is* stable *if for each $\varepsilon > 0$ and $t_1 \geq t_0$ there exists $\delta > 0$ such that $[\phi \in C(t_1), \|\phi\| < \delta, t \geq t_1]$ imply that $|x(t, t_1, \phi)| < \varepsilon$.*

(b) *The zero solution of (4.2.1) is* uniformly stable *if it is stable and if δ is independent of $t_1 \geq t_0$.*

(c) *The zero solution of (4.2.1) is* asymptotically stable *if it is stable and if for each $t_1 \geq t_0$ there is an $\eta > 0$ such that $[\phi \in C(t_1), \|\phi\| < \eta]$ imply that $x(t, t_1, \phi) \to 0$ as $t \to \infty$. (If this is true for every $\eta > 0$, then $x = 0$ is asymptotically stable in the large or globally asymptotically stable.)*

(d) *The zero solution of (4.2.1) is* uniformly asymptotically stable *if it is uniformly stable and if there is an $\eta > 0$ such that for each $\gamma > 0$ there exists $S > 0$ such that $[t_1 \geq t_0, \phi \in C(t_1), \|\phi\| < \eta, t \geq t_1 + S]$ imply that $|x(t, t_1, \phi)| < \gamma$. (If this is true for all $\eta > 0$, then $x = 0$ is uniformly asymptotically stable in the large.)*

Before embarking on abstractions, let us consider some concrete cases.

Example 4.2.6. Consider the scalar equation

$$x' = -ax + bx(t - r)$$

for a, b, and r positive constants with $a \geq b$. We treat the delay term as a perturbation of the asymptotically stable equation $x' = -ax$. Let $t_1 \in R$, $\phi \in C(t_1)$, $x(t) = x(t, t_1, \phi)$ and define the functional

$$V(x_t) = x^2 + b \int_{t-r}^{t} x^2(s) \, ds.$$

Then
$$V'(x_t) = 2xx' + b[x^2 - x^2(t-r)]$$
$$\leq [-2a + 2b]x^2 \leq 0.$$

Hence, an integration yields

$$x^2(x) \leq V(x_t) \leq V(\phi_{t_1})$$
$$= \phi^2(t_1) + b \int_{t_1-r}^{t_1} \phi^2(s)\,ds$$
$$\leq (1 + br)\|\phi_{t_1}\|^2.$$

It is evident that

(a) $x = 0$ is uniformly stable,

(b) this stability is independent of r.

If we ask that $a > b$, then

(c) $x(t)$, $x(t-r)$, and $x'(t)$ are all bounded,

(d) $x^2(t) \in L^1[0, \infty)$ because

$$V'(x_t) \leq -(a-b)x^2,$$

(e) by (c) and (d) it follows that $x(t) \to 0$ as $t \to \infty$ so we have asymptotic stability,

(f) and if we take

$$V(x_t) = x^2 + k \int_{t-r}^{t} x^2(s)\,ds$$

with $b < k < a$, then we obtain

$$V'(x_t) \leq -\mu\{x^2 + [x'(t)]^2\}, \qquad \mu > 0.$$

Exercise 4.2.1. With just the information here, argue that the zero solution is uniformly asymptotically stable in Example 4.2.6. Use the fact that if $|x(t)|$ is small for t in any interval of length r, then $|x(t)|$ remains small. Since

$$V'(x_t) \leq -\mu\{x^2 + [x'(t)]^2\}$$

argue that for each $\varepsilon > 0$ there exists $\delta > 0$ such that $|x(t)| \geq \varepsilon$ implies that $V'(x_t) \leq -\delta|x'(t)|$. This says that the solution has finite arc length outside any neighborhood of $x = 0$.

Example 4.2.7. Consider the scalar equation

$$x'' + ax' + bx(t - r) = 0$$

with a, b, and r positive. Write it as the system

$$x' = y,$$

$$y' = -ay - bx + \int_{-r}^{0} by(t + s) \, ds$$

and use the functional

$$V(x_t, y_t) = y^2 + bx^2 + \gamma \int_{-r}^{0} \int_{t+s}^{t} y^2(u) \, du \, ds$$

to obtain

$$V'(x_t, y_t) = -2ay^2 + \int_{-r}^{0} 2byy(t + s) \, ds$$

$$+ \gamma \int_{-r}^{0} [y^2(t) - y^2(t + s)] \, ds$$

$$\leq \int_{-r}^{0} \{[-(2a/r) + b + \gamma]y^2 + (b - \gamma)y^2(t + s)\} \, ds.$$

If we take $\gamma = b$, then we require

$$-a + br < 0$$

in order to make $V'(x_t, y_t) \leq 0$. In other words, if $a > 0$ and $b > 0$, then for r sufficiently small we have V positive definite and $V' \leq 0$.

This example contrasts with the previous one in that the delay term causes boundedness and the size of the delay is crucial.

Exercise 4.2.2. In Example 4.2.7 let $a > 0$, $b > 0$, and $-a + br < 0$.

(a) Prove that the zero solution is uniformly stable.

(b) Choose $\gamma > b$ and $(-2a/r) + \gamma + b < 0$. Prove that the zero solution is

  (i) asymptotically stable.
  (ii) uniformly asymptotically stable.

Example 4.2.8. Consider the scalar equation

$$x' = -ax + bx(t - r)$$

with a, b, and r positive constants and $a \geq b$. Define a function (not a functional) by

$$V(x) = x^2$$

and obtain

$$\begin{aligned}
V'(x) &= -2ax^2 + 2bxx(t-r) \\
&\leq -2ax^2 + bx^2 + bx^2(t-r) \\
&\leq -bx^2 + bx^2(t-r).
\end{aligned}$$

An integration yields

$$\begin{aligned}
V(x(t)) &\leq V(x(t_0)) - b \int_{t_0}^t x^2(s)\,ds + b \int_{t_0}^t x^2(s-r)\,ds \\
&\leq V(x(t_0)) - b \int_{t_0}^t x^2(s)\,ds + b \int_{t_0-r}^{t-r} x^2(u)\,du \\
&= V(x(t_0)) + b \int_{t_0-r}^{t_0} \phi^2(u)\,du - \int_{t-r}^t bx^2(u)\,du.
\end{aligned}$$

In this example we have begun with what is known as a Razumikhin technique, the art of using a Liapunov function instead of functional. But we avoid the usual geometric argument and supply a purely analytic argument which is easier to follow.

Exercise 4.2.3. Use the Liapunov function of Example 4.2.8 to obtain the results given in Example 4.2.6 using the Liapunov functional.

Example 4.2.9. Consider the scalar equation

$$x' = -ax + b \int_{t-r}^t x(u)\,du$$

with $a > 0$ and $r > 0$, and with a, b, and r constant. Use the change of variable $u = v + t$ to write the equation as

$$x' = -ax + b \int_{-r}^0 x(v+t)\,dv.$$

Define

$$V(x_t) = x^2 + \gamma \int_{-r}^0 \int_{t+v}^t x^2(u)\,du\,dv$$

and obtain

$$\begin{aligned}
V'(x_t) = &-2ax^2 + 2bx \int_{-r}^0 x(v+t)\,dv \\
&+ \gamma \int_{-r}^0 [x^2(t) - x^2(t+v)]\,dv.
\end{aligned}$$

The analysis is then the same as in Example 4.2.7 and Exercise 4.2.2.

Example 4.2.10. Let a, b, and r be positive constants and let $g(x) = \arctan x$. Consider the scalar equation

$$x'' + ax' + bg(x(t-r)) = 0$$

which we write as the system

$$x' = y,$$

$$y' = -ay - bg(x) + (d/dt) \int_{t-r}^{t} bg(x(u)) \, du.$$

Define

$$V(x_t, y_t) = \left[y + ax - \int_{t-r}^{t} bg(x(u)) \, du \right]^2$$

$$+ \left[y - \int_{t-r}^{t} bg(x(u)) \, du \right]^2 + 4b \int_{0}^{x} g(s) \, ds$$

so that

$$V'(x_t, y_t) = 2 \left[y + ax - \int_{t-r}^{t} bg(x(u)) \, du \right] \left[- bg(x) \right]$$

$$+ 2 \left[y - \int_{t-r}^{t} bg(x(u)) du \right] \left[- ay - bg(x) \right] + 4bg(x)y$$

$$\leq -2a \left[bxg(x) + y^2 \right] + ab\pi r |y| + rb^2 \pi^2$$

$$\leq -\left(xg(x) + y^2 \right) + M$$

for positive constants γ and M.

Exercise 4.2.4. Follow the ideas in the proof of Theorem 4.1.16 to show that solutions are uniform bounded and uniform ultimate bounded for bound B.

Example 4.2.11. Consider again the scalar equation

$$x' = -ax + b \int_{t-r}^{t} x(s) \, ds$$

with a, b, and r positive constants and $a > br$. Pursue the Razumikhin technique once more and define

$$V(x(t)) = |x|$$

which yields

$$V'(x_t) \leq -a|x| + b \int_{t-r}^{t} |x(s)| \, ds$$

$$= -a|x| + b \int_{-r}^{0} |x(u+t)| \, du \, .$$

Notice that $V = V(x(t))$, but $V' = V(x_t)$. Take $t_0 = 0$ and integrate to obtain

$$|x(t)| = V(x(t)) \leq V(x(0)) - a \int_{0}^{t} |x(s)| \, ds$$

$$+ b \int_{0}^{t} \int_{-r}^{0} |x(u+v)| \, du \, dv$$

$$= V(x(0)) - a \int_{0}^{t} |x(s)| \, ds + b \int_{-r}^{0} \int_{0}^{t} |x(u+v)| \, dv \, du$$

$$= V(x(0)) - a \int_{0}^{t} |x(s)| \, ds + b \int_{-r}^{0} \int_{u}^{u+t} |x(w)| \, dw \, du$$

$$\leq V(x(0)) - a \int_{0}^{t} |x(s)| \, ds + b \int_{-r}^{t} r|x(w)| \, dw$$

$$\leq V(x(0)) + br \int_{-r}^{0} |\phi(u)| \, du$$

$$\leq (1 + br^2) \|\phi_0\| < \epsilon$$

if $\|\phi_0\| < \epsilon^2/(1 + br^2) = \delta$.

Example 4.2.12. Consider the scalar equation

$$x' = -ax + bx(t-r) + p(t)$$

in which a, b and r are positive constants, $a > b$, p is continuous, $|p(t)| \leq M$ for some $M > 0$. Define

$$V(x_t) = |x| + \int_{t-r}^{t} b|x(u)| \, du$$

so that

$$V'(t_t) \leq -a|x| + b|x(t-r)| + |p(t)|$$

$$+ b|x| - b|x(t-r)|$$

$$\leq -(a-b)|x| + M \, .$$

Let $t \geq t_0$ be any value for which

$$V(x_t) \geq V(x_s) \quad \text{for} \quad t_0 \leq s \leq t.$$

If $t \geq t_0 + r$, then integrate

$$V'(x_t) \leq (a \quad b)|x| \mid M$$

from $t - r$ to t obtaining

$$V(x_t) - V(x_{t-r}) < -(a-b) \int_{t-r}^{t} |x(u)| \, du + Mr \,.$$

Thus, the left side is nonnegative at that t and we have

$$(*) \qquad \int_{t-r}^{t} |x(u)| \, du \leq Mr/(a-b) \,.$$

Thus, $V'(x_t) \geq 0$ so

$$(**) \qquad |x(t)| \leq M/(a-b) \,.$$

Hence, at any such maximum of $V(x_t)$ we have from $(*)$ and $(**)$ that

$$|x(t)| \leq V(x_t) \leq M(1+r)/(a-b) \,.$$

Now for $t_0 \leq t \leq t_0 + r$ we have

$$V(x_t) \leq V(x_{t_0}) + Mr \leq \|\phi\|(1+br) + Mr \,.$$

Thus, for $t \geq t_0$ we have

$$|x(t)| \leq V(x_t) \leq \|\phi\|(1+br) + Mr + \left[M(1+r)/(a-b) \right] \,.$$

Exercise 4.2.5. Consider the scalar equation

$$x' = -ax + bx(t - r(t))$$

for a and b positive constants, $r(t) > 0$, r differentiable, $1 - r'(t) \geq \gamma > 0$. Define $V(x(t)) = |x|$, find V', integrate V', and show that $|x(t)|$ is bounded.

Exercise 4.2.6. Consider the scalar equation

$$x' = -ax + b \int_{t-r}^{t} x(u) \, du$$

for a and b positive constants, $r(t)$ continuous and positive with $r(t) \leq r$, r a positive constant. Define $V(x(t)) = |x|$, obtain

$$V'(x_t) \leq -a|x| + b \int_{t-r(t)}^{t} |x(u)| \, du$$

$$\leq -a|x| + b \int_{t-r}^{t} |x(u)| \, du \,,$$

integrate and obtain a condition for boundedness of solutions.

We turn now to the matter of general theorems.

Definition 4.2.4. *Let $V(t, \psi_t)$ be a scalar-valued functional defined and continuous when $t \geq t_0$ and $\psi_t \in C(t)$. Then the derivative of V with respect to (4.2.1) is*

$$V'_{(4.2.1)}(t, \psi_t) = \limsup_{\Delta t \to 0^+} \left[V(t + \Delta t, \psi^*_{t+\Delta t}) - V(t, \psi_t) \right] / \Delta t$$

where

$$\psi^*(s) = \begin{cases} \psi(s) & \text{for } t_0 - \alpha \leq s \leq t, \\ \psi(t) + F(t, \psi_t)(s - t) & \text{for } t \leq s \leq t + \Delta t. \end{cases}$$

The next result shows that the derivative of V with respect to (4.2.1) is actually the derivative along a solution of (4.2.1). The result is by Driver (1962). This particular formulation asks that solutions be unique, but that is not necessary as may be seen in Yoshizawa (1966).

Theorem 4.2.3. *Let V be a scalar-valued functional as in Def. 4.2.4 and V locally Lipschitz in ψ. Suppose that F is locally Lipschitz in ψ. Then for each $t \geq t_0$ and every $\psi_t \in C(t)$, if $x(s, t, \psi_t)$ is the unique solution of (4.2.1) with initial function ψ_t on $[t - \alpha, t]$, then*

$$V'_{(4.2.1)}(t, \psi_t) = \limsup_{\Delta t \to 0^+} \left[V\big(t + \Delta t, x_{t+\Delta t}(t, \psi_t)\big) - V(t, \psi_t) \right] / \Delta t.$$

Proof. Given t and ψ_t, the solution $x(s, t, \psi)$ is defined on some interval $[t - \alpha, t + h]$ for some $h > 0$. It will suffice to show that

$$V\big(t + \Delta t, x_{t+\Delta t}(t, \psi)\big) - V(t + \Delta t, \psi^*_{t+\Delta t}) = o(\Delta t)$$

as $\Delta t \to 0^+$. Choose $h_1 \in (0, h)$ so small that both $x(s, t, \psi)$ and $\psi^*(s) \in Q$, where Q is some compact set, for $t - \alpha \leq s \leq t + h_1$. Let K be the Lipschitz constant for $V(t, \psi_t)$ associated with $[t - \alpha, t + h_1]$ and Q. Then for $0 < \Delta t < h_1$, we have

$$\left| V\big(t + \Delta t, x_{t+\Delta t}(t, \psi)\big) - V(t + \Delta t, \psi^*_{t+\Delta t} \right|$$
$$\leq K \sup_{t \leq s \leq t + \Delta t} \left| x(s, t, \psi) - \psi(t) - F(t, \psi_t)(s - t) \right|$$
$$\leq K \sup_{t \leq s \leq t + \Delta t} \left| F(\bar{t}, x_t(s, t, \psi) - F(t, \psi_t) \right| \Delta t$$

where $t < \bar{t} < s$. Because $F(t, x_t)$ is continuous, this quantity is $o(\Delta t)$ and the proof is complete.

The next result, given by Driver (1962), allows us to infer that if

$$V'_{(4.2.1)}(t, \psi_t) \leq \omega(t, V)$$

then $V(t, \psi_t)$ is bounded by the maximal solution of the initial value problem

$$r' = \omega(t, r), \qquad r(t_0) = V(t_0, \psi_{t_0}).$$

Theorem 4.2.4. *Let $\omega(t, r)$ be a continuous, nonnegative scalar function of t and r for $t_0 \leq t < \beta$, $r \geq 0$. Let $v(t)$ be any continuous nonnegative scalar function for $t_0 - \alpha \leq t < \beta$ such that*

$$\limsup_{\Delta t \to 0^+} \frac{v(t + \Delta t) - v(t)}{\Delta t} \leq \omega(t, v(t))$$

at those $t \in [t_0, \beta)$ at which

$$v(s) \leq v(t) \qquad \text{for all} \qquad s \in [t_0 - \alpha, t].$$

Let $r_0 \geq \max_{t_0 - \alpha \leq s \leq t_0} v(s)$ be given, and suppose the maximal continuous solution $r(t)$ of

$$r'(t) = \omega(t, r(t)), \qquad r(t_0) = r_0$$

exists for $t_0 \leq t < \beta$. Then

$$v(t) \leq r(t) \qquad \text{for} \qquad t_0 \leq t < \beta.$$

Proof. Choose any $\overline{\beta} \in (t_0, \beta)$. Then, according to Kamke (1930, p. 83), $r(t)$ can be represented as

$$r(t) = \lim_{\epsilon \to 0^+} r(t, \epsilon) \qquad \text{for} \qquad t_0 \leq t < \overline{\beta},$$

where, for each fixed sufficiently small $\epsilon > 0$, $r(t, \epsilon)$ is any solution of

$$r'(t, \epsilon) = \omega(t, r(t, \epsilon)) + \epsilon \qquad \text{for} \qquad t_0 \leq t \leq \overline{\beta}$$

with $r(t_0, \epsilon) = r_0$.

By way of contradiction, suppose that, for some such $\epsilon > 0$, there exists a $t \in (t_0, \overline{\beta})$ such that $v(t) > r(t, \epsilon)$. Let

$$t_1 = \sup \left\{ t \in [t_0, \overline{\beta}) : v(s) \leq r(s, \epsilon) \qquad \text{for all} \qquad s \in [t_0, t] \right\}.$$

It follows that $t_0 \leq t_1 < \overline{\beta}$, and by continuity of $v(t)$ and $r(t, \epsilon)$, $v(t_1) = r(t_1, \epsilon)$. Because $r(t, \epsilon)$ has a positive derivative,

$$v(s) \leq r(t_1, \epsilon) = v(t_1) \qquad \text{for all} \qquad s \in [t_0 - \alpha, t_1],$$

and therefore

$$\limsup_{\Delta t \to 0^+} \frac{v(t_1 + \Delta t) - v(t_1)}{\Delta t} \leq \omega(t_1, v(t_1)).$$

But $v(t_1 + \Delta t) > r(t_1 + \Delta t, \epsilon)$ for certain small $\Delta t > 0$. Hence

$$\limsup_{\Delta t \to 0^+} \frac{v(t_1 + \Delta t) - v(t_1)}{\Delta t} \geq r'(t_1, \epsilon)$$
$$= \omega(t_1, r(t_1, \epsilon)) + \epsilon$$
$$= \omega(t_1, v(t_1)) + \epsilon$$

which is a contradiction. Thus $v(t) \leq r(t, \epsilon)$ for all small $\epsilon > 0$ and all $t \in [t_0, \overline{\beta})$, so $v(t) \leq r(t)$ for all $t \in [t_0, \beta)$. The proof is complete.

We now introduce the standard results on stability and boundedness using Liapunov functionals. Recall that a *wedge* is a continuous function

$$W : [0, \infty) \to [0, \infty), \qquad W(0) = 0, \quad W(r) > 0$$

for $r > 0$, W is strictly increasing, and we now ask that

$$W(r) \to \infty \qquad \text{as} \qquad r \to \infty.$$

Wedges are denoted by W or W_i.

Recall the basic results on Liapunov theory for

$$x' = f(t, x).$$

(i) If there exists $V(t, x)$ with

$$W(|x|) \leq V(t, x), \qquad V(t, 0) = 0, \qquad \text{and} \qquad V'(t, x) \leq 0$$

then $x = 0$ is stable.

(ii) If, in addition to (i),
$$V(t, x) \leq W_1(|x|),$$

then $x = 0$ is uniformly stable.

(iii) Let $f(t, x)$ be bounded for x bounded. If

$$W(|x|) \leq V(t, x), \qquad V(t, 0) = 0, \qquad \text{and} \qquad V'(t, x) \leq -W_2(|x|),$$

then $x = 0$ is asymptotically stable.

(iv) If, in addition to (i) and (ii),

$$V'(t, x) \leq -W_2(|x|)$$

then $x = 0$ is uniformly asymptotically stable.

(v) If there exists $V(t, x)$ with

$$W(|x|) \leq V(t, x) \leq W_2(|x|)$$

and

$$V'(t, x) \leq 0 \qquad \text{for} \qquad |x| \geq m, \ M > 0,$$

then solutions are uniform bounded.

(vi) If, in addition to (v)

$$V'(t, x) \leq -C < 0 \qquad \text{for} \qquad |x| \geq M,$$

then solutions are uniform ultimate bounded for bound B.

If we properly interpret the norm in the wedge

$$V(t, x_t) \leq W(|x_t|)$$

then the properties (i)–(vi) carry over exactly for (4.2.1) and Liapunov functionals.

Theorem 4.2.5. *Suppose that for (4.2.1) and for some $D > 0$ there is a scalar functional $V(t, \psi_t)$ defined, continuous in (t, ψ_t), and locally Lipschitz in ψ_t when $t_0 \leq t < \infty$ and $\psi_t \in C(t)$ with $\|\psi_t\| < D$. Suppose also that $V(t, 0) = 0$ and that $W_1(|\psi(t)|) \leq V(t, \psi_t)$.*

(a) *If $V'_{(4.2.1)}(t, \psi_t) \leq 0$ for $t_0 \leq t < \infty$ and $\|\psi_t\| < D$, then the zero solution of (4.2.1) is stable.*

(b) *If, in addition to (a), $V(t, \psi_t) \leq W_2(\|\psi_t\|)$, then $x = 0$ is uniformly stable.*

(c) *If there is an $M > 0$ with $|F(t, \psi_t)| \leq M$ for $t_0 \leq t < \infty$ and $\|\psi_t\| < D$, and if*

$$V'(t, \psi_t) \leq -W_2(\psi(t)|),$$

then $x = 0$ is asymptotically stable.

Proof. For (a), let $t_1 \geq t_0$ and $\epsilon > 0$ be given with $\epsilon < D$. Since $V(t, \psi)$ is continuous and $V(t_1, 0) = 0$ there exists $\delta > 0$ such that $\phi \in C(t_1)$ with $\|\phi_{t_1}\| < \delta$ implies that $V(t_1, \phi_{t_1}) < W_1(\epsilon)$. Since $V'(t, x_1(t_1, \phi_{t_1})) \leq 0$ we have

$$W_1\big(|x(t, t_1, \phi_{t_1})|\big) \leq V(t, x_t, (t_1, \phi_{t_1})) \leq V(t_1, \phi_{t_1})$$

$$< W_1(\epsilon)$$

so that

$$|x(t, t_1, \phi_{t_1})| < \epsilon$$

as required.

To prove (b), let $\epsilon > 0$ be given with $\epsilon < D$ and find $\delta < 0$ with $W_2(\delta) < W_1(\epsilon)$. If $t_1 \geq t_0$ and $\phi_{t_1} \in C(t_1)$ with $\|\phi_{t_1}\| < \delta$, then

$$W_1\big(|x(t, t_1, \phi_{t_1})|\big) \leq V(t, x_t(t_1, \phi_{t_1}))$$
$$\leq V(t_1, \phi_{t_1}) \leq W_2(\|\phi_{t_1}\|) < W_2(\delta) < W_1(\epsilon)$$

so that

$$|x(t, t_1, \phi_{t_1})| < \epsilon \quad \text{for} \quad t \geq t_1,$$

as required.

To prove (c), let $t_1 \geq t_0$ be given and let $0 < \epsilon < D$. Find δ as in (a). We take $\eta = \delta$. Let $\phi_{t_1} \in C(t_1)$ with $\|\phi_{t_1}\| < \delta$. Write $x(t) = x(t, t_1, \phi_{t_1})$. If $x(t) \nrightarrow 0$ as $t \to \infty$, then there is an $\epsilon_1 > 0$ and a sequence $\{t_n\} \to \infty$ with $|x(t_n)| \geq \epsilon_1$. Since $|F(t, x_t)| \leq M$, there is a $T > 0$ with $|x(t)| \geq \epsilon_1/2$ for $t_n \leq t \leq t_n + T$. From this and $V'(t, x_t) \leq -W_2(|x(t)|)$ we obtain

$$0 \leq V(t, x_t) \leq V(t_1, \phi_{t_1}) - \int_{t_1}^{t} W_2(|x(s)|)\, ds$$
$$\leq V(t_1, \phi_{t_1}) - \sum_{i=2}^{n} \int_{t_i}^{t_i + T} W_2(|x(s)|)\, ds$$
$$\leq V(t_1, \phi_{t_1}) - \sum_{i=2}^{n} \int_{t_i}^{t_i + T} W_2(\epsilon_1/2)\, ds$$
$$= V(t_1, \phi_{t_1}) - (n-1)TW_2(\epsilon_1/2) \to -\infty$$

as $n \to \infty$, a contradiction.

One would readily believe that if, in addition to Theorem 4.2.5(a,b),

$$V'_{(4.2.1)}(t, \psi_t) \leq -W_3(|\psi(t)|)$$

then $x = 0$ is uniformly asymptotically stable. But, at this writing, no one has been able to prove it. Burton (1978) has proved it true when

$$V(t, \psi_t) \leq W_2(|\psi(t)|) + W_4(\|\psi_t\|)$$

where $\|\psi_t\|$ is an L^2-norm (as is the case in most applications) and Wen (1982) has extended Burton's result to include Razumikhin techniques. The idea was carried much deeper by Kato (1980b). The standard gambit is to ask that the right side of (4.2.1) be bounded for x_t bounded. The proof of U.A.S. under either the L^2-norm or bounded $F(t, x_t)$ is very long. We will prove the result under bounded $F(t, x_t)$ and offer the other result without proof since a proof is readily available in Burton (1978,1983a) and is very similar to that of Theorem 4.4.3.

Theorem 4.2.6. *Let the conditions of Theorem 4.2.5 hold and suppose that if $x : [t_0 - \alpha, \infty) \to R^n$ is a bounded continuous function, then $F(t, x_t)$ is bounded on $[t_0, \infty)$. If, in addition to Theorem 4.2.5(a,b),*

$$V'_{(4.2.1)}(t, \psi_t) \leq -W_3(|\psi(t)|)$$

then $x = 0$ is uniformly asymptotically stable.

Proof. Let $\epsilon = \min[1, D/2]$ and find the δ of uniform stability for this ϵ. Take $\eta = \delta$. Next, let $\gamma > 0$ be given. We will find $S > 0$ such that $[t_1 \geq t_0, \ \phi \in C(t_1), \ \|\phi\| < \eta, \ t \geq t_1 + S]$ implies $|x(t, t_1, \phi)| < \gamma$. Let $x(t) = x(t, t_1, \phi)$.

Now, for this $\gamma > 0$, find a δ of uniform stability so that $[t_2 \geq t_0, \ \|\phi_{t_2}\| < \delta, \ t \geq t_2]$ imply that $|x(t, t_2, \phi_{t_2})| < \gamma$. Denote this new δ by μ to avoid confusion with the previous δ.

We now assemble the necessary facts for the proof. Since

$$W_1(|x(t)|) \leq V(t, x_t) \leq W_2(\|x_t\|),$$
$$V'(t, x_t) \leq -W_3(|x(t)|),$$

and

$$\|\phi_{t_1}\| < \eta$$

we have, upon integration,

(i) $$V(t, x_t) \leq V(t_1, \phi_{t_1}) \leq W_2(\|\phi_{t_1}\|) < W_2(\eta).$$

Next, if $|x(t)| \geq \mu/2$ on an interval $[t_2, t_3]$, then $V'(t, x_t) \leq -W_3(\mu/2)$ so

$$0 \leq V(t_3, x_{t_3}) \leq V(t_2, x_{t_2}) - W_3(\mu/2)(t_3 - t_2)$$
$$< W_2(\eta) - W_3(\mu/2)(t_3 - t_2)$$

which implies that

(ii) $$t_3 - t_2 < W_2(\eta)/W_3(\mu/2).$$

As noted before, if $|x(t)| < \mu$ on an interval $[t_4, t_5]$ with $t_5 - t_4 \geq \alpha$, then

(iii) $$|x(t)| < \gamma \quad \text{for} \quad t \geq t_5.$$

The final fact we need is that if $|x(t_6)| \leq \mu/2$ and $|x(t_7)| \geq \mu$ with $t_6 < t_7$, then since $F(t, x_t)$ is a bounded function of t, there is a $T > 0$ with

(iv) $$t_7 - t_6 > T.$$

Now, (iv) implies that

(v) $V(t, x_t)$ decreases $TW_3(\mu/2)$ units on $[t_6, t_7]$.

Thus, find an integer N with

(vi) $NTW_3(\mu/2) > W_2(\eta)$.

From (ii) we see that there is a point s_i on every interval of length $W_2(\eta)/W_3(\mu/2)$ at which $|x(s_i)| \leq \mu/2$. From (iii) we see that there is a point S_i on each t interval of length α with $|x(S_i)| \geq \mu$, otherwise $|x(t)|$ remains smaller than γ. And from (iv) we see that T time units pass between s_i and S_i.

Hence, on each interval of length

$$\alpha + W_2(\eta)/W_3(\mu/2) \overset{\text{def}}{=} U$$

$V(t, x_t)$ decreases $TW_3(\mu/2)$ units. Thus, $S = NU$ suffices and we have $|x(t, t_1, \phi)| < \gamma$ for $t \geq t_1 + S$. The proof is complete.

As mentioned earlier, a proof of the next result may be found in Burton (1978, 1983a). For $\phi \in C(t)$ we denote

$$\|\phi_t\| = \left[\sum_{i=1}^{n} \int_{t-\alpha}^{t} \phi_i^2(s)\, ds \right]^{1/2}$$

where $\phi(t) = \big(\phi_1(t), \ldots \phi_n(t)\big)$.

Theorem 4.2.7. *Suppose that for (4.2.1) and for some $D > 0$ there is a scalar functional $V(t, \psi_t)$ defined, continuous in (t, ψ_t), and locally Lipschitz in ψ_t, when $t_0 \leq t < \infty$ and $\psi_t \in C(t)$ with $\|\psi_t\| < D$. If*

(a) $W(|\phi(t)|) \leq V(t, \phi_t) \leq W_1(|\phi(t)|) + W_2(\|\phi_t\|)$

and

(b) $V'_{(4.2.1)}(t, \phi_t) \leq -W_3(|\phi(t)|)$,

then the zero solution of (4.2.1) is uniformly asymptotically stable.

The prototype V in applications is

$$V(t, x_t) = x^T A(t)x + \int_{t-r(t)}^{t} x^T(s)B(s)x(s)\, ds$$

in which A and B are $n \times n$ matrices. If we denote the second term by $Z(t, x_t)$, then we note that

$$(d/dt)Z(t, x_t) = x^T(t)B(t)x(t)$$
$$- x^T(t - r(t))B(t)x(t - r(t))(1 - r'(t))$$

so that if $B(t)$ and $r'(t)$ are bounded, then Z is Lipschitz in t for any bounded function x. It turns out that we only need a one-sided Lipschitz condition on Z. In the following result a right-hand Lipschitz condition is asked, but it may be replaced by a left-hand one:

$$t_2 > t_1 \quad \text{implies} \quad Z(t_2, x_{t_2}) - Z(t_1, x_{t_1}) \geq K(t_1 - t_2).$$

Results of the following type are developed and used in Busenberg and Cooke (1984), Yoshizawa (1984), and Burton (1979a).

Theorem 4.2.8. *Let $V(t, \phi_t)$ satisfy the continuity assumptions of Theorem 4.2.7 with $V \geq 0$. Suppose there is a functional $Z(t, \phi_t)$ and wedges such that*

$$W(|\phi(t)|) + Z(t, \phi_t) \leq V(t, \phi_t) \leq W_1(|\phi(t)|) + Z(t, \phi_t)$$

and

$$V'_{(4.2.1)}(t, \phi_t) \leq -W_2(|\phi(t)|).$$

If for each bounded continuous function $x : [t_0 - \alpha, \infty) \to R^n$, there exists $K > 0$ such that $t_0 \leq t_1 < t_2$ implies

$$Z(t_2, x_{t_2}) - Z(t_1, x_{t_1}) \leq K(t_2 - t_1),$$

then each bounded solution of (4.2.1) converges to zero.

Proof. If the theorem is false, then there is a bounded solution $x(t)$ on $[t_0, \infty)$, and $\epsilon > 0$, and a sequence $\{\overline{t_n}\}$ with $|x(\overline{t_n})| \geq \epsilon$. As $V'(t, x_t) \leq -W_2(|x(t)|)$ and $V \geq 0$, an integration from t_0 to t shows that there is a sequence $\{t_n^*\}$ with $x(t_n^*) \to 0$. Now $V'(t, x_t) \leq 0$ and so $V(t, x_t) \to L$.

Choose $\gamma > 0$ so that $W_1(\gamma) < W(\epsilon)/4$. Then choose sequences $\{t_n\}$ and $\{T_n\}$ with $|x(t_n)| = \epsilon$, $|x(T_n)| = \gamma$, and $\gamma \leq |x(t)| \leq \epsilon$ for $t_n \leq t \leq T_n$, and $t_n \to \infty$ as $n \to \infty$. We have

$$(*) \qquad W(\epsilon) + Z(t_n, x_t) \leq V(t_n, x_t) \leq W_1(\epsilon) + Z(t_n, x_t)$$

and

$$(**) \qquad W(\gamma) + Z(T_n, x_t) \leq V(T_n, x_t) \leq W_1(\gamma) + Z(T_n, x_t).$$

As $V(t, x_t) \to L$, then for large n (say $n \geq 1$ by renumbering), we have from $(*)$ the relation

$(*)'$ $Z(t_n, x_t) \leq V(t_n, x_t) - W(\epsilon) \leq L + (W(\epsilon)/4) - W(\epsilon)$

and from $(**)$ the relation

$(**)'$ $Z(T_n, x_t) \geq V(T_n, x_t) - W_1(\gamma) \geq L - W_1(\gamma) \geq L - W(\epsilon)/4$.

From these relations we obtain

$$Z(T_n, x_t) - Z(t_n, x_t) \geq L - W(\epsilon)/4 - L + \frac{3}{4}W(\epsilon) = W(\epsilon)/2.$$

To this we add the right-hand Lipschitz condition on Z to obtain

$$W(\epsilon)/2 \leq Z(T_n, x_{T_n}) - Z(t_n, x_{t_n}) \leq K(T_n - t_n)$$

or

$$T_n - t_n \geq W(\epsilon)/2K \stackrel{\text{def}}{=} T.$$

Thus, on intervals $[t_n, T_n]$ we have $V'(t, x_t) \leq -W_2(\gamma)$. As $T_n - t_n \geq T$, an integration yields $V(t, x_t) \to -\infty$ as $t \to \infty$, a contradiction. This completes the proof.

Corollary. *Let the conditions of Theorem 4.2.8 hold. If $W(s) \to \infty$ as $s \to \infty$ and if $Z(t, \phi_t)$ is bounded from below, then all solutions of (4.2.1) are bounded. If $Z(t, \phi_t) \geq 0$ and if $V(t, \phi_t) \leq W_3(\|\phi_t\|)$, then $x = 0$ is asymptotically stable.*

Proof. If $Z(t, \phi_t) \geq -M$ for some $M > 0$, then

$$W(|x(t)|) - M \leq V(t, x_t) \leq V(t_0, \phi_{t_0})$$

yields $x(t, t_0, \phi)$ bounded.
    If $Z(t, \phi_t) \geq 0$, then

$$W(|x(t)|) \leq V(t, x_t) \leq V(t_0, \phi_{t_0}) \leq W_3(\|\phi_{t_0}\|)$$

from which stability follows.

Example 4.2.13. Consider the scalar equation

$$x' = -\big[a + (t \sin t)^2\big]x + bx\big(t - r(t)\big)$$

with $a > 0$ and constant, b constant, $-M \leq r'(t) \leq \gamma < 1$ for some $M > 0$ and some γ satisfying $0 < \gamma < 1$. Define

$$V(t, x_t) = x^2 + 2k \int_{t-r(t)}^{t} x^2(s)\, ds$$

for $k > 0$, k to be determined. We have

$$W(s) = W_1(s) = s^2$$

and

$$Z(t, x_t) = 2k \int_{t-r(t)}^{t} x^2(s)\, ds.$$

Clearly, Z is Lipschitz in t for bounded x. One may note that neither here nor in the last theorem do we actually need $r(t)$ bounded. We find that

$$V'(t, x_t) \le -2ax^2(t) + 2bx(t)x\big(t - r(t)\big)$$
$$+ 2k\big[x^2(t) - x^2\big(t - r(t)\big)\big(1 - r'(t)\big)\big]$$
$$\le -2k(1 - \gamma)x^2\big(t - r(t)\big) + 2bx\big(t - r(t)\big)x(t)$$
$$- 2(a - k)x^2(t).$$

We complete the square, select $k = a/2$, and ask that $a^2(1 - \gamma) > b^2$ to ensure that $V'(t, x_t) \le -\mu x^2(t)$ for some $\mu > 0$. Since $V' \le 0$, $W(|x(t)|) \le V(t, x_t)$, and $W(s) \to \infty$ as $s \to \infty$ we see that solutions are bounded and 0 is asymptotically stable. If $r(t) \ge \alpha$ then

$$V(t, x_t) \le 2[1 + k]\alpha \|x_t\|^2$$

and $x = 0$ is uniformly stable. Moreover, the upper wedge on V can be replaced with an L^2-norm so that Theorem 4.2.7 will yield uniform asymptotic stability.

In this result we see that V is bounded by a "good" part $W_1(|x|)$ and a "bad" part $Z(t, x_t)$. Hatvani (1984) has considered a similar idea for ordinary differential equations arising in mechanical problems. His Liapunov function is $V(t, x) = V_1(t, x) + V_2(t, x)$ in which V_1 is well behaved and V' is negative. He is able to conclude asymptotic stability without asking x' bounded for x bounded.

Recently, Yoshizawa (1984) improved these results and showed that in this same example with the same Liapunov functional one may delete the requirement $-M \le r'(t)$.

Cooke (1984) has applied Theorem 4.2.8 and its corollary to a number of interesting examples with numerous delays. We present two of those examples here.

Example 4.2.14. Consider the nonlinear scalar equation

$$x'(t) = b(t)x^3\big(t - r(t)\big) - c(t)x^3(t).$$

We take

$$V(t, x_t) = x^4(t) + \int_{t-r(t)}^{t} K(s)x^6(s)\, ds$$

and calculate

$$V'(t, x_t) = 4x^3(t)x'(t) + K(t)x^6(t) - [1 - r'(t)] K(h(t))x^6(h(t))$$

where $h(t) = t - r(t)$. Thus

$$V'(t, x_t) = -4c(t)x^6(t) + 4b(t)x^3(h(t))$$
$$+ K(t)x^6(t) - [1 - r'(t)] K(h(t))x^6(h(t)).$$

We assume that $r(t) \leq t$, $r'(t) \leq \gamma < 1$, and let h^{-1} be the inverse of h. We choose, for example,

$$K(t) = a^{-1}b^2(h^{-1}(t))$$

and then since $1 - r'(t) \geq 1 - \gamma$ we get

$$V'(t, x_t) = \left[-4c(t) + a^{-1}b^2(h^{-1}(t)) \right] x^6(t)$$
$$+ 4b(t)x^3(t)x^3(h(t)) - a^{-1}(1 - \gamma)b^2(t)x^6(h(t)).$$

Assume that there are constants $q > 0$ and $a > 0$ such that

$$4c(t) - (4a/(1 - \gamma)) - a^{-1}b^2(h^{-1}(t)) \geq q.$$

Then

$$V'(t, x_t) \leq -qx^6(t) - \left[2\left(\frac{a}{1 - \gamma} \right)^{1/2} x^3(t) - \left(\frac{1 - \gamma}{a} \right)^{1/2} b(t)x^3(h(t)) \right]^2$$

$$\leq -qx^6(t).$$

Thus, in Theorem 4.2.8 take $W_2(|x|) = qx^6$, $W(|x|) = W_1(|x|) = x^4$, and $Z(t, x_t) = \int_{t-r(t)}^t K(s)x^6(s)\,ds$. Then

$$Z(t_2, x_{t_2}) - Z(t_1, x_{t_1}) \leq \|x\|^6 a^{-1} \left(\int_{t_1}^{t_2} b^2(h^{-1}(s))\,ds \right).$$

We require

$$\int_{t_1}^{t_2} b(h^{-1}(s))^2\,ds = \int_{h^{-1}(t_1)}^{h^{-1}(t_2)} b^2(u)(1 - r'(u))\,du$$

$$\leq \gamma(t_2 - t_1)$$

in order to ensure the Lipschitz condition. By Theorem 4.2.8 all solutions tend to zero, and the zero solution is asymptotically stable.

Example 4.2.15. Consider the system

$$x'(t) = -C(t)x(t) + B(t)x\big(t - r(t)\big)$$

where C and B are $n \times n$ matrices and x is an n-vector. We now take a functional of the form

$$V(t, x_t) = x(t)^T Dx(t) + \int_{t-r(t)}^t x^T(s)K(s)x(s)\, ds$$

where $x^T(t)$ is the transpose of $x(t)$, and D and K are $n \times n$ matrices to be chosen below. Then

$$V'(t, x_t) = x'(t)^T Dx(t) + x^T(t)Dx'(t) + x^T(t)K(t)x(t)$$
$$- \big[1 - r'(t)\big]x^T\big(t - r(t)\big)K\big(t - r(t)\big)x\big(t - r(t)\big).$$

We define

$$Z(t, x_t) = \int_{t-r(t)}^t x^T(s)K(s)x(s)\, ds$$

and $W_1(x) = W(x) = x^T Dx$, and take D to be a positive definite symmetric matrix. As before, we let $h(t) = t - r(t)$, and write x for $x(t)$ and $x(h)$ for $x(h(t))$. Then we obtain

$$V'(t, x_t) = -x^T(C^T D + D^T C - K)x + x^T DBx(h)$$
$$+ x^T(h)B^T D^T x - (1 - r')x^T(h)K(h)x(h).$$

We now choose for K the real symmetric matrix

$$K(t) = \big[B\big(h^{-1}(t)\big)\big]^T \big[B\big(h^{-1}(t)\big)\big],$$

and assume that $r'(t) \le \gamma < 1$. Since $K(h) = B^T(t)B(t)$, we obtain

$$V'(t, x_t) \le -x^T\big[C^T D + D^T C - B^T(h^{-1})B(h^{-1})\big]x$$
$$+ x^T DBx(h) + x^T(h)B^T D^T x - (1 - \gamma)x^T(h)B^T Bx(h).$$

If we assume that D can be chosen to satisfy

$$C^T D + D^T C - B^T(h^{-1})B(h^{-1}) - D^T D \ge qI$$

for some $q > 0$, where I is the identity matrix, then

$$V'(t, x_t) \le -qx^T x - x^T D^T Dx + x^T DBx(h) + x^T(h)B^T D^T x$$
$$- (1 - \gamma)x^T(h)B^T Bx(h)$$
$$= -qx^T x - \big[\beta^{1/2}Dx - \beta^{1/2}Bx(h)\big]^T\big[\beta^{-1/2}Dx - \beta^{1/2}Bx(h)\big]$$

where $\beta = 1 - \gamma$. Thus, $V' \leq -qx^T x$, and so we take $W_2(|x|) = qx^2$. Furthermore, we observe that

$$\int_{t_1}^{t_2} x^T B^T(h^{-1}) B(h^{-1}) x \, ds = \int_{t_1}^{t_2} |B(h^{-1})x|^2 \, ds$$

$$\leq |x|^2 \int_{t_1}^{t_2} \left| B\big(h^{-1}(s)\big) \right|^2 ds$$

$$\leq |x|^2 \int_{h^{-1}(t_1)}^{h^{-1}(t_2)} |B(u)|^2 h'(u) \, du.$$

We summarize as follows.

Proposition. *Assume that $B(t)$ and $C(t)$ are continuous $n \times n$ matrices, $r(t)$ is continuous and differentiable, $r(t) \leq t$ and $r'(t) \leq \gamma < 1$. Let $h(t) = t - r(t)$ and let $h^{-1}(t)$ be the inverse function. Further, assume the following.*

(i) *There is a constant μ such that $t_0 \leq t_1 < t_2$,*

$$\int_{h^{-1}(t_1)}^{h^{-1}(t_2)} |B(u)|^2 \big(1 - r'(u)\big) \, du \leq \mu(t_2 - t_1).$$

(ii) *There exists a constant $q > 0$ and an $n \times n$ positive definite symmetric matrix D such that*

$$C^T(t)D + DC(t) - \big[B\big(h^{-1}(t)\big)\big]^T \big[B\big(h^{-1}(t)\big)\big] - D^2 \geq qI.$$

Then all solutions of the equation in Example 4.2.15 tend to zero, and the zero solution is asymptotically stable.

The technique of Theorem 4.2.8 can also be used to prove uniform asymptotic stability, as we now show. This result was obtained in Burton (1979a). The proof there seems not particularly clear at one point and the following one seems much more straightforward.

Theorem 4.2.9. *Let V and $Z : [t_0, \infty) \times C_H \to [0, \infty)$ with V continuous and locally Lipschitz in x_t. Suppose that whenever $\phi \in C_H$ and $t_0 \leq t_1 < t_2$ then*

$$Z(t_2, \phi) - Z(t_1, \phi) \leq K(t_2 - t_1)$$

for some $K > 0$, $K = K(H)$. If

$$W_1(|x|) + Z(t, x_t) \leq V(t, x_t) \leq W_2(|x|) + Z(t, x_t)$$

$$\leq W_3(\|x_t\|)$$

and

$$V'_{(4.2.1)}(t, x_t) \leq -W_4(|x|),$$

then the zero solution of (4.2.1) *is uniformly asymptotically stable.*

Proof. Since

$$W_1(|x|) \leq V(t, x_t) \leq W_3(\|x_t\|)$$

and

$$V'_{(4.2.1)}(t, x_t) \leq 0$$

the zero solution is uniformly stable. Let $\epsilon = H$ and find the δ of uniform stability. Let $\delta = \eta$ and let $\gamma > 0$ be given. We must find $T > 0$ such that $[t_1 \geq t_0, \ \phi \in C_\eta(t_1), \ t \geq t_1 + T]$ imply that $|x(t, t_1, \phi)| < \gamma$.

For this $\gamma > 0$ find μ of uniform stability so that if a solution $x(t)$ satisfies $|x(t)| < \mu$ on an interval of length α, then $|x(t)| < \gamma$ for ever after.

Consider the intervals

$$I_1 = [t_1, t_1 + \alpha], \quad I_2 = [t_1 + \alpha, t_1 + 2\alpha], \dots$$

and suppose that in I_i there is a t_i with $|x(t_i)| \geq \mu$. Find $m > 1$ such that

$$W_2(\mu/m) < W_1(\mu)/2.$$

Because $0 \leq V(t, x_t) \leq W_3(\|x\|)$ and $V'(t, x_t) \leq -W_4(|x|)$ there is a $T_1 > 0$ such that $|x(t)| \geq \mu/m$ fails at some point in every interval of length T_1. Find a sequence $\{S_n\}$ such that $t_n < S_n$, $|x(S_n)| = \mu/m$, and $|x(t)| \geq \mu/m$ for $t_n \leq t \leq S_n$. How large is $S_n - t_n$?

We have

$$W_1(|x|) \leq V(t, x_t) - Z(t, x_t) \leq W_2(|x|),$$

$$W_1(\mu) \leq V(t_i, x_{t_i}) - Z(t_i, x_{t_i}),$$

and

$$V(S_i, x_{S_i}) - Z(S_i, x_{S_i}) \leq W_2(\mu/m) < W_1(\mu)/2.$$

Thus, on each interval $[t_i, S_i]$ the function $V(t, x_t) - Z(t, x_t)$ decreases by at least

$$W_1(\mu)2 \stackrel{\text{def}}{=} \beta.$$

If, on $[t_i, S_i]$, $V(t, x_t)$ has not decreased by $\beta/2$, then $Z(t, x_t)$ has increased by $\beta/2$. But

$$\beta/2 \leq Z(S_i, x_{S_i}) - Z(t_i, x_{t_i}) \leq K(S_i - t_i)$$

implies that

$$S_i - t_i \geq \beta/2K.$$

Now, on the interval $[t_i, S_i]$ we have

$$V'(t, x_t) \leq -W_4(|x|) \leq -W_4(\mu/m)$$

so that V decreases by at least

$$W_4(\mu/m)\beta/2K \,.$$

In summary, then, on each interval $[t_i, S_i]$ it is the case that V decreases by at least

$$\theta \overset{\text{def}}{=} \min \left[W_4(\mu/m)\beta/2K, \ \beta/2 \right].$$

Now, consider intervals

$$J_1 = [t_1, \ t_1 + T_1 + \alpha], \quad J_2 = [t_1 + T_1 + \alpha, \ t_1 + 2(T_1 + \alpha)], \ \ldots$$

There is a point t_i in each of them and, indeed, an interval $[t_i, S_i]$ in each of them. Thus, on each J_i it is true that V decreases at least θ units, unless $|x(t)|$ remains smaller than μ. We then have for $t > S_n$ that

$$0 \leq V(t, x_t) \leq V(t_1, \phi) - \int_{t_1}^{t} W_4(|x(s)|) \, ds$$

$$\leq W_3(\eta) - n\theta < 0$$

if $n > W_3(\eta)/\theta$. Hence, we pick an integer $N > W_3(\eta)/\theta$ and

$$T = N(T_1 + \alpha) \quad \text{so that} \quad t \geq t_1 + T$$

implies $|x(t, t_1, \phi)| < \gamma$. This completes the proof.

The L^2-norm used in Theorem 4.2.7 is also very effective in proving boundedness. Note that W_3 is used twice. We would like to use different wedges and we can do so arguing as in the proof of Theorem 4.4.3. The following result is from Burton and Zhang (1985).

Theorem 4.2.10. *Let $V(t, \phi_t)$ satisfy the continuity conditions of Theorem 4.2.7. If $D = \infty$ and if*

(a) $W(|\phi(t)|) \leq V(t, \phi_t) \leq W_1(|\phi(t)|) + W_2\left(\int_{t-\alpha}^{t} W_3(|\phi(s)|) \, ds\right)$

and

(b) $V'_{(4.2.1)}(t, \phi_t) \leq -W_3(|\phi(t)|) + M$

for some $M > 0$, then solutions of (4.2.1) are uniform bounded and uniform ultimate bounded for bound B.

Proof. Let $B_1 > 0$ be given, $t_1 \geq t_0$, $\phi \in C(t_1)$, $\|\phi\| \leq B_1$, and let $x(t) = x(t, t_1, \phi)$. Integrate (b) from $t - \alpha$ to t (with $t \geq t_1 + \alpha$) and obtain

$$\int_{t-\alpha}^{t} V'(s, x_s)\, ds \leq -\int_{t-\alpha}^{t} W_3(|x(s)|)\, ds + M\alpha$$

so that

$$(*) \qquad \int_{t-\alpha}^{t} W_3(|x(s)|)\, ds \leq V(t - \alpha, x_{t-\alpha}) - V(t, x_t) + M\alpha.$$

Now, consider $V(s) = V(s, x_s)$ on any interval $[t_1, L]$ for any $L > t_1 + \alpha$. Since $V(s)$ is continuous, it has a maximum at some $\bar{t} \in [t_1, L]$. Suppose $\bar{t} \leq t_1 + \alpha$; then

$$V(t) \leq V(\bar{t}) \leq V(t_1) + M(\bar{t} - t_1)$$
$$\leq W_1(B_1) + W_2(\alpha W_3(B_1)) + M\alpha$$

and thus

$$|x(t)| \leq W^{-1}[W_1(B_1) + W_2(\alpha W_3(B_1)) + M\alpha] \overset{\text{def}}{=} B_2^*.$$

If $\bar{t} \in [t_1 + \alpha,\ L]$, then $V(\bar{t} - \alpha) - V(\bar{t}) \leq 0$ so that

$$\int_{\bar{t}-\alpha}^{\bar{t}} W_3(|x(s)|)\, ds \leq V(\bar{t} - \alpha) - V(\bar{t}) + M\alpha \leq M\alpha.$$

We note that for such $\bar{t}$, $V'(\bar{t}) \geq 0$ and hence $|x(\bar{t})| \leq W_3^{-1}(M)$. Thus

$$W(|x(t)|) \leq V(t) \leq V(\bar{t}) \leq W_1(W_3^{-1}(M)) + W_2(M\alpha)$$

for $t \in [t_1, L]$ and, therefore,

$$|x(t)| \leq W^{-1}[W_1(W_3^{-1}(M)) + W_2(M\alpha)] \overset{\text{def}}{=} B_2^{**}.$$

Since L is arbitrary, $B_2 = \max[B_2^*, B_2^{**}]$. This completes the proof of uniform boundedness.

We now show the uniform ultimate boundedness. For a given $L > 3M$ there is a $U > 0$ with $V'(t, x_t) \leq -L$ if $|x(t)| \geq U$. Using the uniform boundedness find U^* and U^{**} such that $[t_1 \geq t_0,\ \|\phi\| \leq U,\ t \geq t_1]$ imply that $|x(t, t_1, \phi)| < U^*$, while $[t_1 \geq t_0,\ \|\phi\| \leq U^*,\ t \geq t_1]$ imply that $|x(t, t_1, \phi)| < U^{**}$.

Moreover, we can choose U^* so large that

$$5M\alpha < W_2^{-1}[W(U^*) - W_1(U)] \overset{\text{def}}{=} \gamma.$$

Next, we can find a wedge W_4 with $V(t,\phi) \leq W_4(\|\phi\|)$. Since $V'(t,x_t) \leq -L$ if $|x(t)| \geq U$, for each $B_3 > 0$ if $S = W_4(B_3)/L$, then there is a $t \in [t_1,\ t_1 + S]$ such that $|x(t,t_1,\phi)| \leq U$ if $\|\phi\| \leq B_3$.

Now, given $B_3 > 0$ consider an arbitrary solution $x(t) = x(t,t_1,\phi)$ with $t_1 \geq t_0$ and $\|\phi\| \leq B_3$. Note that on each interval $[p,\ p+\alpha]$ with $p \geq t_1$, if $|x(t)| \leq U^*$ on that whole interval, then $|x(t)| < U^{**}$ for all $t \geq p + \alpha$. We now determine how long this situation can fail to occur. Let us consider the intervals

$$I_1 = [t_1,\ t_1 + \alpha]\,, \quad I_2 = [t_1 + \alpha,\ t_1 + 2\alpha]\,,\ldots$$

Case 1   If $|x(t)| \geq U$ on an interval I_j, then V decreases on I_j by at least $L\alpha > 3M\alpha$.

Case 2   If $|x(t)|$ does not remain larger than U on I_j, then there is a $t_j \in I_j$ with $|x(t_j)| = U$, a T_j with $|x(T_j)| = U^*$, and $U \leq |x(t)| \leq U^*$ for $t_j \leq t \leq T_j$. (It may be that $T_j \in I_{j+1}$.) But if $T_j - t_j > \alpha$, then $|x(t)| < U^{**}$ for $t \geq T_j$. Thus, if $T_j - t_j \leq \alpha$, then

$$V(t_j, x_{t_j}) \leq W_1(U) + W_2 \left(\int_{t_j - \alpha}^{t_j} W_3(|x(s)|)\, ds \right)$$

and $V'(t,x_t) \leq 0$ on $[t_j, T_j]$ so that

$$W(U^*) \leq V(T_j, x_{T_j}) \leq V(t_j, x_{t_j})$$

$$\leq W_1(U) + W_2 \left(\int_{t_j - \alpha}^{t_j} W_3(|x(s)|)\, ds \right)$$

or

$$\gamma = W_2^{-1}\left[W(U^*) - W_1(U)\right] \leq \int_{t_j - \alpha}^{t_j} W_3(|x(s)|)\, ds\,.$$

But

$$V'(t, x_t) \leq -W_3(|x|) + M$$

so we see that on $[t_j - \alpha,\ t_j]$ V decreased by an amount

$$\gamma - M\alpha > 4M\alpha\,.$$

Now the interval $[t_j - \alpha,\ t_j]$ could intersect I_{j-1}. Thus on $I_{j-1} \cup I_j \cup I_{j+1}$, V increases at most $3M\alpha$ units, but has decreased by at least $4M\delta$ units.

We conclude that over three consecutive intervals I_{j-1}, I_j, I_{j+1} V decreases by at least $M\alpha$. Since

$$V(t_1, \phi) \leq W_4(B_3), \quad \text{if} \quad N > W_4(B_4)/M\alpha,$$

then for
$$t \geq t_1 + (3N\alpha + S)$$

we have $|x(t, t_1, \phi)| < U^{**} \overset{\text{def}}{=} B$, where B is the number for uniform ultimate boundedness. This completes the proof.

In the proof of the next result we denote the arclength of a function
$$x : [a, b] \to R^n \quad \text{by} \quad |x[a, b]|$$

and the Euclidean norm by
$$|\cdot|.$$

This result is essentially Theorems 8.3.4 and 8.3.5 of Burton (1983a) and was originally proved in Burton (1982).

Theorem 4.2.11. *Let $V(t, \phi_t)$ satisfy the continuity conditions of Theorem 4.2.7 and let $D = \infty$. If there are positive constants U, β, μ, and c with*

(a) $V'_{(4.2.1)}(t, x_t) \leq -\mu|F(t, x_t)| - c$ *if* $|x(t)| \geq U$,

(b) $V'_{(4.2.1)}(t, x_t) \leq \beta$ *if* $|x(t)| < U$,

and

(c) $0 \leq V(t, x_t) \leq W_2(\|x_t\|)$,

then solutions of (4.2.1) *are uniform bounded and uniform ultimate bounded for bound B.*

Proof. To show uniform boundedness let $\mu R_1 = \beta\alpha$ define R_1 and let $R > R_1$ be given. We will show that $\left[t_1 \geq t_0, \ \phi \in C(t_1), \ \|\phi\| < R + U \right]$ implies that
$$|x(t)| \leq R + U + \left[W_2(R + U) + \beta\alpha \right]/\mu,$$

where $x(t) = x(t, t_1, \phi)$.

Either $V(t, x_t) < W_2(R + U) + \beta\alpha \overset{\text{def}}{=} s_1$ for $t \geq t_1$ or there is a first t_2 with $V(t_2, x_{t_2}) = s_1$. Clearly, $|x(t_2)| < U$ because V can grow only when $|x(t)| \leq U$.

Now $V(t, x_t)$ can grow no more than $\beta\alpha$ on $[t_2 - \alpha, \ t_2]$ and $V(t, x_t) \leq W_2(\|x_t\|)$, so there is a $t^* \in [t_2 - \alpha, \ t_2]$ with $W_2(|x(t^*)|) \geq W_2(R + U)$, yielding $|x(t^*)| \geq R + U$. As t increases from t^* to t_2, $|x(t)|$ goes from $|x(t^*)| \geq R + U$ to $|x(t)| = U$, so V decreases by at least μR, since $V'(t, x_t) \leq -\mu|x'(t)|$. Thus
$$V(t_2, x_{t_2}) \leq V(t^*, x_{t^*}) - \mu R + \beta\alpha$$
$$< V(t^*, x_{t^*}),$$

a contradiction to $V(t_2, x_{t_2})$ being the maximum on $[t_1, t_2]$.

We then have $V(t, x_t) < s_1$ for $t \geq t_1$ whenever $\|\phi_{t_1}\| < R + U$, and if $|x(t)| \geq U$ on any interval $[a, b]$, then $t \in [a, b]$ yields

$$0 \leq V(t, x_t) \leq s_1 - \mu|x[a, t]|,$$

so that $|x[a, t]| \leq s_1/\mu$, and hence, $|x(t)| \leq R + U + [s_1/\mu]$, which is the uniform boundedness.

Now, if $|x(t)| \geq U$ on $[a, b]$, then for $a \leq t \leq b$ we have

$$0 \leq V(t, x_t) \leq V(a, x_a) - \mu|x[a, t]|.$$

Also, $|x(t)| \geq U$ on $[a, b]$ yields $V'(t, x_t) < -c$, so that

$$0 \leq V(t, x_t) \leq V(a, x_a) - c(t - a)$$

implies that there is a sequence $\{r_n\} \to \infty$ with $|x(r_n)| \leq U$.

To find a lower bound on V, notice that if $x(t)$ is any solution of (4.2.1) with $|x(t_1)| = R + U$ for any $R > 0$ and any t_1, then $|x(t)|$ moves to U and $V(t, x_t)$ decreases at least μR. In other words,

$$V(t_1, x_{t_1}) \geq \mu R = \mu(|x(t_1)| - U)$$

which means that $\mu|x|$ is very nearly a lower wedge for V.

Recall from the proof of uniform boundedness just given that if $R > R_1 = \beta\alpha/\mu$, $t_1 \geq t_0$, and $\|x_{t_1}\| < R + U$, then

$$\mu(|x(t)| - U) \leq V(t, x_t) \leq W_2(R + U) + \beta\alpha \overset{\text{def}}{=} s_1,$$

so that

$$|x(t)| \leq [s_1 + \mu U]/\mu \overset{\text{def}}{=} f(R).$$

Thus, let $R_2 > R$ be fixed and let $R_3 > R_2$ be given. We will find $T > 0$ such that $\|x_{t_1}\| < R_3 + U$ and $t \geq t_1 + T$ implies $|x(t)| < f(R_2) \overset{\text{def}}{=} B$.

On any interval $[t_2 - \alpha, \ t_2]$ either

(a) $\|x_{t_2}\| < R_2 + U$ or

(b) there exists $t_3 \in [t_2 - \alpha, \ t_2]$ with $|x(t_3)| \geq R_2 + U$

and either

(b$_1$) $|x(t)| \geq U$ on $[t_2 - \alpha, \ t_2]$ or

(b$_2$) there exists $t_4 \in [t_2 - \alpha, \ t_2]$ with $|x(t_4)| < U$.

If (a) holds, then $|x(t)| < f(R_2)$ for $t \geq t_2$.

If (b_1) holds, then $V' \leq -c \leq 0$ on $[t_2 - \alpha, t_2]$, so that V decreases at least $c\alpha$ units.

If (b_2) holds, then V decreases at least μR_2 units, but may increase at most $\beta\alpha$ units while $|x(t)| < U$; hence, the net decrease under (b) is at least $\mu R_2 - \beta\alpha \overset{\text{def}}{=} c_1 > 0$.

Let $d = \min[c\alpha, c_1]$. Then for each positive integer n, on the interval $[t_1, \, t_1 + n\alpha]$ either (a) is satisfied at some t_2 resulting in $|x(t)| < f(R_2)$ in the future, or V decreases at least nd units. In the latter case, we can take n so large that for $t = t_1 + n\alpha$ we have

$$|x(t)| < [V(t, x_t) + \mu U]/\mu$$
$$< \{[W_2(R + U) + \beta\alpha + \mu U]/\mu\} - nd/\mu$$
$$< R_2 + U$$

so that for $t \geq t_1 + (n+1)\alpha$ we have $|x(t)| < f(R_2)$. Choosing $T = (n+1)\alpha$ completes the proof of uniform ultimate boundedness.

We are now ready to review properties (i)–(v) given just before Theorem 4.2.5. The problem of extending these to (4.2.1) is almost complete; in fact, for properties (i)–(iv) there is perfect unity between

$$x' = F(t, x)$$

and

$$x' = F(t, x_t).$$

We summarize the results as follows.

Theorem 4.2.12. *Let $V(t, x_t)$ satisfy the continuity properties of Theorem 4.2.5.*

(i) *If*
$$W_1(|x|) \leq V(t, x_t),$$
$$V'_{(4.2.1)}(t, x_t) \leq 0 \quad \text{and} \quad V(t, 0) = 0,$$

then $x = 0$ is stable.

(ii) *If*
$$W_1(|x|) \leq V(t, x_t) \leq W_2(\|x_t\|)$$

and
$$V'_{(4.2.1)}(t, x_t) \leq 0,$$

then $x = 0$ is uniformly stable.

(iii) *If $F(t, x_t)$ is bounded for x_t bounded and if*

$$W_1(|x|) \leq V(t, x_t),$$
$$V(t, 0) = 0 \quad and \quad V'_{(4.2.1)}(t, x_t) \leq -W_3(|x|)$$

then $x = 0$ is asymptotically stable.

(iv) *If*

$$W_1(|x|) \leq V(t, x_t) \leq W_2(|x|) + W_3(\|x_t\|)$$

and

$$V'(t, x_t) \leq -W_4(|x|)$$

then $x = 0$ is uniformly asymptotically stable.

(v) *If $D = \infty$,*

$$W_1(|x|) \leq V(t, x_t) \leq W_2(|x|) + W_3\left(\int_{t-\alpha}^{t} W_4(|x(s)|)\, ds\right)$$

and

$$V'(t, x_t) \leq -W_4(|x|) + M$$

then solutions of (4.2.1) are uniform bounded and uniform ultimate bounded for bound B.

Example 4.2.16. Consider the scalar equation

$$x' = -x + \int_{t-\alpha}^{t} C(t, s)x(s)\, ds + f(t)$$

with C continuous for $0 \leq s \leq t < \infty$, f continuous on $[0, \infty)$, and $|f(t)| \leq M$ for some $M > 0$. Define

$$V(t, x_t) = |x| + K \int_{t-\alpha}^{t} \int_{t}^{\infty} |C(u, s)|\, du\, |x(s)|\, ds$$

so that

$$V'(t, x_t) \leq \left[-1 + K \int_{t}^{\infty} |C(u, t)|\, du\right] |x|$$
$$+ (-K + 1) \int_{t-\alpha}^{t} |C(t, s)|\, |x(s)|\, ds + M.$$

If we suppose that there is $K > 1$ with

$$-1 + K \int_{t}^{\infty} |C(u, t)|\, du \leq -\gamma \quad some \quad \gamma > 0,$$

and with
$$\int_{t-\alpha}^{t} \int_{t}^{\infty} |C(u,s)| \, du \, ds \leq P, \quad \text{some} \quad P > 0,$$

then the conditions of Theorem 4.2.10 hold and solutions are uniform bounded and uniform ultimate bounded for bound B.

Exercise 4.2.7. Determine conditions on C and f in Example 4.2.16 so that Theorem 4.2.2 will imply that there is a T-periodic solution.

When there is a T-periodic solution we are always interested in whether or not it is unique and if all solutions converge to it. Such problems are extensively discussed in Burton and Townsend (1971) and the references therein. Recently Murakami (1984a) and Yoshizawa (1984) have obtained interesting results in that direction for delay equations.

4.3 Volterra Equations with Infinite Delay

4.3.1 Introduction

We consider a system of Volterra integrodifferential equations

$$(4.3.1) \qquad x' = h(t,x) + \int_{-\infty}^{t} q(t,s,x(s)) \, ds$$

in which $h : (-\infty, \infty) \times R^n \to R^n$, $q : (-\infty, \infty) \times (-\infty, \infty) \times R^n \to R^n$, h and q are continuous. This equation was discussed under (3.5.3) in Section 3.5 and results on existence and uniqueness were obtained. We are interested in boundedness, stability, and the existence of periodic solutions.

In this section all of our solutions will start at $t_0 = 0$. To specify a solution we require a continuous initial function $\phi : (-\infty, 0] \to R^n$ and obtain a solution $x(t, 0, \phi)$ on some interval $(-\infty, \beta)$, $\beta > 0$, where $x(t, 0, \phi) = \phi(t)$ for $t \leq 0$ and where $x(t, 0, \phi)$ satisfies (4.3.1) for $0 \leq t < \beta$.

If $\phi : (-\infty, 0] \to R^n$ is continuous and if $\int_{-\infty}^{0} q(t, s, \phi(s)) \, ds \overset{\text{def}}{=} \Phi(t)$ is continuous for $t \geq 0$, then there is a solution $x(t, 0, \phi)$ of (4.3.1) on some interval $(-\infty, \beta)$, $\beta > 0$; if h and q satisfy local Lipschitz conditions in x, then the solution is unique; if the solution remains bounded, then $\beta = \infty$. But we have learned nothing yet about continual dependence of solutions on ϕ.

A study of the existence of a periodic solution of (4.3.1) leads us, quite naturally, into a variety of interesting problems concerning properties of solutions of (4.3.1). One of the first things we discover is that we need to use unbounded initial functions. Next, we find that we need to construct a compact subset of initial functions, and this leads us to consider locally convex spaces or Banach spaces with a weighted norm. We then find that the two problems are related; the unbounded initial functions form a weight for the norm, and are discovered from the properties of $\int_{-\infty}^{0} q(t, s, x(s))\, ds$.

4.3.2   Periodic Solutions

Most investigations into the existence of periodic solutions of differential equations require that one be able to verify that $x(t + T)$ is a solution whenever $x(t)$ is a solution. It is easy to verify that a sufficient condition for that property to hold for (4.3.1) is that $h(t + T, x) = h(t, x)$ and $q(t + T, s + T, x) = q(t, s, x)$.

The basic idea is to find a set S of initial functions and define a mapping $P : S \to S$ by $\phi \in S$ implies

$$P\phi = x(t + T, 0, \phi) \quad \text{for} \quad -\infty < t \le 0.$$

Thus, if P has a fixed point ϕ, then $x(t + T, 0, \phi)$ is a solution (because $x(t, 0, \phi)$ is a solution) which has initial function ϕ (because ϕ is a fixed point of P) and, by uniqueness,

$$x(t, 0, \phi) = x(t + T, 0, \phi).$$

We now indicate that we are going to have to significantly change the approach which we used for finite delay equations when we deal with (4.3.1). We imagine that we extend the definitions of uniform boundedness and uniform ultimate boundedness used for finite delay equations to (4.3.1) by simply allowing the domain of ϕ to be $(-\infty, 0]$ instead of $[-\alpha, 0]$. We ask that for each $B_1 > 0$ there exists $B_2 > 0$ such that $\phi : (-\infty, 0] \to R^n$ with

$$\|\phi\| \overset{\text{def}}{=} \sup_{-\infty < t \le 0} |\phi(t)| \le B_1$$

implies that $|x(t, 0, \phi)| < B_2$. Also, we ask that there exists $B > 0$ such that for each $B_3 > 0$ there exists $K > 0$ such that $\|\phi\| \le B_3$ and $t \ge K$ implies that $|x(t, 0, \phi)| < B$. We would try taking

$$S = \{\phi : (-\infty, 0] \to R^n \,|\, \|\phi\| \le B \text{ and}$$

$$|\phi(u) - \phi(v)| \le L|u - v|\}.$$

Notice that if $B_2 > B_1$, then neither our mapping P nor any of its iterates can ever map S into S because the "hump" is always outside S. Moreover, S is not compact. See Fig. 4.16.

On the other hand, if S becomes large as $t \to -\infty$, then the "hump" can be absorbed inside S as we shift the solution $x(t, 0, \phi)$ to the left using $P\phi, P^2\phi, \dots$. See Fig. 4.17.

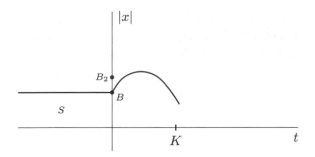

Fig. 4.16

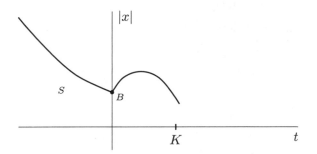

Fig. 4.17

But, is it reasonable to expect to be able to use unbounded initial functions in (4.3.1)? The interesting point here is that if we can admit all bounded initial functions, then we can also admit a large class of unbounded initial functions.

Recall that what we need for existence and uniqueness is that if $\phi : (-\infty, 0] \to R^n$ is continuous then

$$\Phi(t) = \int_{-\infty}^{0} q(t, s, \phi(s)) \, ds$$

is continuous. It is instructive to see what this requires in the scalar linear convolution case where $q(t, s, x) = C(t - s)x(s)$. Then

$$\Phi(t) = \int_{-\infty}^{0} C(t - s)\phi(s) \, ds$$

and, if we are to allow constant functions ϕ, then we are asking that

$$\Phi(t) = \phi_0 \int_{-\infty}^{0} C(t - s) \, ds = \phi_0 \int_{t}^{\infty} C(u) \, du$$

be continuous, or say $C \in L^1[0, \infty)$. However, it is shown in Burton and Grimmer (1973) that if $C \in L^1[0, \infty)$ then there is a continuous increasing function $g : [0, \infty) \to [1, \infty)$ with $g(0) = 1$ and $g(r) \to \infty$ as $r \to \infty$ such that

$$\int_0^\infty |C(u)| \, g(u) \, du < \infty \, .$$

Thus, if we allow bounded initial functions then we may allow initial functions ϕ with $|\phi(s)| \le \gamma g(-s)$ for any $\gamma > 0$.

An examination of this example will show that we are asking that (4.3.1) have a fading memory. The equation representing a real world situation should remember its past, but the memory should fade with time. One such formulation will now be given. It is convenient for that formulation to have the following symbols.

Notation. Let Y denote the set of continuous functions $\phi : (-\infty, 0] \to R^n$:

(a) $|\phi(t)|$ denote the Euclidean length of $\phi(t)$,

(b) $\|\phi\| = \sup_{-\infty < t \le 0} |\phi(t)|$, if it exists,

(c) if $g : (-\infty, 0] \to (0, \infty)$ is continuous, then

$$|\phi|_g = \sup_{-\infty < t \le 0} |\phi(t)/g(t)|$$

if it exists.

Definition 4.3.1. *Equation* (4.3.1) *is said to have a* fading memory *if for each $\epsilon > 0$ and for each $B > 0$, there exists $K > 0$ such that $[\phi \in Y, \|\phi\| \le B, \, t \ge 0]$ imply that*

$$\int_{-\infty}^{-K} |q(t, s, \phi(s))| \, ds < \epsilon \, .$$

Theorem 4.3.1. *If* (4.3.1) *has a fading memory, then for every $B > 0$ there is a continuous nonincreasing function $g : (-\infty, 0] \to [B, \infty)$ with $g(0) = B$ and $g(r) \to \infty$ as $r \to -\infty$ such that $[\phi \in Y$ and $|\phi|_g \le 1]$ imply that $\int_{-\infty}^0 q(t, s, \phi(s)) \, ds$ is continuous for $t \ge 0$.*

Proof. Let $B > 0$ be given. Take $\epsilon_n = 1/n^2$, $B_n = B + n$, and use Definition 4.3.1 to find K_n such that $[\phi \in Y, \|\phi\| \le B_n + 1, \, t \ge 0]$ imply that $\int_{-\infty}^{-K_n} |q(t, s, \phi(s))| \, ds \le \epsilon_n$. Let $K_n < K_{n+1}$ and g be a continuous piecewise linear function with $g(0) = B$, $g(-K_1) = B+1, \ldots, g(-K_n) = B+n$.

Now, let $\phi \in Y$ with $|\phi(t)| \leq g(t)$ on $(-\infty, 0]$. Then

$$\int_{-\infty}^{0} |q(t, s, \phi(s))| \, ds$$

$$= \int_{-K_1}^{0} |q(t, s, \phi(s))| \, ds + \sum_{n=1}^{\infty} \int_{-K_{n+1}}^{-K_n} |q(t, s, \phi(s))| \, ds$$

$$\leq \int_{-K_1}^{0} |q(t, s, \phi(s))| \, ds + \sum_{n=1}^{\infty} \int_{-\infty}^{-K_n} |q(t, s, \phi_n(s))| \, ds$$

$$\leq \int_{-K_1}^{0} |q(t, s, \phi(s))| \, ds + \sum_{n=1}^{\infty} (1/n^2)$$

where $\phi_n(t) = \phi(t)$ on $[-K_{n+1}, -K_n]$ and $\phi_n(t) = \phi(-K_{n+1})$ for $t \leq -K_{n+1}$. This shows that for fixed t, then $\int_{-\infty}^{0} |q(t, s, \phi(s))| \, ds$ converges. To see that it is continuous for $t > 0$, let $t_0 > 0$ be fixed, and let $\epsilon > 0$ be given. Note that for any $K > 0$, $\int_{-K}^{0} q(t, s, \phi(s)) \, ds$ is continuous for $t \geq 0$. We seek $\delta > 0$ such that $|t - t_0| < \delta$ implies that

$$\left| \int_{-\infty}^{0} q(t, s, \phi(s)) \, ds - \int_{-\infty}^{0} q(t_0, s, \phi(s)) \, ds \right| < \epsilon .$$

Find $j > 0$ with

$$\sum_{n=j}^{\infty} (1/n^2) < \epsilon/3 .$$

Then find K with

$$\int_{-\infty}^{-K} |q(t, s, \phi(s))| \, ds < \sum_{n=j}^{\infty} (1/n^2) .$$

Proceed as with the estimate for the existence of $\int_{-\infty}^{0} |q(t, s, \phi(s))| \, ds$ and conclude that for this K, there is a $\delta > 0$ such that $|t - t_0| < \delta$ implies that

$$\left| \int_{-K}^{0} q(t, s, \phi(s)) \, ds - \int_{-K}^{0} q(t_0, s, \phi(s)) \, ds \right| < \epsilon/3 .$$

Then for $|t - t_0| < \delta$ we have

$$\left| \int_{-\infty}^{0} q(t, s, \phi(s)) \, ds - \int_{-\infty}^{0} q(t_0, s, \phi(s)) \, ds \right|$$

$$\leq \left| \int_{-K}^{0} q(t, s, \phi(s)) \, ds - \int_{-K}^{0} q(t_0, s, \phi(s)) \, ds \right|$$

$$+ \left| \int_{-\infty}^{-K} q(t, s, \phi(s)) \, ds \right| + \left| \int_{-\infty}^{-K} q(t_0, s, \phi(s)) \, ds \right|$$

$$< (\epsilon/3) + (\epsilon/3) + (\epsilon/3),$$

as required.

We now prepare to construct a set S of initial functions so that the operator P can map S into S. We will ask for a single function g such that $Bg(t)$ satisfies the conditions of Theorem 4.3.1.

Basic Assumption for (4.3.1). *There is a continuous function* $g :$ $(-\infty, 0] \to [0, \infty)$, $g(0) = 1$, $g(r) \to \infty$ *as* $r \to -\infty$, *and* g *is decreasing such that* $[\phi \in Y, \ |\phi(s)| \leq \gamma g(s)$ *for some* $\gamma > 0$ *and* $-\infty < s \leq 0$, *and* $t \geq 0]$ *imply that* $\int_{-\infty}^{0} q(t, s, \phi(s)) \, ds$ *is continuous.*

For this g, we denote by

$$(X, | \cdot |_g)$$

the Banach space of continuous $\phi : (-\infty, 0] \to R^n$ for which

$$|\phi|_g = \sup_{-\infty < t \leq 0} |\phi(t)/g(t)|$$

exist.

Definition 4.3.2. *Solutions of (4.3.1) are* g-*uniform bounded at* $t = 0$ *if for each* $B_1 > 0$ *there exists* $B_2 > 0$ *such that* $[\phi \in X, \ |\phi|_g \leq B_1, \ t \geq 0]$ *imply that* $|x(t, 0, \phi)| < B_2$.

Definition 4.3.3. *Solutions of (4.3.1) are* g-*uniform ultimate bounded for bound* B *at* $t = 0$ *if for each* $B_3 > 0$ *there is a* $K > 0$ *such that* $[\phi \in X, \ |\phi|_g \leq B_3, \ t \geq K]$ *imply that* $|x(t, 0, \phi)| < B$.

We want to select a subset S of initial functions and define a map $P : S \to S$ by $\phi \in S$ implies

$$P\phi = x(t + T, 0, \phi) \quad \text{for} \quad -\infty < t \leq 0$$

and prove that there is a fixed point so that there will be a periodic solution. The fixed-point theorems ask that S be compact. Exercise 3.1.1 describes our situation precisely and the natural space is (Y, ρ) where

$$\rho(\phi, \psi) = \sum_{n=1}^{\infty} 2^{-n} \rho_n(\phi, \psi) / \{1 + \rho_n(\phi, \psi)\}$$

described in Example 3.1.5.

The space (Y, ρ) is a locally convex topological vector space and the fixed-point theorem for it is the Schauder-Tychonov theorem which calls for the mapping $P : S \to S$ to be continuous in (Y, ρ); that is, if $\phi \in S$, then given $\epsilon > 0$ there exists $\delta > 0$ such that $\psi \in Y$ and $\rho(\phi, \psi) < \delta$ implies $\rho(P\phi, P\psi) < \epsilon$. And this is a very strong type of continuity.

Definition 4.3.4. *Let $\overline{\rho}$ be any metric on Y. Solutions of (4.3.1) are continuous in ϕ relative to a set $M \subset Y$ and $\overline{\rho}$ if (for each $\phi \in M$, for each $\epsilon > 0$, and for each $J > 0$) there exists $\delta > 0$ such that $[\psi \in M, \overline{\rho}(\phi, \psi) < \delta]$ imply that*

$$\overline{\rho}(P\phi, P\psi) < \epsilon$$

where

$$P\phi = x(t + j, 0, \phi) \quad for \quad -\infty < t \leq 0.$$

Theorem 4.3.2. *Let solutions of (4.3.1) be g-uniform bounded, let (4.3.1) have a fading memory, let h satisfy a local Lipschitz condition in x, and let q satisfy a Lipschitz condition of the following type. For each $H > 0$, each $J > 0$ there exists $M > 0$ such that $|x_i| \leq H$ and $-\infty \leq s \leq t \leq J$ imply that*

$$\left| q(t, s, x_1) - q(t, s, x_2) \right| \leq M |x_1 - x_2|.$$

If S is any bounded (sup norm) subset of Y, then solutions of (4.3.1) are continuous in ϕ relative to S and ρ.

Proof. Let $B_1 > 0$ be given and find $B_2 > 0$ such that $[\phi \in Y, |\phi|_g \leq B_1, t \geq 0]$ imply that $|x(t, 0, \phi)| < B_2$. Define $S = \{\phi \in Y \mid \|\phi\| \leq B_1\}$ and $H = B_2$. Let $J > 0$, $\epsilon > 0$, and $\phi \in S$ be given. It will suffice to find $\delta > 0$ such that $[\psi \in S, \rho(\phi, \psi) < \delta]$ imply $|x(t, 0, \phi) - x(t, 0, \psi)| < \epsilon$ for $0 \leq t \leq J$. For $\epsilon_1 > 0$ satisfying $2\epsilon_1 J e^{M(1+J)J} < \epsilon/2$ find $D > 0$ such that $[\phi \in S, t \geq 0]$ imply $\int_{-\infty}^{-D} |q(t, s, \phi(s))| \, ds < \epsilon_1$. Here, M is the Lipschitz constant for h and q when $|x_i| \leq H$ and $-\infty \leq s \leq t \leq J$. Let $\psi \in S$ and $\rho(\phi, \psi) < \delta$ so that $|\phi(t) - \psi(t)| < k\delta$ for some $k > 0$ when $-D \leq t \leq 0$ and $k\delta(1 + MJD)e^{M(1+J)J} < \epsilon/2$ Let $x_1(t) = x(t, 0, \phi)$ and

$x_2(t) = x(t, 0, \psi)$. Then

$$x_1'(t) - x_2'(t) = h(t, x_1) - h(t, x_2)$$
$$+ \int_{-\infty}^t \left[q(t, s, x_1(s)) - q(t, s, x_2(s)) \right] ds$$

so that

$$x_1(t) - x_2(t) = \phi(0) - \psi(0)$$
$$+ \int_0^t \left[h(s, x_1(s)) - h(s, x_2(s)) \right] ds$$
$$+ \int_0^t \int_{-\infty}^u \left[q(u, s, x_1(s)) - q(u, s, x_2(s)) \right] ds \, du .$$

Then

$$|x_1(t) - x_2(t)| \le k\delta + M \int_0^t |x_1(s) - x_2(s)| \, ds$$
$$+ \int_0^t \int_{-\infty}^{-D} |q(u, s, \phi(s)) - q(u, s, \psi(s))| \, ds \, du$$
$$+ \int_0^t \int_{-D}^u M|x_1(s) - x_2(s)| \, ds \, du$$
$$\le k\delta + M \int_0^t |x_1(s) - x_2(s)| \, ds + 2\epsilon_1 J$$
$$+ \int_0^t \int_{-D}^0 M|\phi(s) - \psi(s)| \, ds \, du$$
$$+ \int_0^t \int_0^u M|x_1(s) - x_2(s)| \, ds \, du$$
$$\le k\delta + 2\epsilon_1 J + k\delta M D J$$
$$+ M \int_0^t |x_1(s) - x_2(s)| \, ds$$
$$+ M \int_0^t (t - s) |x_1(s) - x_2(s)| \, ds$$
$$\le k\delta + 2\epsilon_1 J + k\delta M D J$$
$$+ M(1 + J) \int_0^t |x_1(s) - x_2(s)| \, ds .$$

By Gronwall's inequality

$$|x_1(t) - x_2(t)| \leq \left[k\delta(1 + MDJ) + 2\epsilon_1 J\right] e^{M(1+J)J}$$

$$< \epsilon,$$

as required.

Remark 4.3.1. Theorem 4.3.2 can be proved in the same way when (Y, ρ) is replaced by $(Y, | \cdot |_g)$. Moreover, the fading memory definition can be altered to ask that $[\phi \in Y, |\phi|_g \leq B, t \geq 0]$ imply that

$$\int_{-\infty}^{-K} |q(t, s, \phi(s))| \, ds < \epsilon.$$

Theorem 4.3.2 can also be proved with that change. We will be needing those changes in the coming theorems. But Theorem 4.3.2 is simply one type of continual dependence result and so our future theorems do not rely on it, but rather hypothesize the continual dependence property. The general question of continual dependence is treated by Haddock (1984), Hale and Kato (1978), Kaminogo (1978), Kappel and Schappacher (1980); and Sawano (1979).

Theorem 4.3.3. *Let the following conditions hold.*

(a) *If $\phi \in X$ and $|\phi|_g \leq \gamma$ for any $\gamma > 0$, then there is a unique solution $x(t, 0, \phi)$ of (4.3.1) on $[0, \infty)$.*

(b) *Solutions of (4.3.1) are g-uniform bounded and g-uniform ultimate bounded for bound B at $t = 0$.*

(c) *For each $\gamma > 0$ there is an $L > 0$ such that $|\phi(t)| \leq \gamma$ on $(-\infty, 0]$ implies that $|x'(t, 0, \phi)| \leq L$ on $[0, \infty)$.*

(d) *For each $\gamma > 0$, if*

$$U = \{\phi \in X \mid |\phi(t)| \leq \gamma \text{ on } (-\infty, 0]\},$$

then solutions of (4.3.1) depend continuously on ϕ in U relative to (X, ρ).

(e) *If $x(t)$ is a solution of (4.3.1) on $[0, \infty)$, so is $x(t + T)$.*

Under these conditions, (4.3.1) has an mT-periodic solution for some positive integer m.

Proof. For the number B in Def. 4.3.3, let $B_1 = B$ in Def. 4.3.2 and find $B_2 = B_2(B)$. Also, find $K = K(B)$. If $|\phi|_g \leq B$, then $|x(t, 0, \phi)| \leq B_2$ for $t \geq 0$ and $|x(t, 0, \phi)| < B$ for $t \geq K$.

Determine a number $H > 0$ with $Bg(-H) = B_2$; then determine a positive integer m with $mT > K + H$. Define

$$S_1 = \{\phi \in X \mid |\phi|_g \leq B\}$$

and, with $L = L(B_2)$ in (c), let

$$S = \{\phi \in S_1 \mid |\phi(t)| \leq Bg(-H),$$
$$|\phi(u) - \phi(v)| \leq L|u - v|\}.$$

See Fig. 4.18

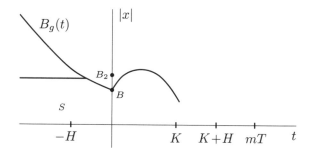

Fig. 4.18

The set S is compact in (X, ρ), as we saw in Example 3.1.2. It is convex, as we now show. Let $\phi_1, \phi_2 \in S$ and $0 \leq k \leq 1$. For

$$\phi = k\phi_1 + (1 - k)\phi_2$$

we have

$$|\phi(u) - \phi(v)| \leq k|\phi_1(u) - \phi_1(v)| + (1 - k)|\phi_2(u) - \phi_2(v)|$$
$$\leq L|u - v|.$$

Also, for $t \leq -H$ we have

$$|\phi(t)| = |k\phi_1(t) + (1 - k)\phi_2(t)|$$
$$\leq k|\phi_1(t)| + (1 - k)|\phi_2(t)|$$
$$\leq kBg(-H) + (1 - k)Bg(-H)$$
$$= Bg(-H).$$

When $-H < t \le 0$, then

$$|\phi(t)| \le kBg(t) + (1-k)Bg(t) = Bg(t),$$

so S is convex.

Define a mapping $P : S \to S$ by $\phi \in S$ implies

$$P\phi = x(t + mT, 0, \phi) \quad \text{for} \quad -\infty < t \le 0$$

where $mT > K + H$ and m is an integer. By (c) we see that for $\psi = P\phi$ then $|\psi(u) - \psi(v)| \le L|u - v|$. By choice of m, K, H, B, and B_2, we also see that $P\phi \in S$ since $P\phi$ is just the solution through ϕ shifted to the left mT units. Moreover, P is continuous by (d). By the Schauder-Tychonov theorem P has a fixed point: $P\phi = \phi$. That is,

$$x(t + mT, 0, \phi) \equiv \phi(t) \quad \text{on} \quad (-\infty, 0].$$

By (e), $x(t + mT, 0, \phi)$ is a solution of (4.3.1) for $t \ge 0$ and it has ϕ as its initial function. By uniqueness, $x(t + mT, 0, \phi) = x(t, 0, \phi)$. This completes the proof.

The space (X, ρ) is the natural space for the problem because of its compactness properties. The difficulty here is that we would like to use an asymptotic fixed-point theorem to conclude that there is a T-periodic solution. But the appropriate asymptotic fixed-point theorems seem to require a Banach space. Browder (1970) has numerous general asymptotic fixed-point theorems but they do not seem to fit our situation. Too often the fixed-point theorems require that the initial function set be open and they depend on the map smoothing the functions. Both properties fail here because open sets are simply too large in these spaces and when the delay is infinite, then the mapping can never smooth the whole function ϕ.

Horn's theorem seems to be just right for the occasion; moreover, the function g which we introduced to allow unbounded initial functions gives us the Banach space

$$(X, |\cdot|_g)$$

and is well suited to Horn's theorem. The next result is a special case of a theorem of Arino *et al.* (1985).

Theorem 4.3.4. *Let the conditions of Theorem 4.3.3 hold with* (d) *modified to require continuity in* $(X, |\cdot|_g)$. *Then* (4.3.1) *has a T-periodic solution.*

Proof. For B in Def. 4.3.3, use Def. 4.3.2 to find $B_1 > 0$ such that

$$|x(t, 0, \phi)| < B_1 \quad \text{if} \quad [\phi \in X, \ |\phi|_g \le B, \ t \ge 0].$$

Find K_1 using Def. 4.3.3 so that

$$|x(t,0,\phi)| < B \quad \text{if} \quad \left[\phi \in X, \ |\phi|_g \le B, \ t \ge K_1\right].$$

In the same way, find B_2 and K_2 such that $|x(t,0,\phi)| < B_2$ for $t \ge 0$ and $|x(t,0,\phi)| < B$ if $t \ge K_2$ and $|\phi|_g \le B_1 + 1$. See Fig. 4.19.

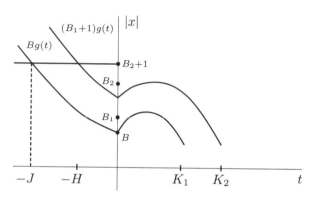

Fig. 4.19

Find $J > 0$ with $Bg(-J) = B_2 + 1$ and find an integer m with $mT > J + K_2$. Then find $B_3 > 0$ such that $\left[\phi \in X, \ |\phi|_g \le (B_2 + 1), \ t \ge 0\right]$ imply that $|x(t,0,\phi)| < B_3$. Finally, find $L > 0$ such that $\left[\phi \in X, \ |\phi(s)| \le B_3 \text{ on } (-\infty, 0], \text{ and } t \ge 0\right]$ imply that $|x'(t,0,\phi)| \le L$.

Define

$$S_0 = \left\{\phi \in X \mid |\phi(t)| \le \min[Bg(t), B_2 + 1]\right.$$

$$\text{on } (-\infty, 0],$$

$$\left.|\phi(u) - \phi(v)| \le L|u - v|\right\},$$

$$\widetilde{S_1} = \left\{\phi \in X \mid |\phi|_g < B_1 + 1\right\},$$

$$S_2 = \left\{\phi \in X \mid |\phi(t)| \le B_2 + 1\right.$$

$$\text{on } (-\infty, 0],$$

$$\left.|\phi(u) - \phi(v)| \le L|u - v|\right\},$$

and

$$S_1 = \widetilde{S_1} \cap S_2.$$

Then S_0 and S_2 are convex and compact in $(X, |\cdot|_g)$. Also, S_1 is convex and open in S_2 because $\widetilde{S_1}$ is open.

Define a mapping $P : S_2 \to X$ by $\phi \in S_2$ implies

$$P\phi = x(t + T, 0, \phi) \quad \text{for} \quad -\infty < t \le 0.$$

Thus, P translates ϕ to the left T units, along with $x(t, 0, \phi)$ on $[0, T]$. Now $x(t + T, 0, \phi)$ is also a solution of (4.3.1) for $t \geq 0$ and its initial function is $P\phi$. Hence,

$$P(P\phi) = x(t + T, 0, P\phi) \quad \text{for} \quad -\infty < t \leq 0.$$

or

$$P^2\phi = x(t + 2T, 0, \phi) \quad \text{for} \quad -\infty < t \leq 0.$$

In general,

$$P^k\phi = x(t + kT, 0, \phi) \quad \text{for} \quad -\infty < t \leq 0.$$

By the construction,

$$P^j S_0 \subset S_1 \quad \text{for} \quad 0 \leq j < \infty$$

since $|x'(t, 0, \phi)| \leq L$. Also, for $j > m$ we have

$$P^j S_1 \subset S_0$$

because $mT > K_2 + J$ and $S_1 \subset S_2$ so that $\phi \in S_1$ is L-Lipschitz. By Horn's theorem P has a fixed point $\phi \in S_0$. Hence, $P\phi = \phi$ so that

$$x(t, 0, \phi) = x(t, 0, P\phi) = x(t + T, 0, \phi)$$

and the proof is complete.

If we are willing to strengthen conditions (b)–(d) slightly, then we can avoid the constructions and offer a proof virtually indistinguishable from the one for ordinary differential equations. Recall that for

$$x' = F(t, x)$$

we ask that;

(a) For each (t_0, x_0) there is a solution $x(t, t_0, x_0)$ on $[t_0, \infty)$ which is continuous in x_0.

(b) Solutions are uniform bounded and uniform ultimate bounded for bound B.

(c) For each $H > 0$ there exists $L > 0$ such that $|x_0| \leq H$ implies $|x'(t, t_0, x_0)| \leq L$ for $t \geq t_0$.

(d) If $x(t)$ is a solution so is $x(t + T)$.

We then conclude that there is a T-periodic solution.

The proof uses Horn's theorem and sets

$$S_0 = \{x \in R^n \mid |x| \leq B\},$$
$$S_2 = \{x \in R^n \mid |x| \leq B_3\},$$

and

$$S_1 = \{x \in R^n \mid |x| < B_2\} \cap S_2$$

with a mapping $P : S_2 \to R^n$.

The sets S_0 and S_2 are concentric compact balls in the initial condition space R^n. What can be interpreted as concentric compact balls in $(X, |\cdot|_g)$? One answer might be the pair of sets

$$S_0 = \{\phi \in X \mid |\phi(t)| \leq \gamma\sqrt{g(t)}$$
$$\text{on } (-\infty, 0],$$
$$|\phi(u) - \phi(v)| \leq L|u - v|\}$$

and

$$S_2 = \{\phi \in X \mid |\phi(t)| \leq \gamma\sqrt{g(t - H)}$$
$$\text{on } (-\infty, 0],$$
$$|\phi(u) - \phi(v)| \leq L|u - v|\}.$$

With this interpretation, the theorem and proof of the existence of a T-periodic solution of (4.3.1) becomes very similar to that for $x' = F(t, x)$. The next result is found in Burton (1985b). To make S_2 compact we ask that g be Lipschitz.

Theorem 4.3.5. *Let the following conditions hold for (4.3.1).*

(a) *If $\phi \in X$, then $x(t, 0, \phi)$ exists on $[0, \infty)$ and is continuous in ϕ relative to X and $|\cdot|_g$.*

(b) *Solutions of (4.3.1) are g-uniform bounded and g-uniform ultimate bounded for bound B at $t = 0$.*

(c) *For each $H > 0$ there exists $L > 0$ such that $\phi \in X$ and $|\phi(t)| \leq B\sqrt{g(t - H)}$ on $(-\infty, 0]$ imply that $|x'(t, 0, \phi)| \leq L$ on $[0, \infty)$.*

(d) *If $x(t)$ is a solution (4.3.1) so is $x(t + T)$.*

Then (4.3.1) has a T-periodic solution.

Proof. Given $B > 0$ there is a $B_2 > 0$ such that $\left[\phi \in X, \ |\phi|_g \leq B, \ t \geq 0\right]$ imply that $|x(t, 0, \phi)| < B_2$ and there is a $K_1 > 0$ with $|x(t, 0, \phi)| < B$ if $t \geq K_1$. For the $B_2 > 0$ there exists $B_3 > 0$ and $K > 0$ such that $\left[\phi \in X, \ |\phi|_g \leq B_2, \ t \geq 0\right]$ imply that $|x(t, 0, \phi)| < B_3$ and $\left[\phi \in X, \ |\phi|_g \leq B_2, \ t \geq K\right]$ imply that $|x(t, 0, \phi)| < B$.

Find $H > 0$ with $B\sqrt{g(-H)} = B_3$. Use (c) to find L such that $|\phi(t)| \leq B\sqrt{g(t - H)}$ on $(-\infty, 0]$ implies that $|x'(t, 0, \phi| \leq L$. Define

$$S_0 = \left\{\phi \in X \mid |\phi(t)| \leq B\sqrt{g(t)}\right.$$
$$\text{on } (-\infty, 0],$$
$$\left. |\phi(u) - \phi(v)| \leq L|u - v|\right\}$$

$$S_2 = \left\{\phi \in X \mid |\phi(t)| \leq B\sqrt{g(t - H)}\right.$$
$$\text{on } (-\infty, 0],$$
$$\left. |\phi(u) - \phi(v)| \leq L|u - v|\right\}.$$

and

$$S_1 = \left\{\phi \in X \mid |\phi|_g < B_2\right\} \cap S_2.$$

See Fig. 4.20.

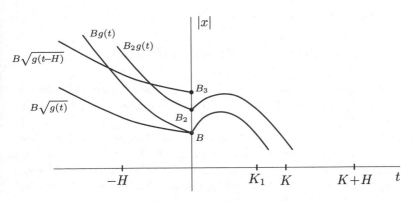

Fig. 4.20

For $\phi \in X$, define

$$P\phi = x(t + T, 0, \phi) \quad \text{for} \quad -\infty < t \leq 0$$

and note that

$$P^j\phi = x(t + jT, 0, \phi) \quad \text{for} \quad -\infty < t \leq 0.$$

By the choices of B_2 and B_3, if m is an integer with $mT > K + H$, then

$$P^j S_0 \subset S_1 \quad \text{for all} \qquad j \,,$$

$$P^j S_1 \subset S_0 \quad \text{for} \qquad j \geq m \,,$$

and

$$P^j S_1 \subset S_2 \quad \text{for all} \qquad j \,.$$

By Example 3.1.7 we see that S_0 and S_2 are compact, while S_0, S_1, and S_2 are convex with S_1 open in S_2. By Horn's theorem P has a fixed point, say ϕ; then

$$x(t + T, 0, \phi) = x(t, 0, \phi) \,,$$

and the proof is complete.

In the last theorem and proof we attempted to reduce the concepts and techniques for (4.3.1) to those for an ordinary differential equation $x' = F(t, x)$. The next idea is to try to reduce the problem of showing the existence of a periodic solution of (4.3.1) to that of showing the existence of a periodic solution of

(4.2.1) $$x' = F(t, x_t) \,.$$

In particular, we consider the family of systems

(4.3.2) $$y' = h(t, y) + \int_{t-kT}^{t} q(t, s, y(s)) \, ds \quad k = 1, 2, \dots .$$

We prove that when each member of the family has a T-periodic solution, then the sequence of these periodic solutions converges to a T-periodic solution of (4.3.1).

For (4.3.2) we require a continuous initial function $\phi : [-kT, 0] \to R^n$ and denote by $y(t, \phi, k)$ the solution of (4.3.2) with $y(t, \phi, k) = \phi(t)$ for $-KT \leq t \leq 0$. Conditions for the existence of a solution for (4.3.1) more than suffice for the existence of a solution of (4.3.2). We now list five assumptions which are used to prove the desired result. The first assumption reduces our basic assumptions for (4.3.1).

Assumption 4.3.1. *Let h and q be continuous and satisfy a local Lipschitz condition in x. Suppose also that for bounded and continuous $\phi : (-\infty, 0] \to R^n$ the function*

$$\int_{-\infty}^{0} q(t, s, \phi(s)) \, ds$$

is continuous for $t \geq 0$.

Assumption 4.3.2. *Let $(Z, \|\cdot\|)$ denote the Banach space of continuous T-periodic functions from $(-\infty, \infty) \to R^n$ with the supremum norm. Assume that for $\eta \in Z$ and $0 \le t \le T$ the operator defined formally by*

(4.3.3)
$$(P\eta)(t) = \eta(0) + \int_0^t h(u, \eta(u))\, du$$
$$+ \int_0^t \int_{-\infty}^u q(u, s, \eta(s))\, ds\, du$$

is uniformly continuous at each $\eta \in Z$ in the following sense: For each $\eta \in Z$ and each $\epsilon > 0$ there exists $\delta > 0$ such that $\psi \in Z$ and $\|\psi - \eta\| < \delta$ imply that
$$\left|(P\eta)(t) - (P\psi)(t)\right| < \epsilon \quad \text{for} \quad 0 \le t \le T.$$

The next lemma gives a sufficient condition for this continuity of P.

Lemma 4.3.1. *Suppose that for each $M > 0$ there exists a continuous function $L_M : [0, \infty) \to [0, \infty)$ such that when $\phi, \psi : (-\infty, 0] \to R^n$ are continuous and satisfy $|\phi(s)| \le M$ and $|\psi(s)| \le M$ for $s \le t$, then*

(a) $\quad \left|q(t, s, \phi(s)) - q(t, s, \psi(s))\right| \le L_M(t - s)|\phi(s) - \psi(s)|$

and

(b) $\quad \int_{-\infty}^t L_M(t - s)\, ds < \infty.$

Then P is continuous.

The lemma is proved by writing $(P\phi)(t) - (P\psi)(t)$ and using (a), (b), and the local Lipschitz condition to form a simple estimate.

The assumptions of the lemma are readily fulfilled for the equation

$$x' = h(t, x) + \int_{-\infty}^t C(t - s)g(x(s))\, ds$$

with h and g locally Lipschitz and $C \in L^1[0, \infty)$, and the lemma is easily extended to include

$$x' = h(t, x) + \int_{-\infty}^t \sum_{i=1}^k C_i(t - s)g_i(x(s))\, ds$$

in which each g_i is locally Lipschitz and $C_i \in L^1[0, \infty)$.

Assumption 4.3.3. *Assume that*

(4.3.4)   *for each k, (4.3.2) has a T-periodic solution $y_k(t)$,*

and

(4.3.5)   *there is a number $B > 0$ with $|y_k(t)| \leq B$ for all k and all t.*

Assumption 4.3.3 is, of course, satisfied if the conditions of Theorem 4.2.2 hold.

Our next assumption is a special form of the fading memory property.

Assumption 4.3.4. *There exists $\{\epsilon_k\} \to 0$ as $k \to \infty$ such that if $\eta \in Z$ with $\|\eta\| \leq B$ and if m is defined by*

$$(4.3.6) \qquad m(u, k, \eta) = \int_{-\infty}^{u-kT} q(u, s, \eta(s)) \, ds \quad 0 \leq u \leq T,$$

then

$$(4.3.7) \qquad |m(u, k, \eta)| \leq \epsilon_k.$$

Assumption 4.3.5. *If $\eta \in Z$ then*

$$h(t, \eta(t)) + \int_{t-kT}^{t} q(t, s, \eta(s)) \, ds$$

is T-periodic.

The next result is found in Burton (1985c).

Theorem 4.3.6. *Let Assumptions 4.3.1–4.3.5 hold. Then 4.3.1 has a T-periodic solution.*

Proof. Consider the sequence $\{y_k\}$ in Assumption 4.3.3. We have $|y_k(t)| \leq B$ and, by Assumption 4.3.5, it is seen that $|y_k'(t)| \leq M$. Thus, the sequence is equicontinuous. There is then a subsequence, say $\{y_k\}$ again, converging uniformly on $(-\infty, \infty)$ to a T-periodic function ϕ. Write $P(y_k(t))$ as

$$P(y_k(t)) = y_k(0) + \int_0^t h(u, y_k(u)) \, du$$

$$+ \int_0^t \int_{u-kT}^{u} q(u, s, y_k(s)) \, ds \, du + \int_0^t \int_{-\infty}^{u-kT} q(u, s, y_k(s)) \, ds \, du$$

$$= y_k(t) + \int_0^t \int_{-\infty}^{u-kT} q(u, s, y_k(s)) \, ds \, du.$$

Now, P is uniformly continuous (as described in Assumption 4.3.2), $y_k(t)$ converges uniformly to ϕ, and $\left| \int_{-\infty}^{u-kT} q(u,s,y_k(s))\,ds \right| \leq \epsilon_k$. Hence, if we let $k \to \infty$ and have $0 \leq t \leq T$, then we obtain

$$\lim_{k \to \infty} P(y_k(t)) = \lim_{k \to \infty} y_k(t) + \lim_{k \to \infty} \int_0^t \int_{-\infty}^{u-kT} q(u,s,y_k(s))\,ds\,du\,.$$

And

$$\lim_{k \to \infty} P(y_k(t)) = P\left(\lim_{k \to \infty} y_k(t) \right) = P(\phi(t))\,,$$

while

$$\left| \int_0^t \int_{-\infty}^{u-kT} q(u,s,y_k(s))\,ds\,du \right| \leq T\epsilon_k$$

so that $P(\phi(t)) = \phi(t)$, or from (4.3.3) that

$$\phi(t) = \phi(0) + \int_0^t h(u,\phi(u))\,du + \int_0^t \int_{-\infty}^u q(u,s,\phi(s))\,ds\,du$$

and

$$\phi'(t) = h(t,\phi(t)) + \int_{-\infty}^t q(t,s,\phi(s))\,ds\,.$$

Thus, ϕ is a T-periodic solution of (4.3.1) and the proof is complete.

Furumochi (1982) shows the existence of periodic solutions of equations with infinite delay using a Razumikhin technique and differential inequalities. Other periodic results for infinite delay equations are found in Leitman and Mizel (1978) and Grimmer (1979). Hale (1977) has extensive periodic material for bounded delays.

4.3.3 Boundedness and Liapunov Functionals

We now consider a series of examples and techniques for showing uniform boundedness and uniform ultimate boundedness. Our first two results are found in Burton (1985d).

Consider the scalar equation

$$(4.3.8) \qquad x' = -x^3 + \int_{-\infty}^t C(t-s)v(s,x(s))\,ds + f(t)$$

where C, f, and v are continuous, $|C(t)| \leq \mu(1+t)^{-5}$, $v(s+T,x) = v(s,x)$, $|v(t,x)| \leq \alpha + \beta x^2$, $|f(t)| \leq M$ for some $M > 0$, and v is locally Lipschitz in x.

Proposition 4.3.1. *Solutions of* (4.3.8) *are g-uniform bounded for all positive constants* α, β, M, *and* μ.

Proof. Let $g(t) = 1 + |t|$ on $(-\infty, 0]$ so that if $|\phi(t)| \leq \gamma g(t)$, then

$$\left| \int_{-\infty}^{0} C(t-s)v(s, \phi(s))\, ds \right| \leq \int_{-\infty}^{0} \mu(1 + t - s)^{-5}\left[\alpha + \beta\gamma^2(|s| + 1)^2\right] ds$$

$$\leq \int_{-\infty}^{0} \mu(1-s)^{-5}\left[\alpha + \beta\gamma^2(|s| + 1)^2\right] ds$$

$$\leq K \int_{-\infty}^{0} (1-s)^{-5}(1 + s^2)\, ds$$

which is finite. We will therefore have existence and uniqueness of solutions.

Now for $|\phi(t)| \leq \gamma g(t)$ on $(-\infty, 0]$ we consider a solution $x(t) = x(t, 0, \phi)$ and define a Liapunov functional

$$V(t, x(\cdot)) = |x| + \beta \int_{-\infty}^{t} \int_{t}^{\infty} |C(u-s)|\, du\, x^2(s)\, ds$$

so that

$$V(0, \phi(\cdot)) \leq |\phi(0)| + \beta \int_{-\infty}^{0} \int_{0}^{\infty} \mu(1 + u - s)^{-5}\, du\, \gamma^2(1 + |s|)^2\, ds$$

$$\leq \gamma + \beta\mu\gamma^2 \int_{-\infty}^{0} \int_{0}^{\infty} (1 + u - s)^{-5}\, du\, (1 + |s|)^2\, ds$$

$$\leq \gamma + \beta\mu\gamma^2 \int_{-\infty}^{0} \frac{1}{4}(1-s)^{-4}(1 + |s|)^2\, ds$$

$$= \gamma + (\beta\mu\gamma^2/4) \int_{-\infty}^{0} (1-s)^{-2}\, ds$$

$$= \gamma + (\beta\mu\gamma^2/4) \stackrel{\text{def}}{=} V_0.$$

We compute the derivative of V along the solutions $x(t)$ and obtain

$$V'_{(4.3.8)}(t, x(\cdot)) \leq -|x^3| + \int_{-\infty}^{t} |C(t-s)|\, |v(s, x(s)|\, ds$$

$$+ \beta \int_{t}^{\infty} |C(u-t)|\, du\, x^2 - \beta \int_{-\infty}^{t} |C(t-s)|x^2(s)\, ds + |f(t)|$$

$$\leq -|x^3| + \beta \int_t^\infty |C(u-t)|\, du\, x^2 + \int_{-\infty}^t |C(t-s)|\, (\alpha + \beta x^2(s))\, ds$$

$$- \beta \int_{-\infty}^t |C(t-s)| x^2(s)\, ds + |f(t)|$$

$$\leq -|x^3| + \beta \int_0^\infty |C(v)|\, dv\, x^2 + \alpha \int_{-\infty}^t |C(t-s)|\, ds + |f(t)|$$

$$\leq -|x^3| + C_1 x^2 + C_2 \leq -\delta x^2 + K$$

for some positive numbers δ and K.

In summary, we have

(4.3.9) $$|x| \leq V(t, x(\cdot)) \leq |x| + \int_{-\infty}^t \Phi(t-s) x^2(s)\, ds$$

and

(4.3.10) $$V'(t, x(\cdot)) \leq -\delta x^2 + K$$

where $\Phi(t-s) = \beta \int_t^\infty |C(u-s)|\, du = \beta \int_{t-s}^\infty |C(v)|\, dv$. The seasoned investigator of stability theory of ordinary differential equations will readily believe that (4.3.9) and (4.3.10) imply very strong boundedness results; but implications of (4.3.9) and (4.3.10) are not yet fully understood and it turns out that conclusions from them still demand much work. We illustrate one approach.

Let $0 \leq s \leq t < \infty$ and write $V(t) = V(t, x(\cdot))$ so that

$$\left[V'(s) \leq -\delta x^2(s) + K\right] \Phi(t-s)$$

which yields

$$\int_0^t V'(s)\Phi(t-s)\, ds \leq -\delta \int_0^t x^2(s)\Phi(t-s)\, ds + K \int_0^t \Phi(t-s)\, ds\,.$$

The last term is bounded by a positive constant Q. If we integrate $V'\Phi$ by parts we may obtain

$$V(t)\Phi(0) - V(0)\Phi(t) + \int_0^t V(s)\Phi'(t-s)\, ds$$

$$\leq -\delta \int_0^t x^2(s)\Phi(t-s)\, ds + Q$$

or

$$\delta \int_0^t \Phi(t-s)x^2(s)\, ds \leq Q + V(0)\Phi(t) - V(t)\Phi(0)$$

$$- \int_0^t V(t)\Phi'(t-s)\, ds\,.$$

Recall that $\Phi' < 0$ and let t be selected so that $V(t) \geq \max_{0 \leq s \leq t} V(s)$. Then

$$\delta \int_0^t \Phi(t-s)x^2(s)\,ds \leq Q + V(0)\Phi(t) - V(t)\Phi(0)$$

$$- V(t) \int_0^t \Phi'(t-s)\,ds$$

$$= Q + V(0)\Phi(t) - V(t)\Phi(0)$$

$$+ V(t)\big[\Phi(0) - \Phi(t)\big]$$

$$\leq Q + V(0)\Phi(t)$$

$$\leq Q + V_0(\mu\beta/4) = Q + \big[\gamma + (\beta\mu\gamma^2/4)\big]\mu\beta/4$$

where Q is independent of γ. Thus, if t is any value for which $V(t)$ is the maximum on $[0, t]$, then $\delta \int_0^t \Phi(t-s)x^2(s)\,ds \leq [Q + V_0(\mu\beta/4)]$. But there is a $U > 0$ with $V' < 0$ if $|x| \geq U$; hence, either $t = 0$ (and the result is trivial) or the t occurs with $|x(t)| < U$ yielding

$$|x(t)| \leq V(t) \leq U + \int_{-\infty}^t \Phi(t-s)x^2(s)\,ds$$

$$\leq U + \int_{-\infty}^0 \Phi(t-s)\gamma^2 g^2(s)\,ds + \int_0^t \Phi(t-s)x^2(s)\,ds$$

$$\leq U + \int_{-\infty}^0 \Phi(t-s)\gamma^2 g^2(s)\,ds + \big[Q + V_0(\mu\beta/4)\big]/\delta$$

$$\overset{\text{def}}{=} B_1(\gamma)$$

so that solutions are g-uniform bounded.

The g-uniform ultimate boundedness will follow from Theorem 4.4.5. But when V is of a certain form, then the Laplace transform yields a nice variation of parameters inequality, as we had in Section 1.5.

Proposition 4.3.2. *Solutions of* (4.3.8) *are g-uniform ultimate bounded for all positive α, β, and M when $\int_{-\infty}^t |C(t-s)|\,ds \leq p < 2$.*

Proof. Define a new functional

$$W(t, x(\cdot)) = x^4 + K \int_{-\infty}^t \Phi(t-s)x^4(s)\,ds$$

where

$$\Phi(t-s) = \int_{t-s}^\infty |C(v)|\,dv\,.$$

If $|\phi(t)| \leq \gamma g(t) = \gamma[1 + |t|]^{1/8}$ on $(-\infty, 0]$, then let $x(t) = x(t, 0, \phi)$ and write $W(t) = V(t, x(\cdot))$ so that

$$W'_{(4.3.8)}(t) \leq -4x^6 + 4|x|^3 \int_{-\infty}^{t} |C(t-s)| \, |v(s, x(s))| \, ds$$

$$+ K\Phi(0)x^4 - K \int_{-\infty}^{t} |C(t-s)|x^4(s) \, ds + 4|x^3| \, |f(t)|$$

$$\leq -4x^6 + 2 \int_{-\infty}^{t} |C(t-s)| \big(x^6 + v^2(s, x(s))\big) \, ds$$

$$+ K\Phi(0)x^4 - K \int_{-\infty}^{t} |C(t-s)|x^4(s) \, ds + 4|x^3| \, |f(t)|$$

$$\leq -4x^6 + 2px^6 + 2 \int_{-\infty}^{t} |C(t-s)|(\alpha + \beta x^2(s))^2 \, ds$$

$$+ K\Phi(0)x^4 - K \int_{-\infty}^{t} |C(t-s)|x^4(s) \, ds + 4|x^3| \, |f(t)|$$

$$\leq (-4 + 2p)x^6 + L_1 \int_{-\infty}^{t} |C(t-s)| \, ds$$

$$+ L_2 \int_{-\infty}^{t} |C(t-s)|x^4(s) \, ds + K\Phi(0)x^4$$

$$- K \int_{-\infty}^{t} |C(t-s)|x^4(s) \, ds + 4|x^3| \, |f(t)|$$

$$\leq -\delta^4 + H$$

where δ and H are positive numbers, L_1 and L_2 are positive, and $K = L_2$.

In summary, we now have

(4.3.11) $$W(t, x(\cdot)) = x^4 + K \int_{-\infty}^{t} \Phi(t-s)x^4(s) \, ds$$

and

(4.3.12) $$W'_{(4.3.8)}(t, x(\cdot)) \leq -\delta x^4 + H \, .$$

Because of the very special form of this pair, (4.3.11) and (4.3.12), we will be able to obtain a "variation of parameters formula" for $x^4(t)$ which shows the required boundedness.

The integral equation

(4.3.13) $$y^4(t) + \int_{0}^{t} [\delta + K\Phi(t-s)]y^4(s) \, ds = 1$$

will serve as the "unforced" equation. It is trivial to verify that (4.3.13) has a solution on $[0, \infty)$ which is positive, $y^4(t) \in L^1[0, \infty)$, and $y^4(t) \to 0$ as $t \to \infty$.

Define a positive continuous function $\eta(t)$ from (4.3.12) by

$$(4.3.12)' \qquad\qquad W'(t) = -\eta(t) - \delta x^4(t) + H.$$

We will take Laplace transforms of (4.3.11), (4.3.12)', and (4.3.13) letting L denote the transform and $*$ denote the convolution integral from 0 to t. In (4.3.13) we have

$$L(y^4) + (\delta/s)L(y^4) + KL(\Phi)L(y^4) = 1/s$$

so that

$$L(y^4) = 1/\{1 + KL(\Phi) + (\delta/s)\}s\,.$$

The transform of (4.3.12)' is

$$sL(W) = W(0) - L(\eta) - \delta L(x^4) + (H/s)$$

or

$$L(W) = \big[W(0) - L(\eta) - \delta L(x^4) + (H/s)\big]/s\,.$$

And the transform of (4.3.11) is

$$L(W) = L(x^4) + KL(\Phi)L(x^4) + kL\left(\int_{-\infty}^0 \Phi(t-s)\phi^4(s)\right)ds$$

so that

$$L(x^4) + KL(\Phi)L(x^4) + KL\left(\int_{-\infty}^0 \Phi(t-s)\phi^4(s)\,ds\right)$$
$$= \big[W(0) - L(\eta) - \delta L(x^4) + (H/s)\big]/s\,.$$

If we solve for $L(x^4)$ we have

$$L(x^4) = \left\{ \big[W(0) - L(\eta) + (H/s)\big]/s \right.$$

$$\left. - KL\left(\int_{-\infty}^0 \Phi(t-s)\phi^4(s)\,ds\right)\right\}\bigg/\big[1 + KL(\Phi) + (\delta/s)\big]$$
$$= W(0)L(y^4) + L(H)L(y^4) - L(\eta)L(y^4)$$
$$- KsL\left(\int_{-\infty}^0 \Phi(t-s)\phi^4(s)\,ds\right)L(y^4)$$
$$= W(0)L(y^4) + L(H)L(y^4) - L(\eta)L(y^4)$$
$$- K\left[L\left(\int_{-\infty}^0 \Phi'(t-s)\phi^4(s)\,ds\right) + \int_{-\infty}^0 \Phi(-s)\phi^4(s)\,ds\right]L(y^4)\,.$$

Now $\Phi(t-s) = \int_{t-s}^{\infty} |C(v)|\, dv$ so $|\phi'(t-s)| = +|C(t-s)| \leq \mu(1+t-s)^{-5}$ and $|\phi(t)| \leq \gamma g(t) \leq \gamma[1+|t|]^{1/8}$ so that

$$\int_{-\infty}^{0} |\Phi'(t-s)\phi^4(s)|\, ds \leq \mu\gamma^4 \int_{-\infty}^{0} (1+t-s)^{-5}(1-s)^{1/2}\, ds$$

$$\leq \mu\gamma^4 \int_{-\infty}^{0} (1+t-s)^{-4}\, ds = \mu\gamma^4 (1+t-s)^{-3}/3 \Big|_{-\infty}^{0}$$

$$= \mu\gamma^4 (1+t)^{-3}/3 \to 0$$

as $t \to \infty$. Hence

$$L\left(\int_{-\infty}^{0} \Phi'(t-s)\phi^4(s)\, ds \right) L(y^4) = L(\lambda(t))\mu\gamma^4$$

where $\lambda(t)$ is the convolution of an L^1 function $(y^4(t))$ with a function tending to zero $\left(\int_{-\infty}^{0} \Phi'(t-s)\phi^4(s)\, ds \right)$, and so $\lambda(t) \to 0$ as $t \to \infty$. Also, $\int_{-\infty}^{0} \Phi(-s)\phi^4(s)\, ds = P_1$ where P_1 is a constant and $|P_1| \leq \mu\gamma^4 P$, for P a fixed constant independent of γ. Thus,

$$x^4 = \left[W(0) - KP_1 \right]y^4 - \eta * y^4 - K\mu\gamma^4\lambda(t) + H \int_{0}^{t} y^4(s)\, ds$$

and as $\eta \geq 0$ we have

$$x^4(t) \leq \left[W(0) + K\mu\gamma^4 P \right]y^4(t) + K\mu\gamma^4\lambda(t) + H \int_{0}^{t} y^4(s)\, ds.$$

As λ and y^4 are fixed functions independent of γ, definitions of the required boundedness are satisfied. The proof is complete.

Exercise 4.3.1. Suppose there is a functional V for (4.3.1) satisfying

$$W_1(|x|) \leq V(t, x(\cdot))$$

$$\leq W_2(|x|) + W_3\left(\int_{-\infty}^{t} \Phi(t-s)W_4(|x(s)|)\, ds \right)$$

and

$$V'_{(4.3.1)}(t, x(\cdot)) \leq -W_4(|x|) + K$$

for $\Phi \geq 0$, Φ continuous, $\Phi \in L^1[0, \infty)$, and K is positive. Follow the proof of Proposition 4.3.1 to show that solutions of (4.3.1) are g-uniform bounded for some g.

Exercise 4.3.2. Suppose there is a functional V for (4.3.1) satisfying

$$V(t, x(\cdot)) = W(|x|) + \int_{-\infty}^{t} \Phi(t - s)W(|x(s)|)\, ds$$

and

$$V'_{(4.3.1)}(t, x(\cdot)) \leq -W(|x|) + K$$

for K and Φ as in Exercise 4.3.1. Follow the proof of Proposition 4.3.2 to show that solutions of (4.3.1) are g-uniform ultimate bounded. Is a condition like $\int_{-\infty}^{t} |\Phi(t - s)|\, ds \leq p \leq 2$ needed in either this exercise or in Proposition 4.3.2?

Exercise 4.3.3. In some of these simple examples one may dispense with g-uniform ultimate boundedness and with g itself. Consider (4.3.8) again and look at

$$M = \{\phi : (-\infty, 0] \to R^n \mid \text{ is continuous and }$$

$$|\phi(s)| \leq H \text{ for } -\infty < s \leq 0\}$$

with H a positive constant. Let $V(x) = |x|$ and note that for large H if we consider $x(t) = x(t, 0, \phi)$ for $\phi \in M$ then

$$V'_{(4.3.8)}(x) < 0$$

for any t for which $|x(t)| = H$. Put an appropriate Lipschitz condition on M, choose an appropriate space, and choose a fixed-point theorem that will yield a fixed point of the map $P : M \to M$ defined by

$$P\phi = x(t + T, 0, \phi) \quad \text{for} \quad -\infty < t \leq 0.$$

State the continuity condition for P.

The next three propositions are found in Arino *et al.* (1985).
Consider the scalar equation

$$(4.3.13) \qquad x' = -x^3 + \int_{-\infty}^{t} q(t, s, x(s))\, ds + \cos t$$

where q is continuous, locally Lipschitz in x,

$$|q(t, s, x)| \leq C(t - s)x^3,$$

and where

(a) $C : [0, \infty) \to [0, \infty)$ is continuous,

(b) $\int_0^\infty C(t)\, dt = M < 1$,

and

(c) $\int_t^\infty C(u)\, du \le KC(t)$ for some $K > 0$ and all $t > 0$.

Proposition 4.3.3. *If* (a), (b), *and* (c) *hold then solutions of* (4.3.13) *are g-uniform bounded and g-uniform ultimate bounded for bound B at t = 0. And* (4.3.13) *has a 2π-periodic solution.*

Proof. A suitable function g can be found for which

$$\int_0^\infty C(t)g^3(-t)\, dt = L < \infty.$$

Define

$$V(t, x(\cdot)) = |x| + J \int_{-\infty}^t \int_{t-s}^\infty C(u)\, du\, |x^3(s)|\, ds$$

where $J > 1$ and $JM = \alpha < 1$. We then have

$$V'_{(4.3.13)}(t, x(\cdot)) \le [-1 + \alpha]\, |x|^3$$

$$- \beta \int_{-\infty}^t \int_{t-s}^\infty C(u)\, du\, |x^3(s)|\, ds + 1$$

where $\beta = (J - 1)/K$. Thus,

$$V'(t, x(\cdot)) \le [-1 + \alpha]\, |x|$$

$$- (\beta/J)J \int_{-\infty}^t \int_{t-s}^\infty C(u)\, du\, |x^3(s)|\, ds + 2$$

and, for $\gamma = \min\{1 - \alpha, \beta/J\}$, then

$$V'(t, x(\cdot)) \le -\gamma V(t, x(\cdot)) + 2.$$

An integration yields

$$|x(t, 0, \phi)| \le V(0, \phi(\cdot))e^{-\gamma t} + \int_0^t 2e^{-\gamma(t-s)}\, ds$$

from which the required boundedness properties follow and, hence, the existence of a periodic solution.

Consider the scalar equation

$$(4.3.14) \qquad x' = -h(x) + \int_{-\infty}^t C(t - s)q(x(s))\, ds + f(t)$$

where

(a) $h : (-\infty, \infty) \to (-\infty, \infty)$, $q : (-\infty, \infty) \to (-\infty, \infty)$, $f : (-\infty, \infty) \to (-\infty, \infty)$, and $C : [0, \infty) \to (-\infty, \infty)$ are continuous with h and q possessing continuous derivatives;

(b) $h'(x) > 0$ for all $x \neq 0$, $h(0) = 0$ and $|q'(x)| \leq \inf_{x \in R} h'(x)$ for all x;

(c) $\int_0^\infty |C(t)| \, dt < 1$;

(d) $\int_0^\infty t |C(t)| \, dt < \infty$;

(e) $f(t + T) = f(t)$ for all t and some $T > 0$.

Proposition 4.3.4. *If (a)–(e) hold, then (4.3.14) has at most one T-periodic solution.*

Proof. Let $x(t)$ and $p(t)$ be T-periodic solutions of (4.3.14) and set $y(t) = x(t) - p(t)$. Now, $y(t)$ is T-periodic,

$$h(x(t)) - h(p(t)) = h'(\xi(t))(x(t) - p(t))$$

and

$$q(x(t)) - q(p(t)) = q'(\eta(t))(x(t) - p(t))$$

for certain functions ξ and η. Define $z(t) = h'(\xi(t))$ and $r(t) = q'(\eta(t))$. Then there is a constant $K > 0$ such that

$$0 \leq |r(t)| \leq z(t) \leq K \quad \text{for all} \quad t \in (-\infty, \infty),$$

and y satisfies the equation

$$(4.3.14) \qquad y' = -z(t)y + \int_{-\infty}^t C(t - s)r(s)y(s) \, ds.$$

By assumption, (4.3.14) has a (nontrivial) T-periodic solution, and that is one we now discuss. Define a functional

$$V(t, y(\cdot)) = |y| + \int_{-\infty}^t \int_{t-s}^\infty |C(u)| \, du \, z(s)|y(s)| \, ds$$

and obtain

$$V'_{(4.3.14)}(t, y(\cdot)) \leq - \left(1 - \int_0^\infty |C(u)| \, du \right) z(t)|y|.$$

An integration shows that the only periodic function y having this property is the zero function. This completes the proof.

We now consider the scalar equation

$$(4.3.15) \qquad x' = \int_{-\infty}^{t} C(t-s)x(s)\,ds + f(t,x)$$

where

(a) $C : [0,\infty) \to (-\infty,0)$ is continuous and $f : (-\infty,\infty) \times (-\infty,\infty) \to (-\infty,\infty)$ is continuous and locally Lipschitz in x;

(b) $\int_0^\infty |C(t)|\,dt < \infty$;

(c) $|f(t,x)| \le M$ for all (t,x) and some $M > 0$;

(d) $\int_0^t \int_0^{t_0} C(u+v)\,du\,dv \le N$ for all $t, t_0 \ge 0$ and some $N \ge 0$;

(e) there is a function G with $G'(t) = C(t)$ and $G(0) > 0$; and

(f) $\int_0^\infty |G(u)|\,du < 1$.

Proposition 4.3.5. *Let* (a)–(f) *hold. Then there is a function g such that solutions of (4.3.15) are g-uniform and g-uniform ultimate bounded for bound B.*

Proof. Let us first consider the reduced equation

$$(4.3.16) \qquad x' = \int_0^t C(t-s)x(s)\,ds$$

and show that the solution $z(t)$ satisfying $z(0) = 1$ has the properties: $z(t) \to 0$ as $t \to \infty$ and $\int_0^\infty |z(t)|\,dt < \infty$. To this end, we write (4.3.16) as

$$(4.3.17) \qquad x' = -G(0)x + (d/dt)\int_0^t G(t-s)x(s)\,ds.$$

Define a functional

$$V(t,x(\cdot)) = \left[x - \int_0^t G(t-s)x(s)\,ds \right]^2$$
$$+ G(0)\int_0^t \int_t^\infty |G(u-s)|\,du\,x^2(s)\,ds.$$

A calculation will show that

$$V'_{(4.3.17)}(t,x(\cdot)) \le -\beta x^2$$

for some $\beta > 0$. Next, define a new functional

$$U(t, x(\cdot)) = x^2 + \int_0^t \int_t^\infty |C(u - s)|\, du\, x^2(s)\, ds$$

and obtain

$$U'_{(4.3.16)}(t, x(\cdot)) \leq Jx^2$$

for some $J > 0$. [See Burton (1983a, pp. 135–136) for details of these computations.] But solutions of (4.3.16) and (4.3.17) are the same so the functional

$$W(t, x(\cdot)) = (\beta/2J)U(t, x(\cdot)) + V(t, x(\cdot))$$

satisfies

$$W'_{(4.3.17)}(t, x(\cdot)) \leq -\beta x^2/2\,.$$

Thus, all solutions of (4.3.17) are bounded, $\int_0^\infty x^2(t)\, dt < \infty$, and since $\int_0^\infty |C(t)|\, dt < \infty$, we have $x'(t)$ bounded. Hence $x(t) \to 0$ as $t \to \infty$. It follows from a result in Burton (1983a, p. 56) that $z(t) \in L^1[0, \infty)$.

We now write (4.3.15) as

$$x' = \int_0^t C(t - s)x(s)\, ds + f(t, x) + \int_{-\infty}^0 C(t - s)\phi(s)\, ds\,.$$

Since $C \in L^1[0, \infty)$, a function g of the type required does exist and, when $|\phi(t)| \leq \gamma g(t)$, we apply the variation of parameters formula and obtain

(4.3.18)
$$x(t) = z(t)\phi(0) + \int_0^t z(t - u)\left[f(u, x(u))\right.$$
$$\left. + \int_{-\infty}^0 C(u - s)\phi(s)\, ds\right] du\,.$$

Now $|f(u, x(u))| \leq M$ and

$$\int_{-\infty}^0 |C(t - s)|\, |\phi(s)|\, ds \leq \gamma \int_{-\infty}^0 |C(t - s)|g(s)\, ds$$

$$= \gamma \int_t^\infty |C(u)|g(t - u)\, du \leq \gamma \int_t^\infty |C(u)|g(t - u)\, du$$

$$\to 0 \quad \text{as} \quad t \to \infty\,.$$

Hence, the last integral in (4.3.18) tends to zero as $t \to \infty$ and we have g-uniform and g-uniform ultimate boundedness, as required.

Consider the scalar second-order equation

(4.3.19) $x'' + f(x)x' + \int_{-\infty}^{t} h(t-s)v(x(s))\, ds = p(t)$

in which

(a) $h : [0, \infty) \to (-\infty, \infty)$ is continuous and $L^1[0, \infty)$;

(b) there is an $H : [0, \infty) \to [0, \infty)$ with $H'(t) = -h(t)$, and $\int_0^\infty H(u)\, du < \infty$;

(c) $v : (-\infty, \infty) \to [-L, L]$ for some $L > 0$, $xv(x) > 0$ if $x \neq 0$, and v is continuous;

(d) $p : (-\infty, \infty) \to [-J, J]$ for some $J > 0$ and p is continuous.

Equation (4.3.19) can be expressed as the system

$$x' = y\,,$$

(4.3.20)
$$y' = -f(x)y - H(0)v(x)$$
$$+ (d/dt) \int_{-\infty}^{t} H(t-s)v(x(s))\, ds + p(t)\,.$$

If we define the functional

$$W(t, x(\cdot), y(\cdot)) = \left[y - \int_{-\infty}^{t} H(t-s)v(x(s))\, ds \right]^2$$

$$+ 4H(0)V(x) + \left[y + F(x) - \int_{-\infty}^{t} H(t-s)v(x(s))\, ds \right]^2$$

with $V(x) = \int_0^x v(s)\, ds$ and $F(x) = \int_0^x f(s)\, ds$, then we obtain

$$W'_{(4.3.20)}(t, x(\cdot), y(\cdot)) \le -2\big[f(x)y^2 + H(0)v(x)F(x) \big]$$
$$+ 4yp(t) + 2F(x)p(t)$$

$$+ \big[4H(0)v(x) - 4p(t) + 2f(x)y \big] \int_{-\infty}^{t} H(t-s)v(x(s))\, ds\,.$$

Several interesting results may be obtained from this relation. Note that $|v(x)| \le L$ and $\int_0^\infty H(s)\, ds < \infty$ imply that any continuous initial function can be used. Under proper growth conditions on f and v, we can bound W by

$$W_1(|(x, y)|) \le W(t, x(\cdot), y(\cdot)) \le W_2(|(x, y)|) + K\,,$$

$K > 0$, independent of the initial function ϕ. Thus, if we can make W' negative for large $x^2 + y^2$, then the same arguments as those used for ordinary differential equations (see Theorem 4.1.16) will yield uniform boundedness and uniform ultimate boundedness.

To be definite we give conditions for such boundedness.   Let $\int_0^\infty H(s)\, ds = \overline{H} < \infty$, $|p(t)| \leq 1$, $v(x) = x$ for $|x| \leq 1$, $|v(x)| = 1$ for $|x| \geq 1$, $f(x) = c > 0$, and $H(0) > c$.

Proposition 4.3.6. *Let the preceding conditions hold for* (4.3.20). *Then for $x^2 + y^2$ large we have*

$$W'(t, x(\cdot), y(\cdot)) \leq -\alpha < 0$$

and for any g solutions are g-uniform bounded and g-uniform ultimate bounded.

The proof is left as an exercise.

4.4   Stability of Systems with Unbounded Delays

We now consider a system of Volterra functional differential equations

(4.4.1) $$x' = F(t, x_t)$$

where x_t represents a function from $[\alpha, t] \to R^n$, $-\infty \leq \alpha \leq t_0$, and where F is defined for $t \geq t_0$. The existence theory for this was developed in Chapter 3 with (3.5.6)

When $\alpha = -\infty$ we shall mean that $x : (\alpha, t] \to R^n$. In fact, when $\alpha = -\infty$ and we write that x is a continuous function from $(\alpha, t] \to R^n$, then we mean that $x(t)$ is bounded as $t \to -\infty$ so that for a continuous x, then $\sup_{s \in (\alpha, t]} |x(s)|$ always exists as a finite number.

For any $t \geq t_0$, by

$$\bigl(X(t), \| \cdot \|\bigr)$$

we shall mean the space of continuous bounded functions $\phi : [\alpha, t] \to R^n$ with

$$\|\phi\| = \sup_{\alpha \leq s \leq t} |\phi(s)|$$

and $|\cdot|$ is any norm on R^n. The symbol

$$x_H(t)$$

denotes those $\phi \in X(t)$ with $\|\phi\| \leq H$. An element $\phi \in X(t)$ will often be denoted by ϕ_t so that this notation will correspond to the notation introduced in Section 4.2.1 for the system (4.2.1) with bounded delay. In particular, $\|\phi_t\|$ indicates that $\phi \in X(t)$ and that the supremum is taken over the interval $[\alpha, t]$. We emphasize that ϕ_t is *not* shifted to the interval $[\alpha, 0]$.

It is supposed that $F(t, x_t)$ is a continuous function of t for $t_0 \leq t < \infty$ whenever $x_t \in X_H(t)$ for $t_0 \leq t < \infty$. In addition, it is assumed that whenever $\psi \in X_H(t)$ and $\{\psi^{(n)}\}$ is a sequence in $X_H(t)$ with $\|\psi^{(n)} - \psi\| \to 0$ as $n \to \infty$, then $F(t, \psi^{(n)}) \to F(t, \psi)$ as $n \to \infty$. Here $\{\psi^{(n)}\}$ denotes the sequence of functions ordinarily denoted by $\{\psi_n\}$ which would be confusing since ψ_t has a different meaning. It is supposed that F takes closed bounded sets of $R \times X(t)$ into bounded sets of R^n.

To specify a solutions of (4.4.1) we require a $t_1 \geq t_0$ and $\phi \in X(t_1)$; we then obtain a continuous solution $x(t, t_1, \phi)$ satisfying (4.4.1) on some interval $[t_1, t_1 + \beta)$, $\beta > 0$, and $x(t, t_1, \phi) = \phi(t)$ for $\alpha \leq t \leq t_1$. If there is an $\widetilde{H} < H$ such that $|x(t, t_1, \phi)| \leq \widetilde{H}$ so long as $x(t, t_1, \phi)$ is defined, then $\beta = \infty$.

Stability definitions correspond almost exactly to those for finite delay equations. One could, of course, extend definitions using a weighted norm and obtain statements about g-stability as we did for g-uniform boundedness. Such a task would be simple, but so far there seems to be no reason for doing so. Also, for a general F, unbounded initial functions call for considerable extra care.

Definition 4.4.1. *Let $F(t, 0) = 0$. The zero solution of (4.4.1) is:*

(a) *stable if for each $t_1 \geq t_0$ and $\epsilon > 0$ there exists $\delta > 0$ such that $[\phi \in X_\delta(t_1), t \geq t_1]$ imply that $|x(t, t_1, \phi)| < \epsilon$;*

(b) uniformly stable *if it is stable and δ is independent of t_1;*

(c) asymptotically stable *if it is stable and if for each $t_1 \geq t_0$ there is an $\eta > 0$ such that $\phi \in X_\eta(t_1)$ implies that $x(t, t_1, \phi) \to 0$ as $t \to \infty$;*

(d) uniformly asymptotically stable *if it is uniformly stable and if there is an $\eta > 0$ such that for each $\mu > 0$ there is a $T > 0$ such that $[t_1 \geq t_0, \ \phi \in X_\eta(t_1), \ t \geq t_1 + T]$ imply that $|x(t, t_1, \phi)| < \mu$.*

A functional $V(t, x_t)$ is locally Lipschitz with respect to x if, for every $\gamma \in [t_0, \infty)$ and every $L > 0$, there is a constant $K(\gamma, L)$ such that $\psi, \phi \in X_L(t)$ and $t_0 \leq t \leq \gamma$ imply that

$$|V(t, \phi) - V(t, \psi)| \leq K\|\phi - \psi\|.$$

Theorem 4.4.1. *Let $V(t, x_t)$ be a continuous scalar function which is locally Lipschitz in x_t for $x_t \in X_H(t)$ for some $H > 0$ and all $t \geq t_0$.*

(a) *If*

$$W(|x(t)|) \leq V(t, x_t)$$

$$V(t, 0) = 0, \quad V'_{(4.4.1)}(t, x_t) \leq 0$$

for $t \geq t_0$, then $x = 0$ is stable.

(b) *If, in addition to* (a)

$$V(t, x_t) \leq W_1(\|x_t\|)$$

then $x = 0$ is uniformly stable.

(c) *Let* (a) *hold and suppose there is an $M > 0$ such that $[x_t \in X_H(t), \ t_0 \leq t < \infty]$ imply that $|F(t, x_t)| \leq M$ and*

$$V'_{(4.4.1)}(t, x_t) \leq -W_3(|x(t)|).$$

Then $x = 0$ is asymptotically stable.

The proofs are identical to those for Theorem 4.2.5 and will not be given here.

Remark 4.4.1. The reader is reminded that Theorem 4.2.8 also holds for equations with infinite delay and so properly belongs here.

Theorem 4.2.7 suggests that the upper wedge on V needs an L^2-norm for uniform asymptotic stability for finite delay. This, and examples, suggest that for infinite delay we need a weighted norm. The next two results from Burton and Zhang (1985) illustrate one type of relation; others are found in Burton *et al.* (1985a).

Theorem 4.4.2. *Let V be continuous and locally Lipschitz for $x_t \in X_H(t)$. Suppose there is a continuous function $\Phi : [0, \infty) \to [0, \infty)$ which is $L^1[0, \infty)$ and satisfies $\Phi(t) \to 0$ as $t \to \infty$ with*

(a) $W_1(|x|) \leq V(t, x_t) \leq W_2(|x|) + W_3\left(\int_\alpha^t \Phi(t - s) W_4(|x(s)|) \, ds \right)$

and

(b) $V'_{(4.4.1)}(t, x_t) \leq -W_4(|x|).$

Then the zero solution of (4.4.1) *is uniformly stable and asymptotically stable.*

Proof. Let $\epsilon > 0$ be given. If $t_1 \geq t_0$ and $\phi \in X_\delta(t_1)$ with $\delta < \epsilon$, then for $x(t, t_1, \phi) = x(t)$ we have

$$W_1(|x(t)|) \leq V(t, x_t) \leq V(t_1, \phi)$$

$$\leq W_2(\delta) + W_3\left(W_4(\delta) \int_\alpha^{t_1} \Phi(t_1 - s)\, ds\right)$$

$$= W_2(\delta) + W_3\left(W_4(\delta) \int_0^{t_1 - \alpha} \Phi(u)\, du\right)$$

$$\leq W_2(\delta) + W_3\left(W_4(\delta) \int_0^\infty \Phi(u)\, du\right) < W_1(\epsilon)$$

if δ is small enough. This is uniform stability.

Now, for some $\epsilon > 0$ we find δ of uniform stability and take $\eta = \delta$. We see that for $x(t) = x(t, t_1, \phi)$ with $t_1 \geq t_0$ and $\phi \in x_\eta(t_1)$, then $V'(t, x_t) \leq -W_4(|x|)$ implies that $W_4(|x(t)|) \in L^1[0, \infty)$. Since $\Phi(t) \to 0$ as $t \to \infty$, $\int_0^t \Phi(t - s)W_4(|x(s)|)\, ds \to 0$ as $t \to \infty$. Thus, we have the relations

$$W_1(|x|) \leq V(t, x_t) \leq W_2(|x|) + W_3\left(\int_\alpha^{t_1} \Phi(t - s)W_4(\delta)\, ds\right.$$

$$\left. + \int_{t_1}^t \Phi(t - s)W_4(|x(s)|)\, ds\right)$$

$$\overset{\text{def}}{=} W_2(|x|) + \lambda(t)$$

and

$$V'(t, x_t) \leq -W_4(|x|)$$

where $\lambda(t)$ is positive and tends to zero as $t \to \infty$. Because $V \geq 0$ and $V' \leq -W_4(|x|)$, there is a sequence $\{t_n\} \to \infty$ with $|x(t_n)| \to 0$. Hence, for $t > t_n$ we have

$$W_1(|x(t)|) \leq V(t, x_t) \leq V(t_n, x_{t_n})$$

$$\leq W_2(|x(t_n)|) + \lambda(t_n) \to 0 \quad \text{as} \quad t \to \infty$$

and so $|x(t)| \to 0$ as $t \to \infty$. The proof is complete.

This last result is included because it is simple and extends readily to a uniform boundedness theorem for perturbed equations. With hard work one may both weaken the assumptions and extend the result to uniform asymptotic stability.

Lemma 4.4.1. *Let $\{x_n\}$ be a sequence of continuous functions with continuous derivatives, $x_n : [0, 1] \to [0, 1]$. Let $g : [0, \infty) \to [0, \infty)$ be continuous,*

$g(0) = 0$, $g(r) > 0$ if $r > 0$, and let g be nondecreasing. If there exists $\alpha > 0$ with $\int_0^1 x_n(t)\, dt \geq \alpha$ for all n, then there exists $\beta > 0$ with $\int_0^1 g(x_n(t))\, dt \geq \beta$ for all n.

Proof. Let $M_k = \{t \mid x_k(t) \geq \alpha/2\}$, M_k^c be the complement of M_k on $[0, 1]$, and let m_k be the measure of M_k. Then $m_k \geq \alpha/2$. To see this, if $m_k < \alpha/2$, then

$$\alpha \leq \int_0^1 x_k(t)\, dt = \int_{M_k} x_k(t)\, dt + \int_{M_k^c} x_k(t)\, dt < (\alpha/2) + (\alpha/2)\,.$$

a contradiction. (The first $\alpha/2$ follows from $m_k < \alpha/2$ and $x_k(t) \leq 1$. The second $\alpha/2$ follows from measure $M_k^c \leq 1$ and $x_k(t) \leq \alpha/2$.) Hence, for any k, we have

$$\int_0^1 g(x_k(t))\, dt \geq \int_{M_k} g(x_k(t))\, dt \geq \int_{M_k} g(\alpha/2)\, dt$$

$$= m_k g(\alpha/2) \geq g(\alpha/2)\alpha/2 \stackrel{\text{def}}{=} \beta\,.$$

This completes the proof.

Theorem 4.4.3. *Let V be continuous and locally Lipschitz for $t_0 \leq t < \infty$ and $x_t \in X_H(t)$. Suppose there is a continuous function $\Phi : [0, \infty) \to [0, \infty)$ which is $L^1[0, \infty)$ and bounded. If*

$$W_1(|x|) \leq V(t, x_t) \leq W_2(|x|) + W_3\left(\int_\alpha^t \Phi(t - s)W_4(|x(s)|)\, ds\right)$$

and

$$V'_{(4.4.1)}(t, x_t) \leq -W_5(|x|)\,,$$

then $x = 0$ is uniformly asymptotically stable.

Proof. The proof of Theorem 4.4.2 shows the uniform stability. For $\epsilon = H$ find δ of uniform stability and let $\delta = \eta$.

Let $\gamma > 0$ be given. We must find $T > 0$ such that $[t_1 \geq t_0,\ \phi \in X_\eta(t_1),\ t \geq t_1 + T]$ imply that $|x(t, t_1, \phi)| < \gamma$.

For this $\gamma > 0$, find $\theta > 0$ such that

$$W_2[W_4^{-1}(\theta/J)] + W_3(2\theta) < W_1(\gamma)\,,$$

where we let $\Phi(t) \leq J$ for $t \geq 0$. Now, find $r > 1$ with

$$W_4(\epsilon) \int_r^\infty \Phi(u)\, du < \theta\,.$$

If $\phi \in X_\eta(t_1)$, $t_1 \geq t_0$, $t \geq t_1 + r$, then for $x(t) = x(t, t_1, \phi)$ we have

$$W_1(|x(t)|) \leq V(t, x_t)$$

$$\leq W_2(|x(t)|) + W_3\left(\int_\alpha^t \Phi(t-s)W_4(|x(s)|)\,ds\right)$$

$$\leq W_2(|x(t)|) + W_3\left(\int_\alpha^{t-r} \Phi(t-s)W_4(\epsilon)\,ds\right.$$

$$\left. + \int_{t-r}^t \Phi(t-s)W_4(|x(s)|)\,ds\right)$$

$$\leq W_2(|x(t)|) + W_3\left(W_4(\epsilon)\int_r^{t-\alpha}\Phi(u)\,du\right.$$

$$\left. + \int_{t-r}^t JW_4(|x(s)|)\,ds\right)$$

$$\leq W_2(|x(t)|) + W_3\left(W_4(\epsilon)\int_r^\infty \Phi(u)\,du\right.$$

$$\left. + J\int_{t-r}^t W_4(|x(s)|)\,ds\right)$$

$$\leq W_2(|x(t)|) + W_3\left(\theta + J\int_{t-r}^t W_4(|x(s)|)\,ds\right).$$

We will find a t_2 with $|x(t_2)| < W_4^{-1}(\theta/Jr) \overset{\text{def}}{=} \epsilon_1$ and $\int_{t_2-r}^{t_2} W_4(|x(s)|)\,ds < \theta/J$. This will mean that

$$W_1(|x(t)|) \leq V(t, x_t) \leq V(t_2, x_{t_2})$$

$$< W_2(W_4^{-1}(\theta/Jr)) + W_3(2\theta) < W_1(\gamma)$$

for $t \geq t_2$.

Because $V'(t, x_t) \leq -W_5(|x|)$, there is a $T_2 \geq r$ such that $|x(t)| \geq W_4^{-1}(\theta/Jr) = \epsilon_1$ fails for some value of t on every interval of length T_2. Hence, there exists $\{t_n\} \to \infty$ such that $|x(t_n, t_1, \theta)| < \epsilon_1$. In particular, we choose

$$t_n \in [t_1 + (n-1)T_2,\ t_1 + nT_2] \quad \text{for} \quad n = 3, 4, \ldots.$$

The length of these intervals is independent of t_1 and $\phi \in X_H(t_1)$.

Now, consider the sequence of functions $\{x_k(t)\}$ defined by

$$x_k(t) = x(t, t_1, \phi) \quad \text{for} \quad t_k - r \leq t \leq t_k.$$

Examine those members satisfying

$$(*) \qquad \int_{t_k-r}^{t_k} JW_4(|x(s)|)\,ds \geq \theta$$

so that by Lemma 4.4.1 we have

$$\int_{t_k-r}^{t_k} W_5(|x(s)|)\,ds \geq \beta$$

for some $\beta > 0$.

Next,

$$V(t_1, \phi) \leq W_2(\epsilon) + W_3 \left(\int_\alpha^{t_1} \Phi(t_1 - s) W_4(\delta)\,ds \right)$$

$$\leq W_2(\epsilon) + W_3 \left(W_4(\delta) \int_0^\infty \Phi(u)\,du \right)$$

$$\overset{\text{def}}{=} \mu$$

is a positive number independent of $t_1 \geq t_0$ and independent of $\phi \in X_\eta(t_1)$. Now, for $t > T_1 + r$ and $t > t_{2n}$ we have

$$V'(t, x_t) \leq -W_5(|x|)$$

so that

$$V(t, x_t) \leq V(t_1, \phi) - \int_{t_1}^t W_5(|x(s)|)\,ds$$

$$\leq \mu - \sum_{i=1}^n \int_{t_{2i}-r}^{t_{2i}} W_5(|x(s)|)\,ds \leq \mu - n\beta < 0$$

if $n > \mu/\beta$. (Here we have integrated over alternate intervals to be sure the intervals are disjoint.) Hence, if $n > \mu/\beta$, then t_n fails to exist with $(*)$ holding. We choose n as the smallest integer greater than μ/β and we then have

$$|x(t_n, t_1, \phi)| < \epsilon_1$$

and

$$\int_{t_n-r}^{t_n} JW_4(|x(s)|)\,ds < \theta.$$

Thus, if

$$T = T_1 + r + 2nT_2$$

then $t > t_1 + T$ implies $|x(t, t_1, \phi)| < \gamma$. This completes the proof.

Example 4.4.1. Consider the scalar equation

$$x' = -x^3 + \int_{-\infty}^{t} C(t-s)v(x(s))\,ds$$

with

$$\int_{0}^{\infty} |C(u)|\,du < 1\,,\ \int_{t}^{\infty} |C(u)|\,du \in L^1[0, \infty)\,,\ |v(x)| \le \alpha|x|^3 \text{ for } 0 \le \alpha \le 1\,.$$

Define

$$V(t, x_t) = |x| + \int_{-\infty}^{t} \int_{t-s}^{\infty} |C(u)|\,du\,|x(s)|^3\,ds$$

so that

$$V'(t, x_t) \le -|x|^3 + \int_{-\infty}^{t} |C(t-s)|\,|v(x(s))|\,ds$$

$$+ \int_{0}^{\infty} |C(u)|\,du\,|x|^3 - \int_{-\infty}^{t} |C(t-s)|\,|x(s)|^3\,ds$$

$$\le -\gamma|x|^3$$

for some $\gamma > 0$. The conditions of Theorem 4.4.3 are satisfied and the zero solution is uniformly asymptotically stable.

Example 4.4.2. Consider the scalar equation

$$x' = -G(0)f(x) + (d/dt)\int_{0}^{t} G(t-s)f(x(s))\,ds$$

in which $G(0) > 0$, $xf(x) > 0$ if $x \ne 0$, $|f(x)| \le |x|$, $\int_{0}^{\infty} |G(u)|\,du = L < 1$, and $\int_{t}^{\infty} |G(u)|\,du \in L^1[0, \infty)$. Define a wedge by $W(r) = \sqrt{r}$ if $0 \le r \le 1$ and $W(r) = r$ for $r > 1$. Define a functional

$$V(t, x_t) = \left(x - \int_{0}^{t} G(t-s)f(x(s))\,ds \right)^2$$

$$+ G(0) \int_{0}^{t} \int_{t}^{\infty} |G(u-s)|\,du\,f^2(x(s))\,ds$$

so that

$$V'(t, x_t) = 2 \left(x - \int_0^t G(t-s)f(x(s))\,ds \right) [-G(0)f(x)]$$

$$+ G(0) \int_t^\infty |G(u-t)|\,du\, f^2(x) - G(0) \int_0^t |G(t-s)|f^2(x(s))\,ds$$

$$\le -2G(0)xf(s) + \int_0^t |G(t-s)|G(0) \left[f^2(x(s)) + f^2(x) \right] ds$$

$$+ G(0)Lf^2(x) - G(0) \int_0^t |G(t-s)|f^2(x(s))\,ds$$

$$\le -2G(0)xf(x) + 2G(0)Lf^2(x)$$

$$\le 2G(0) \left[-f^2(x) + Lf^2(x) \right]$$

$$\le -\gamma f^2(x)$$

for some $\gamma > 0$.

For bounds on V we have

$$V(t, x_t) \le x^2 + \int_0^t |G(t-s)| \left(f^2(x(s)) + x^2 \right) ds$$

$$+ \left(\int_0^t G(t-s)f(x(s))\,ds \right)^2$$

$$+ G(0) \int_0^t \int_{t-s}^\infty |G(u)|\,du\, f^2(x(s))\,ds$$

$$\le [1+L]x^2 + \int_0^t |G(t-s)|f^2(x(s))\,ds$$

$$+ \left[\int_0^t |G(t-s)|\,ds \int_0^t |G(t-s)|f^2(x(s))\,ds \right]$$

$$+ G(0) \int_0^t \int_{t-s}^\infty |G(u)|\,du\, f^2(x(s))\,ds$$

$$\le [1+L]x^2 + W \left(\int_0^t \left\{ 2|G(t-s)| + G(0) \int_{t-s}^\infty |G(u)|\,du \right\} \right.$$

$$\left. \times\, f^2(x(s))\,ds \right).$$

The lower bound is obtained as in Proposition 4.3.5.

The upper wedges on V in the proof of Theorem 4.4.3 illustrate exactly the kind of fading memory that a Liapunov functional must have to allow one to conclude asymptotic stability. Something must be said to prevent a solution from spending arbitrarily large time intervals near zero, and later moving far away from zero.

Definition 4.4.2. *A functional $V(t, x_t)$ defined for $t_0 \leq t < \infty$ and $x_t \in X_H(t)$ has a fading memory if for each $\eta > 0$, $\delta > 0$, and $\theta > 0$, there exists $T_1 > 0$ and $r > 0$ such that*

$$\left[t_1 \geq t_0, \ \phi \in X_\delta(t_1), \ |x(t, t_1, \phi)| < \eta \text{ on } [t_1, \infty), \right.$$

and $t \geq t_1 + r$] imply that for $x(t) = x(t, t_1, \phi)$, then

$$V(t, x_t) \leq W_2(|x(t)|) + W_3 \left(\theta + \sup_{t-r \leq s \leq t} |x(s)| \right).$$

This definition simply formalizes the situation observed in the proof of Theorem 4.4.3 and it would, of course, be possible to replace the condition

$$V(t, x_t) \leq W_2(|x|) + W_3 \left(\int_0^t \Phi(t - s) W_4(|x(s)|) \, ds \right)$$

with the condition that V has a fading memory. Such a formulation is left as an exercise.

The technique of Theorem of 4.4.2 is well suited for uniform boundedness. But recall that we have (4.4.1) only defined for bounded initial functions. Thus, we need to state our boundedness definitions accordingly. We say that solutions of (4.4.1) are *uniform bounded* if for each $B_1 > 0$ there is a $B_2 > 0$ such that $[t_1 \geq t_0, \ \phi \in X_{B_1}(t_1), \ t \geq t_1]$ imply that $|x(t, t_1, \phi)| < B_2$. Solutions are *uniform ultimate bounded for bound B* if for each $B_3 > 0$ there is a $K > 0$ such that $[t_1 \geq t_0, \ \phi \in X_{B_3}(t_1), \ t \geq t_1+K]$ imply that $|x(t, t_1, \phi)| < B_2$.

The next result is from Burton and Zhang (1985).

Theorem 4.4.4. *Let V be continuous and locally Lipschitz for $x_t \in X(t)$ and $t_0 \leq t < \infty$. Suppose there is a continuous function $\Phi : [0, \infty) \rightarrow [0, \infty)$ which is $L^1[0, \infty)$ and $\Phi'(t) \leq 0$ with*

(a) $\quad W_1(|x|) \leq V(t, x_t) \leq W_2(|x|) + W_3 \left(\int_\alpha^t \Phi(t - s) W_4(|x(s)|) \, ds \right)$

and

(b) $\quad V'_{(4.4.1)}(t, x_t) \leq -W_4(|x|) + M$

for some $M > 0$.

Then solutions of (4.4.1) are uniform bounded and uniform ultimate bounded for bound B.

Proof. Let $B_1 > 0$ be given and suppose $\phi \in X_{B_1}(t_1)$ for arbitrary $t_1 \geq t_0$. Let $x(t) = x(t, t_1, \phi)$ and $V(t) = V(t, x_t)$. For $t_1 \leq s \leq t < \infty$ we have

$$\left[V'(s) \leq -W_4(|x(s)|) + M \right] \Phi(t - s)$$

so that

$$\int_{t_1}^{t} W_4(|x(s)|) \Phi(t - s) \, ds$$

$$\leq - \int_{t_1}^{t} V'(s) \Phi(t - s) \, ds + M \int_{t_1}^{t} \Phi(t - s) \, ds$$

$$= - \left[V(s) \Phi(t - s) \Big|_{t_1}^{t} + \int_{t_1}^{t} V(s) \Phi'(t - s) \, ds \right] + M \int_{0}^{t - t_1} \Phi(u) \, du$$

$$\leq -V(t) \Phi(0) + V(t_1) \Phi(t - t_1) + V(\tau) \left[\Phi(0) - \Phi(t - t_1) \right] + ML$$

where $t_1 \leq \tau \leq t$ and $L = \int_0^\infty \Phi(u) \, du$.

Suppose there is a t with $V(t) \geq V(s)$ for $t_1 \leq s \leq t$. Either $t = t_1$ or $V'(t) \geq 0$. If $t = t_1$ we have

$$W_1(|x(t)|) \leq V(t, x_t) \leq V(t_1, \phi)$$

$$\leq W_2(B_1) + W_3 \left(\int_{\alpha}^{t_1} \Phi(t - s) W_4(B_1) \, ds \right)$$

$$\leq W_2(B_1) + W_3(W_4(B_1)L)$$

so

$$|x(t)| \leq W_1^{-1} \left[W_2(B_1) + W_3(W_4(B_1)L) \right] \stackrel{\text{def}}{=} B_2^* .$$

If $V'(t) \geq 0$ then $|x(t)| \leq W_4^{-1}(M)$ and

$$\int_{t_1}^{t} W_4(|x(s)|) \Phi(t - s) \, ds \leq V(t_1) \Phi(t - t_1) + ML$$

so that at the maximum of V we have

$$W_1(|x(t)|) \leq V(t, x_t) \leq W_2(W_4^{-1}(M))$$

$$+ W_3 \left(\int_{\alpha}^{t_1} \Phi(t - s) W_4(|\phi(s)|) \, ds + V(t_1) \Phi(t - t_1) + ML \right)$$

$$\leq W_2(W_4^{-1}(M)) + W_3 \left(W_4(B_1)L \right.$$

$$+ \left\{ W_2(B_1) + W_3(W_4(B_1))L \right\} \Phi(t - t_1) + ML)$$

or
$$|x(t)| \leq W_1^{-1}\big[W_2(W_4^{-1}(M)) + W_3\big(W_4(B_1)L$$
$$+ \{W_2(B_1) + W_3(W_4(B_1))L\}\Phi(0) + ML\big)\big]$$
$$\overset{\text{def}}{=} B_2^{**}.$$

We then have
$$|x(t)| \leq B_2 = \max[B_2^*, B_2^{**}],$$

and this is uniform boundedness.

To prove the uniform ultimate boundedness, choose $U > 0$ so that $|x| \geq U$ implies $V'(t, x_t) \leq -W_4(|x|) + M < -1$. Let $B_3 > 0$ be given and find B_4 such that $[t_1 \geq t_0, \phi \in X(t_1), \|\phi\| \leq B_3, t \geq t_1]$ imply that $|x(t, t_1, \phi)| < B_4$. Find $T > t_0$ with $W_4(B_4) \int_T^\infty \Phi(u)\,du < 1$. For $t_1 \geq t_0$ and $x(t) = x(t, t_1, \phi)$ with $\|\phi\| \leq B_3$, if $t \geq t_1 + T$ and if $\int_0^\infty \Phi(u)\,du = J$, then

$$\int_{t-T}^t \Phi(t - s)W_4(|x(s)|)\,ds$$

$$\leq -\int_{t-T}^t V'(s)\Phi(t - s)\,ds + M\int_{t-T}^t \Phi(t - s)\,ds$$

$$\leq -\left[V(s)\Phi(t - s)\Big|_{t-T}^t + \int_{t-T}^t V(s)\Phi'(t - s)\,ds\right] + MJ$$

$$= -V(t)\Phi(0) + V(t - T)\Phi(T) + V(\tau)[\Phi(0) - \Phi(t)] + MJ$$

where $V(\tau)$ is the maximum of $V(s)$ on $[t - T, t]$. This yields

$$(*) \qquad \int_{t-T}^t \Phi(t - s)W_4(|x(s)|)\,ds \leq -V(t)\Phi(0) + V(\tau)\Phi(0) + MJ.$$

Consider the intervals

$$I_1 = [t - T, t], \quad I_2 = [t, t + T], \quad I_3 = [t + T, t + 2T], \ldots$$

and select $t_i \in I_i$ such that $V(t_i)$ is the maximum of $V(s)$ on I_i, unless t_i is the left end point of I_i with $|x(t_i)| > U$; in the exceptional case we determine a point $\overline{t_i} > t_i$ such that $|x(\overline{t_i})| = U$ and $V'(t) < 0$ on $[t_i, \overline{t_i}]$. To accomplish this we may assume T so large that a $\overline{t_i}$ exists on I_i because

$$(**) \qquad V(t) \leq W_2(B_4) + W_3\big(1 + W_4(B_4)J\big)$$

and $V'(t) \leq -1$ if $|x| \geq U$. Now, replace I_i by the interval having $\overline{t_i}$ as its left end point and $t + (i - 1)T$ as its right end point. Call the exceptional interval I_i again and select t_i in I_i with $V(t_i)$ the maximum on I_i.

Now, consider the intervals

$$L_2 = [t_2 - T, t_2] \quad L_3 = [t_3 - T, t_3] \ldots .$$

Note that when $V(t_i)$ is the maximum on I_i, then $V'(t_i) \geq 0$ so $|x(t_i)| \leq U$. Next, consider each i:

Case 1   Suppose $V(t_i) + 1 \geq V(s)$ for all $s \in L_i$.

Case 2   Suppose $V(t_i) + 1 < V(s_i)$ for some $s_i \in L_i$.

Note that in Case 2 we may conclude that $s_1 \in I_{i-1}$ and $V(t_i) + 1 < V(t_{i-1})$. (This is true because $V(t_i)$ is maximum on I_i, so $s_i \in I_{i-1}$; in the exceptional case where we selected $\overline{t_i}$, if s_i is in $(t_i, \overline{t_i})$, then $s_i = t_i$ also qualifies because V decreases on $(t_i, \overline{t_i})$ and $t_i \in I_{i-1}$.) This means that if $V_i = \sup_{t \in I_i} V(t)$, then $V_i + 1 < V_{i-1}$. Because of $(**)$, there is an integer P such that Case 2 can hold on no more than P consecutive intervals.

Thus, on some L_j with $j \leq P$ we have $V(t_j) + 1 \geq V(s)$ for all $s \in L_j$. This means that if $V(\tau)$ is the maximum of V on L_j, then $V(\tau) \leq V(t_j) + 1$ so that from $(*)$ we have (for $t = t_j$)

$$\int_{t-T}^{t} \Phi(t - s) W_4(|x(s)|) \, ds$$
$$\leq -[V(t_j) + 1]\Phi(0) + V(\tau)\Phi(0) + MJ + \Phi(0)$$
$$\leq MJ + \Phi(0).$$

Now, recall that $|x(t_j)| \leq U$ so $V(t_j) \leq W_2(U) + W_3\big(1 + \Phi(0) + MJ\big)$ and at the maximum of V, $V(\tau)$, we have $V(\tau) \leq V(t_j) + 1$.

We claim that

$$V(t) \leq W_2(U) + W_3\big(1 + \Phi(0) + MJ\big) + 1$$

for all $t \geq t_j$; to see this, let t_p be the first $t > t_j$ with $V(t_p) = V(\tau)$. Then notice that $V(t_p)$ is the maximum of V on $[t_p - T, t_p]$. For $t = t_p$ this yields

$$\int_{t-T}^{t} W_4(|x(s)|)\Phi(t_p - s) \, ds$$
$$\leq -V(t_p)\Phi(0) + V(t_p)\Phi(0) + MJ = MJ$$

so that

$$V(t_p) \leq W_2(U) + W_3(1 + MJ),$$

as required. Moreover, if there is a $t > t_p$ with $V(t) = \max_{t_p \leq s \leq t} V(s)$, then the same bound holds.

Hence, for $t \geq t_1 + T + PT$ we have

$$W_1(|x|) \leq V(t) \leq V(t_j) + 1$$
$$\leq W_2(U) + W_3\big(1 + \Phi(0) + MJ\big) + 1$$

or

$$|x(t)| \leq W_1^{-1}\big[W_2(U) + W_3\big(1 + \Phi(0) + MJ\big) + 1\big]$$
$$\overset{\text{def}}{=} B .$$

This completes the proof.

These last three theorems complete the program illustrating unity of Liapunov theory for

(a) ordinary,

(b) finite delay,

and

(c) infinite delay

differential equations relative to

(a) stability,

(b) uniform stability,

(c) uniform asymptotic stability,

and

(d) uniform boundedness and uniform ultimate boundedness.

There are, of course, endless attractive problems remaining. Among these are

(a) parallel results using a Razumikhin technique, and

(b) converse theorems concerning the L^2 and weighted norm upper wedges on V.

The techniques used in the proof of Theorem 4.4.4 will allow us to obtain g-uniform ultimate boundedness results for the system

$$(4.3.1) \qquad x' = h(t, x) + \int_{-\infty}^{t} q(t, s, x(s)) \, ds$$

which were needed in Section 4.3.2 to prove the existence of a T-periodic solution. As an example, this result will complete the material needed in Proposition 4.3.1 for g-uniform ultimate boundedness. The theorem is found in Burton and Zhang (1985).

Theorem 4.4.5. *In* (4.3.1) *let* h *and* q *be continuous and suppose there is a continuous decreasing function* $g : (-\infty, 0] \to [1, \infty)$ *with* $g(r) \to \infty$ *as* $r \to -\infty$, *together with a function* $\Phi : [0, \infty) \to [0, \infty)$ *with* $\Phi \in L^1[0, \infty)$, $\Phi'(t) \leq 0$, *and such that if* $\phi : (-\infty, 0] \to R^n$ *is continuous and* $|\phi|_g \leq \gamma$ *for* $\gamma \in R$, *then*

(i) $\int_{-\infty}^0 q(t, s, \phi(s)) \, ds$ *is continuous for* $t \geq 0$;

(ii) $\int_{-\infty}^0 \Phi(t-s) W_4(\gamma g(s)) \, ds \overset{def}{=} G(t, \gamma)$ *is continuous in* t *for each* $\gamma > 0$;

(iii) $W_1(|x|) \leq V(t, x(\cdot)) \leq W_2(|x|) + W_3\left(\int_{-\infty}^t \Phi(t - s) W_4(|x(s)|) \, ds\right)$ *for every continuous* $x : (-\infty, \infty) \to R^n$ *with* $|x(s)| \leq \gamma g(s)$ *on* $(-\infty, 0]$ *for some* $\gamma > 0$;

(iv) $V'_{(4.3.1)}(t, x(\cdot)) \leq -W_4(|x|) + M$.

Under these conditions solutions of (4.3.1) *are* g-*uniform bounded and* g-*uniform ultimate bounded for bound* B.

Proof. Since h and q are continuous condition (i) shows that there exists a solution $x(t, 0, \phi)$ on an interval $[0, \beta)$. That solution can be continued so long as it remains bounded for t bounded. By (ii) we see that (iii) is defined so long as the solution exists. Using (iv) we see that $W_1(|x(t)|) \leq V(t, x(\cdot)) \leq V(0, \phi) + Mt$ and so $|x(t)|$ is bounded for t bounded. This means that $x(t, 0, \phi)$ is defined on $[0, \infty)$.

To prove g-uniform boundedness we proceed as in the proof of Theorem 4.4.4. Let $B_1 > 0$ be given, $|\phi|_g \leq B_1$, $x(t) = x(t, 0, \phi)$, and let $V(t) = V(t, x(\cdot))$. Either $V(t) \leq V(0)$ on $[0, \infty)$ or there is a $t > 0$ with $V(t) = \max_{0 \leq s \leq t} V(s)$. Then for $0 \leq s \leq t$ and for $V'(s) = dV(s)/ds$ we have

$$\left[V'(s) \leq -W_4(|x(s)|) + M\right]\Phi(t - s)$$

and an integration yields

$$\int_0^t \Phi(t - s) W_4(|x(s)|) \, ds \leq -\int_0^t V'(s)\Phi(t - s) \, ds + MJ$$

$$\leq V(0)\Phi(0) + MJ$$

where $J = \int_0^\infty \Phi(s) \, ds$. Since $V(t)$ is the maximum, $V'(t) \geq 0$ and so if we define U by $-W_4(U) + M = -1$, then $|x(t)| < U$. This yields

$$V(t) \leq W_2(U) + W_3\left(\int_{-\infty}^t \Phi(t - s) W_4(B_1 g(s)) \, ds + V(0)\Phi(0) + MJ\right)$$

$$\leq W_2(U) + W_3\big(G(0, B_1) + V(0)\Phi(0) + MJ\big)$$

$$\overset{def}{=} V_m.$$

Hence, for $0 \leq t < \infty$ we have $V(t) \leq V(0) + V_m$ so that for any $t \geq 0$ we have

$$W_1(|x(t)|) \leq V(t, x(\cdot)) \leq V(0, \psi) + V_m$$

$$\leq W_2(B_1) + W_3(G(0, B_1)) + V_m \overset{\text{def}}{=} W_1(B_2)$$

defining B_2. This proves g-uniform boundedness.

To prove the g-uniform ultimate boundedness we first note that

$$G(t, \gamma) = \int_{-\infty}^{0} \Phi(t - s) W_4(\gamma g(s)) \, ds$$

$$= \int_{t}^{\infty} \Phi(u) W_4(\gamma g(t - u)) \, du$$

$$\leq \int_{t}^{\infty} \Phi(u) W_4(\gamma g(-u)) \, du$$

and this tends to zero as $t \to \infty$.

Now, given $B_3 > 0$, if $|\phi|_g \leq B_3$, then there is a B_4 as a bound on $|x(t, 0, \phi)|$. Then for $T > 0$ and $t > T$ we have

$$\int_{-\infty}^{t} \Phi(t - s) W_4(|x(s, 0, \phi)|) \, ds$$

$$\leq \int_{-\infty}^{0} \Phi(t - s) W_4(B_3 g(s)) \, ds + \int_{0}^{t-T} \Phi(t - s) W_4(B_4) \, ds$$

$$+ \int_{t-T}^{t} \Phi(t - s) W_4(|x(s, 0, \phi)|) \, ds$$

$$\leq G(t, B_3) + \int_{T}^{\infty} \Phi(u) W_4(B_4) \, du$$

$$+ \int_{t-T}^{t} \Phi(t - s) W_4(|x(s, 0, \phi)|) \, ds$$

$$< 1 + \int_{t-T}^{t} \Phi(t - s) W_4(|x(s, 0, \phi)|) \, ds$$

if T (and, hence, t) is sufficiently large. The remainder of the proof of g-uniform ultimate boundedness is now exactly as that for Theorem 4.4.4.

There are many other Liapunov theorems for unbounded delay which we have not considered here. A variety of these are given in Burton (1983a; pp. 227–302). For additional stability theory for infinite delay equations we refer the reader to Corduneanu (1973, 1981, 1982), Gopalsamy (1983), Haddock (1984), Hale and Kato (1978), Hino (1970, 1971a,b), Kaminogo (1978), Kappel and Schappacher (1980), Kato (1978, 1980b), Krisztin (1981), Naito (1976a,b), and Wang (1983).

References

Andronow, A. A., and Chaikin, C. E. (1949). "Theory of Oscillations." Princeton Univ. Press, New Jersey

Antosiewicz, H. A. (1958). A survey of Liapunov's second method. *Ann. Math. Stud.* **41**, 141–156.

Arino, O., Burton, T. A., and Haddock, J. (1985). Periodic solutions of functional differential equations, *Proc. Roy. Soc. Edinburgh Sect. A* **101**, 253-271.

Artstein, Z. (1978). Uniform asymptotic stability via limiting equations. *J. Differential Equations* **27**, 172–189.

Barbashin, E. A. (1968). The construction of Liapunov functions. *Differential Equations* **4**, 1097–1112.

Bellman, R. (1953). "Stability Theory of Differential Equations." McGraw-Hill, New York.

Bellman, R. (1961). "Modern Mathematical Classics: Analysis." Dover, New York.

Bellman, R., and Cooke, K. L. (1963). "Differential-Difference Equations." Academic Press, New York.

Bendixson, I. (1901). Sur les courbes définies par des équations différentielles. *Acta Math.* **24**, 1–88.

Bihari, I. (1957). Researches of the boundedness and stability of the solutions of non-linear differential equations. *Acta Math. Acad. Sci. Hungar.* **VIII**, 261–278.

Billotti, J. E., and LaSalle, J. P. (1971). Dissipative periodic processes. *Bull. Amer. Math. Soc. (N.S.)* **77**, 1082–1088.

Boyce, W. E., and DiPrima, R. C. (1969). "Elementary Differential Equations and Boundary Value Problems." Wiley, New York.

Brauer, F. (1978). Asymptotic stability of a class of integrodifferential equations. *J. Differential Equations* **28**, 180–188.

Brauer, F., and Nohel, J. (1969). "Qualitative Theory of Ordinary Differential Equations." Benjamin, New York.

Braun, M. (1975). "Differential Equations and their Applications." Springer, New York.

Brouwer, L. E. J. (1912). Beweis des ebenen Translationssatzes. *Math. Ann.* **72**, 37–54

Browder, F. E. (1959). On a generalization of the Schauder fixed point theorem. *Duke Math. J.* **26**, 291–303.

Browder, F. E. (1970). Asymptotic fixed point theorems. *Math. Ann.* **185**, 38–60.

Burton, T. A. (1965). The generalized Liénard equation. *SIAM J. Control Optim.* **3**, 223–230.

Burton, T. A. (1969). An extension of Liapunov's direct method. *J. Math. Anal. Appl.* **28**, 545–552.

Burton, T. A. (1970). Correction to "An extension of Liapunov's direct method." *J. Math. Anal. Appl.* **32**, 689–691.

Burton, T. A. (1977). Differential inequalities for Liapunov functions. *Nonlinear Anal.* **1**, 331–338.

Burton, T. A. (1978). Uniform asymptotic stability in functional differential equations. *Proc. Amer. Math. Soc.* **68**, 195–199.

Burton, T. A. (1979a). Stability theory for delay equations. *Funkcial. Ekvac.* **22**, 67–76.

Burton, T. A. (1979b). Stability theory for functional differential equations. *Trans. Amer. Math. Soc.* **255**, 263–275.

Burton, T. A. (ed.) (1981). "Mathematical Biology." Pergamon, New York.

Burton, T. A. (1982). Boundedness in functional differential equations. *Funkcial. Ekvac.* **25**, 51–77.

Burton, T. A. (1883a). "Volterra Integral and Differential Equations." Academic Press, New York.

Burton, T. A. (1983b). Structure of solutions of Volterra equations. *SIAM Rev.* **25**, 343–364.

Burton, T. A. (1984). Periodic solutions of linear Volterra equations. *Funkcial. Ekvac.* **27**, 229–253.

Burton, T. A. (1985a). Periodicity and limiting equations in Volterra systems. *Boll. Un. Mat. Ital. Sect. C.* Series IV -CN. **1**, 31-39.

Burton, T. A. (1985b). Toward unification of periodic theory. *In* "Differential Equations: Qualitative Theory" (Szeged, 1984), *Colloq. Math. Soc. János Bolyai* **47**. North-Holland, Amsterdam and New York, 127-141.

Burton, T. A. (1985c). Periodic solutions of functional differential equations. *J. London Math Soc. (2)* (6) IV, 31–39.

Burton, T. A. (1985d). Periodic solutions of nonlinear Volterra equations. *Funkcial Ekvac.* **27**, 301–317.

Burton, T. A., and Grimmer, R. C. (1973). Oscillation, continuation, and uniqueness of solutions of retarded differential equations. *Trans. Amer. Math. Soc.* **179**, 193–209.

Burton, T. A., and Mahfoud, W. E. (1983). Stability criteria for Volterra equations. *Trans. Amer. Math. Soc.* **279**, 143–174.

Burton, T. A., and Mahfoud, W. E. (1984). Stability by decomposition for Volterra equations. (appeared in Tohoku Math. J. **37**(1985), 489-511.)

Burton, T. A., and Townsend, C. G. (1968). On the generalized Liénard equation with forcing function. *J. Differential Equations* **4**, 620–633.

Burton, T. A., and Townsend, C. G. (1971). Stability regions of the forced Liénard equation. *J. London Math. Soc. (2)* **3**, 393–402.

Burton, T. A., and Zhang, S. (1985). Uniform boundedness in functional differential equations. *Ann. Math. Pure Appl.* (appeared in **CXLV** (1986), 129-158).

Burton, T. A., Huang, Q., and Mahfoud, W. E. (1985a). Liapunov functionals of convolution type. *J. Math. Anal. Appl.* **106**, 249-272.

Burton, T. A., Huang, Q., and Mahfoud, W. E. (1985b). Rate of decay of solutions of Volterra equations, *J. Nonlinear Anal.* **9**, 651–663.

Busenberg, S. V., and Cooke, K. L. (1984). Stability conditions for linear non-autonomous delay differential equations. *Quart. Appl. Math.* **42**, 295–306.

Cartwright, M. L. (1950). Forced oscillations in nonlinear systems. *In* "Contributions to the Theory of Nonlinear Oscillations" (S. Lefshetz, ed.), Vol. 1, pp. 149–241, Princeton Univ. Press., Princeton, New Jersey.

Churchill, R. V. (1958). "Operational Mathematics." McGraw-Hill, New York.

Coleman, C. S. (1976). Combat models. MAA Workshop on Modules in Applied Math., Cornell University. [Also in "Differential Equations Models" (M. Braun, C. S. Coleman, and D. A. Drew, eds.). Springer Publ., New York].

Cooke, K. L. (1976). An epidemic equation with immigration. *Math. Biosci.* **29**, 135–158.

Cooke, K. L. (1984). Stability of nonautonomous delay differntial equations by Liapunov functionals. *In* "Infinite-Dimensional Systems" (F. Kappel and W. Schapppacher, eds.). Springer Lecture Notes 1076. Springer, Berlin.

Cooke, K. L., and Yorke, J. A. (1973). Some equations modelling growth processes and gonorrhea epidemics. *Math. Biosci.* **16**, 75–101.

Corduneanu, C. (1973). "Integral Equations and Stability of Feedback Systems." Academic Press, New York.

Corduneanu, C. (1981). Almost periodic solutions for infinite delay systems. *In* "Spectral Theory of Differential Operators." (I. W. Knowles, and R. T. Lewis, eds.). pp. 99–106. North-Holland Publ., Amsterdam.

Corduneanu, C. (1982). Integrodifferntial equations with almost periodic solutions. *In* "Volterra and Functional Differential Equations" (K. B. Hannsgen, T. L. Herdman, H. W. Stech, and R. L. Wheeler, eds.). Dekker, New York.

Cronin, J. (1964). "Fixed Points and Topological Degree in Nonlinear Analysis." Amer. Math. Soc., Providence, Rhode Island.

Davis, H. T. (1962). "Introduction to Nonlinear Differential and Integral Equations." Dover, New York.

DeVito, C. L. (1978). "Functional Analysis." Academic Press, New York.

Dickinson, A. B. (1972). "Differential Equations: Theory and use in Time and Motion." Addison-Wesley, Reading, Massachusetts.

Driver, R. D. (1962). Existence and stability of solutions of a delay-differential system. *Arch. Rational Mech. Anal.* **10**, 401–426.

Driver, R. D. (1977). "Ordinary and Delay Differential Equations." Springer Publ., New York.

Dunford, N., and Schwartz, J. T. (1964). "Linear Operators," Part I. Wiley (Interscience), New York.

El'sgol'ts, L. E. (1966). "Introduction to the Theory of Differential Equations with Deviating Arguments." Holden-Day, San Francisco.

Epstein, I. J. (1962). Periodic solutions of systems of differential equations. *Proc. Amer. Math. Soc.* **13**, 690–694.

Erhart, J. (1973). Lyapunov theory and perturbations of differential equations. *SIAM J. Math. Anal.* **4**, 417–432.

Feller, W. (1941). On the integral equation of renewal theory. *Ann. Math Statist.* **12**, 243–267.

Finkbeiner, D. T. (1960). "Introduction to Matrices and Linear Transformations." Freeman, San Francisco.

Fulks, W. (1969). "Advanced Calculus." Wiley, New York.

Furumochi, T. (1981). Periodic solutions of periodic functional differential equations. *Funkcial. Ekvac.* **24**, 247–258.

Furumochi, T. (1982). Periodic solutions of functional differential equations with large delays. *Funkcial. Ekvac.* **25**, 33-42.

Gantmacher, F. R. (1960). "Matrix Theory." Vol. II. Chelsea, New York.

Gopalsamy, K. (1981). Time lags in Richardson's arms race model. *J. Social. Biol. Struc.* **4**, 303–317.

Gopalsamy, K. (1983). Stability and decay rates in a class of linear integrodifferential systems. *Funkcial. Ekvac.* **26**, 251–261.

Graef, J. R. (1972). On the generalized Liénard equation with negative damping. *J. Differential Equations* **12**, 34–62.

Grimmer, R. (1979). Existence of periodic solutions of functional differential equations. *J. Math. Anal. Appl.* **72**, 666–673.

Grimmer, R., and Seifert, G. (1975). Stability properties of Volterra integrodifferential equations. *J. Differential Equations* **19**, 142–166.

Grossman, S. I., and Miller, R. K. (1970). Perturbation theory for Volterra integrodifferential systems. *J. Differential Equations* **8**, 457–474.

Grossman, S. L., and Miller, R. K. (1973). Nonlinear Volterra integrodifferential systems with L^1-kernels. *J. Differential Equations* **13**, 551–566.

Haag, J. (1962). "Oscillatory Motions." Wadsworth, Belmont, California.

Haddock, J. R. (1972a). A remark on a stability theorem of Marachkoff. *Proc. Amer. Math. Soc.* **31**, 209–212.

Haddock, J. R. (1972b). Liapunov functions and boundedness and global existence of solutions. *Applicable Anal.* **1**, 321–330.

Haddock, J. R. (1974). On Liapunov functions for nonautonomous systems. *J. Math. Anal. Appl.* **47**, 599–603.

Haddock, J. R. (1984). A friendly space for functional differential equations with infinite delay. *In* "Trends in the Theory and Practice of Non-Linear Analysis" (V. Lakshmikantham, ed.), North-Holland Publ., Amsterdam (appeared (1985), 173-182).

Haddock, J. R., and Terjéki, J. (1983). Liapunov-Razumikhin functions and an invariance principle for functional differential equations. *J. Differential Equations* **48**, 93–122.

Haddock, J. R., Krisztin, T., and Terjéki, J. (1984). Invariance principles for autonomous functional differential equations. *J. Integral Equations* (appeared in **10** (1985), 123-136).

Hahn, W. (1963). "Theory and Application of Liapunov's Direct Method." Prentice-Hall, Englewood Cliffs, New Jersey.

Hahn, W. (1967). "Stability of Motion." Springer Publ., New York.

Hale, J. K. (1963). "Oscillations in Nonlinear Systems." McGraw-Hill, New York.

Hale, J. K. (1965). Sufficient conditions for stability and instability of autonomous functional differential equations. *J. Differential Equations* **1**, 452–482.

Hale, J. K. (1969). "Ordinary Differential Equations." Wiley, New York.

Hale, J. K. (1977). "Theory of Functional Differential Equations." Springer Publ., New York.

Hale, J. K., and Kato, J. (1978). Phase space for retarded equations with infinite delay. *Funkcial Ekvac.* **21**, 11-41.

Hale, J. K., and Lopes, O. (1973). Fixed point theorems and dissipative processes. *J. Differential Equations* **13**, 391–402.

Halmos, P. R. (1958). "Finite Dimensional Vector Spaces," Second Ed. Van Nostrand, New York.

Hartman, P. (1964). "Ordinary Differential Equations." Wiley, New York.

Hatvani, L. (1978). Attractivity theorems for non-autonomous systems of differential equations. *Acta Sci. Math.* **40**, 271–283.

Hatvani, L. (1983). On partial asymptotic stability and instability, II (The method of limiting equations.). *Acta Sci. Math.* **46**, 143–156.

Hatvani, L. (1984). On partial asymptotic stability by energy-like Lyapunov functions. *Acta Sci. Math.* (appeared in **48** (1985), 187-200).

Herdman, T. L. (1980). A note on non-continuable solutions of a delay differential equation. *In* "Differential Equations." (S. Ahmad, M. Keener, A. C. Lazer, eds.), pp. 187–192. Academic Press, New York.

Hill, W. W. (1978). A time lagged arms race model. *J. Peace Sci.* **3**, 55–62.

Hino, Y. (1970). Asymptotic behavior of solutions of some functional differential equations. *Tôhoku Math. J.* (2) **22**, 98–108.

Hino, Y. (1971a). Continuous dependence for some functional differential equations. *Tôhoku Math. J.* (2) **23**, 565–571.

Hino, Y. (1971b). On stability of some functional differential equations. *Funkcial. Ekvac.* **14**, 47–60.

Horn, W. A. (1970). Some fixed point theorems for compact maps and flows in Banach spaces. *Trans. Amer. Math. Soc.* **149**, 391–404.

Jacobson, J. C. (1984). Delays and memory functions in Richardson type arms race models. M.S. thesis, Southern Illinois Univ., Carbondale.

Jones, G. S. (1963). Asymptotic fixed point theorems and periodic systems of functional differential euqations. *Contrib. Differential Equations* **2**, 385–406.

Jones, G. S. (1964). Periodic motions in Banach space and application to functional differential equations. *Contrib. Differential Equations* **3**, 75–106.

Jones, G. S. (1965). Stability and asymptotic fixed-point theory. *Proc. Nat. Acad. Sci.* **53**, 1262–1264.

Jordon, G. S. (1979). Asymptotic stability of a class of integrodifferential systems. *J. Differential Equations* **31**, 359–365.

Kaminogo, T. (1978). Kneser's property and boundary value problems for some retarded functional differential equations. *Tôhoku Math. J.* **30**, 471–486.

Kamke, E. (1930). "Differentialgleichungen reeler Functionen." Leipzig, Akademische Verlagsgellschaft.

Kaplan, J. L., Sorg, M., and Yorke, J. A. (1979). Solutions of $x'(t) = f(x(t), x(t - L))$ have limits when f is an order relation. *Nonlinear Anal.* **3**, 53–58.

Kaplansky, I. (1957). "An Introduction to Differential Algebra." Hermann, Paris.

Kappel, F., and Schappacher, W. (1980). Some considerations to the fundamental theory of infinite delay equations. *J. Differential Equations* **37**, 141–183.

Kato, J. (1978). Stability problem in functional differential equations with infinite delays. *Funkcial. Ekvac.* **21**, 63–80.

Kato, J. (1980a). An autonomous system whose solutions are uniformly ultimately bounded but not uniformly bounded. *Tôhoku Math. J.* **32**, 499–504.

Kato, J. (1980b). Liapunov's second method in functional differential equations. *Tôhoku Math. J.* **32**, 487–497.

Kato, J., and Yoshizawa T. (1981). Remarks on global properties in limiting equations *Funkcial. Ekvac.* **24**, 363–371.

Krasovskii, N. N. (1963). "Stability of Motion." Stanford Univ. Press, Stanford, California.

Krisztin, T. (1981). On the convergence of solutions of functional differential equations. *Acta Sci. Math.* **43**, 45–54.

Lakshmikantham, V., and Leela, S. (1969). "Differential and Integral Inequalities," Vol. I. Academic Press, New York.

Langenhop, C. E. (1973). Differentiability of the distance to a set in Hilbert space. *J. Math. Anal. Appl.* **44**, 620–624.

Langenhop, C. E. (1985). Periodic and almost periodic solutions of Volterra integral differential equations with infinite memory. *J. Differential Equations* **58**, 391–403.

Lefschetz, S. (1965). "Stability of Nonlinear Control Systems," Academic Press, New York.

Leitman, M. J., and Mizel, V. J. (1978). Asymptotic stability and the periodic solutions of $x(t) + \int_{-\infty}^{t} a(t-s)g(s,x(s))\,ds = f(t)$. *J. Math. Anal. Appl.* **66**, 606–625.

Levin, J. (1963). The asymptotic behavior of a Volterra equation. *Proc. Amer. Math. Soc.* **14**, 534–541.

Levin, J., and Nohel, J. (1960). On a system of integrodifferential equations occurring in reactor dynamics. *J. Math. Mechanics* **9**, 347–368.

Levinson, N. (1944). Transformation theory of non-linear diferential equations of second order. *Ann. Math.* (2) **45**, 723–737.

Liénard, A. (1928). Étude des oscillations entretenues. *Rev. Gen. de l' Electricité* **23**, 901–946.

Lotka, A. J. (1907a). Studies of the mode of growth of material aggregates. *Amer. J. Sci.* **24**, 199–216.

Lotka, A. J. (1907b). Relation between birth rates and death rates. *Amer. J. Sci.* **26**, 21–22.

Massera, J. L. (1950). The existence of periodic solutions of systems of differential equations. *Duke Math. J.* **17**, 457–475.

Miller, R. K. (1971). Asymptotic stability properties of linear Volterra integrodifferential equations. *J. Differential Equations* **10**, 485–506.

Minorsky, N. (1962). "Nonlinear Oscillations." Van Nostrand, New York.

Murakami, S. (1984a). Asymptotic behavior of solutions of some differential equations. *J. Math. Anal. Appl.* (appeared in **109** (1985), 534-545).

Murakami, S. (1984b). Perturbation theorems for functional differential equations with infinite delay via limiting equations. (appeared in *J. Differential Equations* **59** (1985), 314-335).

Myshkis, A. D. (1951). General theory of differential equations with retarded argument. *Amer. Math. Soc. Transl.* No. 55.

Naito, T. (1976a). On autonomous linear functional differential equations with infinite retardations. *Tôhoku Math. J.* **21**, 297–315.

Naito, T. (1976b). Adjoint equations of autonomous linear functional differential equations with infinite retardations. *Tôhoku Math J.* **28**, 135–143.

Nemytskii, V. V., and Stepanov, V. V. (1960). "Qualitative Theory of Differential Equations." Princeton Univ. Press, Princeton, New Jersey.

Perron, O. (1930). Die Stabilitatsfrage bei Differential-gleichungungssysteme. *Math. Z.* **32**, 703–728.

Picard, E. (1907). La Mécanique classique et ses approximations successive. *Révista de Scienza* **1**, 4–15.

Reed, M., and Simon, B. (1972). "Methods of Modern Mathematical Physics." Academic Press, New York.

Reissig, R., Sansone, G., and Conti, R. (1963). "Qualitative Theorie Nichtlinearer Differentialgleichungen." Edizioni Cremonese, Roma.

Richardson, L. F. (1960). "Arms and Insecurity." Boxwood, Pittsburgh.

Sansone, G., and Conti, R. (1964). "Non-Linear Differential Equations." Macmillan, New York.

Sawano, K. (1979). Exponential asymptotic stability for functional differential equations with infinite retardations. *Tôhoku Math. J.* **31**, 363–382.

Seifert, G. (1981). Almost periodic solutions for delay-differential equations with infinite delays. *J. Differential Equations* **41**, 416–425.

Shi, S. L. (1980). A concrete example of the existence of 4 limit cycles for plane quadratic systems. *Systems Sci. Sinica* **2**, 153–158.

Smart, D. R. (1980). "Fixed Point Theorems." Cambridge Univ. Press, Cambridge.

Somolinos, A. (1977). Stability of Lurie-type functional equations. *J. Differential Equations* **26**, 191–199.

Somolinos, A. (1978). Periodic solutions of the sunflower equation. *Quart. Appl. Math.* **35**, 465–478.

Spiegel, M. R. (1958). "Applied Differential Equations." Prentice-Hall, New Jersey.

van der Pol, B. (1927). Forced oscillations in a circuit with non-linear resistance (Reception with reactive triode) *Philos. Mag.* **iii** [*In* Bellman (1961)].

van der Pol, B., and van der Mark, J. (1928). The heartbeat considered as a relaxation oscillation, and an electrical model of the heart. *Philos. Mag.* 7th Ser. **6**, 763–775.

Wang, Z. (1983). Comparison method and stability problem in functional differential equations. *Tôhoku Math. J.* **35**, 349–356.

Wang, L., and Wang, M. (1983). On construction of Lyapunov's global asymptotic stable functions of a type of nonlinear third-order systems. *Kexue Tonbao*, Special Issue.

Wen, L. Z. (1982). On the uniform asymptotic stability in functional differential equations. *Proc. Amer. Math. Soc.* **85**, 533–538.

Yanqian, Ye. (1982). Some problems in the qualitative theory of ordinary differential equations. *J. Differential Equations* **46**, 153–164.

Yoshizawa, T. (1963). Asymptotic behavior of solutions of a system of differential equations. *Contrib. Differential Equations* **1**, 371–387.

Yoshizawa, T. (1966). "Stability Theory by Liapunov's Second Method." Math. Soc. Japan, Tokyo.

Yoshizawa, T. (1975). "Stability Theory and the Existence of Periodic and Almost Periodic Solutions." Springer Publ., New York.

Yoshizawa, T. (1984). Asymptotic behavior of solutions of differential equations. (appeared in *Colloquia Mathematica Societatis Janos Bolyai* **47** *Differential Equations: Qualitative Theory II*, Szeged (Hungary), (1987), 1141-1159).

Author Index

Subject Index

A CATALOG OF SELECTED
DOVER BOOKS
IN SCIENCE AND MATHEMATICS

Astronomy

BURNHAM'S CELESTIAL HANDBOOK, Robert Burnham, Jr. Thorough guide to the stars beyond our solar system. Exhaustive treatment. Alphabetical by constellation: Andromeda to Cetus in Vol. 1; Chamaeleon to Orion in Vol. 2; and Pavo to Vulpecula in Vol. 3. Hundreds of illustrations. Index in Vol. 3. 2,000pp. 6⅛ x 9¼.
Vol. I: 23567-X
Vol. II: 23568-8
Vol. III: 23673-0

EXPLORING THE MOON THROUGH BINOCULARS AND SMALL TELESCOPES, Ernest H. Cherrington, Jr. Informative, profusely illustrated guide to locating and identifying craters, rills, seas, mountains, other lunar features. Newly revised and updated with special section of new photos. Over 100 photos and diagrams. 240pp. 8¼ x 11. 24491-1

THE EXTRATERRESTRIAL LIFE DEBATE, 1750–1900, Michael J. Crowe. First detailed, scholarly study in English of the many ideas that developed from 1750 to 1900 regarding the existence of intelligent extraterrestrial life. Examines ideas of Kant, Herschel, Voltaire, Percival Lowell, many other scientists and thinkers. 16 illustrations. 704pp. 5⅜ x 8½. 40675-X

THEORIES OF THE WORLD FROM ANTIQUITY TO THE COPERNICAN REVOLUTION, Michael J. Crowe. Newly revised edition of an accessible, enlightening book recreates the change from an earth-centered to a sun-centered conception of the solar system. 242pp. 5⅜ x 8½. 41444-2

A HISTORY OF ASTRONOMY, A. Pannekoek. Well-balanced, carefully reasoned study covers such topics as Ptolemaic theory, work of Copernicus, Kepler, Newton, Eddington's work on stars, much more. Illustrated. References. 521pp. 5⅜ x 8½.
65994-1

A COMPLETE MANUAL OF AMATEUR ASTRONOMY: Tools and Techniques for Astronomical Observations, P. Clay Sherrod with Thomas L. Koed. Concise, highly readable book discusses: selecting, setting up and maintaining a telescope; amateur studies of the sun; lunar topography and occultations; observations of Mars, Jupiter, Saturn, the minor planets and the stars; an introduction to photoelectric photometry; more. 1981 ed. 124 figures. 26 halftones. 37 tables. 335pp. 6½ x 9¼.
42820-6

AMATEUR ASTRONOMER'S HANDBOOK, J. B. Sidgwick. Timeless, comprehensive coverage of telescopes, mirrors, lenses, mountings, telescope drives, micrometers, spectroscopes, more. 189 illustrations. 576pp. 5⅝ x 8¼. (Available in U.S. only.)
24034-7

STARS AND RELATIVITY, Ya. B. Zel'dovich and I. D. Novikov. Vol. 1 of *Relativistic Astrophysics* by famed Russian scientists. General relativity, properties of matter under astrophysical conditions, stars, and stellar systems. Deep physical insights, clear presentation. 1971 edition. References. 544pp. 5⅝ x 8¼. 69424-0

Chemistry

THE SCEPTICAL CHYMIST: The Classic 1661 Text, Robert Boyle. Boyle defines the term "element," asserting that all natural phenomena can be explained by the motion and organization of primary particles. 1911 ed. viii+232pp. 5⅜ x 8½.
42825-7

RADIOACTIVE SUBSTANCES, Marie Curie. Here is the celebrated scientist's doctoral thesis, the prelude to her receipt of the 1903 Nobel Prize. Curie discusses establishing atomic character of radioactivity found in compounds of uranium and thorium; extraction from pitchblende of polonium and radium; isolation of pure radium chloride; determination of atomic weight of radium; plus electric, photographic, luminous, heat, color effects of radioactivity. ii+94pp. 5⅜ x 8½.
42550-9

CHEMICAL MAGIC, Leonard A. Ford. Second Edition, Revised by E. Winston Grundmeier. Over 100 unusual stunts demonstrating cold fire, dust explosions, much more. Text explains scientific principles and stresses safety precautions. 128pp. 5⅜ x 8½.
67628-5

THE DEVELOPMENT OF MODERN CHEMISTRY, Aaron J. Ihde. Authoritative history of chemistry from ancient Greek theory to 20th-century innovation. Covers major chemists and their discoveries. 209 illustrations. 14 tables. Bibliographies. Indices. Appendices. 851pp. 5⅜ x 8½.
64235-6

CATALYSIS IN CHEMISTRY AND ENZYMOLOGY, William P. Jencks. Exceptionally clear coverage of mechanisms for catalysis, forces in aqueous solution, carbonyl- and acyl-group reactions, practical kinetics, more. 864pp. 5⅜ x 8½.
65460-5

ELEMENTS OF CHEMISTRY, Antoine Lavoisier. Monumental classic by founder of modern chemistry in remarkable reprint of rare 1790 Kerr translation. A must for every student of chemistry or the history of science. 539pp. 5⅜ x 8½.
64624-6

THE HISTORICAL BACKGROUND OF CHEMISTRY, Henry M. Leicester. Evolution of ideas, not individual biography. Concentrates on formulation of a coherent set of chemical laws. 260pp. 5⅜ x 8½.
61053-5

A SHORT HISTORY OF CHEMISTRY, J. R. Partington. Classic exposition explores origins of chemistry, alchemy, early medical chemistry, nature of atmosphere, theory of valency, laws and structure of atomic theory, much more. 428pp. 5⅜ x 8½. (Available in U.S. only.)
65977-1

GENERAL CHEMISTRY, Linus Pauling. Revised 3rd edition of classic first-year text by Nobel laureate. Atomic and molecular structure, quantum mechanics, statistical mechanics, thermodynamics correlated with descriptive chemistry. Problems. 992pp. 5⅜ x 8½.
65622-5

FROM ALCHEMY TO CHEMISTRY, John Read. Broad, humanistic treatment focuses on great figures of chemistry and ideas that revolutionized the science. 50 illustrations. 240pp. 5⅜ x 8½.
28690-8

Engineering

DE RE METALLICA, Georgius Agricola. The famous Hoover translation of greatest treatise on technological chemistry, engineering, geology, mining of early modern times (1556). All 289 original woodcuts. 638pp. 6¾ x 11.   60006-8

FUNDAMENTALS OF ASTRODYNAMICS, Roger Bate et al. Modern approach developed by U.S. Air Force Academy. Designed as a first course. Problems, exercises. Numerous illustrations. 455pp. 5⅜ x 8½.   60061-0

DYNAMICS OF FLUIDS IN POROUS MEDIA, Jacob Bear. For advanced students of ground water hydrology, soil mechanics and physics, drainage and irrigation engineering, and more. 335 illustrations. Exercises, with answers. 784pp. 6⅛ x 9¼.   65675-6

THEORY OF VISCOELASTICITY (Second Edition), Richard M. Christensen. Complete, consistent description of the linear theory of the viscoelastic behavior of materials. Problem-solving techniques discussed. 1982 edition. 29 figures. xiv+364pp. 6⅛ x 9¼.   42880-X

MECHANICS, J. P. Den Hartog. A classic introductory text or refresher. Hundreds of applications and design problems illuminate fundamentals of trusses, loaded beams and cables, etc. 334 answered problems. 462pp. 5⅜ x 8½.   60754-2

MECHANICAL VIBRATIONS, J. P. Den Hartog. Classic textbook offers lucid explanations and illustrative models, applying theories of vibrations to a variety of practical industrial engineering problems. Numerous figures. 233 problems, solutions. Appendix. Index. Preface. 436pp. 5⅜ x 8½.   64785-4

STRENGTH OF MATERIALS, J. P. Den Hartog. Full, clear treatment of basic material (tension, torsion, bending, etc.) plus advanced material on engineering methods, applications. 350 answered problems. 323pp. 5⅜ x 8½.   60755-0

A HISTORY OF MECHANICS, René Dugas. Monumental study of mechanical principles from antiquity to quantum mechanics. Contributions of ancient Greeks, Galileo, Leonardo, Kepler, Lagrange, many others. 671pp. 5⅜ x 8½.   65632-2

STABILITY THEORY AND ITS APPLICATIONS TO STRUCTURAL MECHANICS, Clive L. Dym. Self-contained text focuses on Koiter postbuckling analyses, with mathematical notions of stability of motion. Basing minimum energy principles for static stability upon dynamic concepts of stability of motion, it develops asymptotic buckling and postbuckling analyses from potential energy considerations, with applications to columns, plates, and arches. 1974 ed. 208pp. 5⅜ x 8½.   42541-X

METAL FATIGUE, N. E. Frost, K. J. Marsh, and L. P. Pook. Definitive, clearly written, and well-illustrated volume addresses all aspects of the subject, from the historical development of understanding metal fatigue to vital concepts of the cyclic stress that causes a crack to grow. Includes 7 appendixes. 544pp. 5⅜ x 8½.   40927-9

ROCKETS, Robert Goddard. Two of the most significant publications in the history of rocketry and jet propulsion: "A Method of Reaching Extreme Altitudes" (1919) and "Liquid Propellant Rocket Development" (1936). 128pp. 5⅜ x 8½.   42537-1

STATISTICAL MECHANICS: Principles and Applications, Terrell L. Hill. Standard text covers fundamentals of statistical mechanics, applications to fluctuation theory, imperfect gases, distribution functions, more. 448pp. 5⅜ x 8½.   65390-0

ENGINEERING AND TECHNOLOGY 1650–1750: Illustrations and Texts from Original Sources, Martin Jensen. Highly readable text with more than 200 contemporary drawings and detailed engravings of engineering projects dealing with surveying, leveling, materials, hand tools, lifting equipment, transport and erection, piling, bailing, water supply, hydraulic engineering, and more. Among the specific projects outlined–transporting a 50-ton stone to the Louvre, erecting an obelisk, building timber locks, and dredging canals. 207pp. 8⅜ x 11¼.   42232-1

THE VARIATIONAL PRINCIPLES OF MECHANICS, Cornelius Lanczos. Graduate level coverage of calculus of variations, equations of motion, relativistic mechanics, more. First inexpensive paperbound edition of classic treatise. Index. Bibliography. 418pp. 5⅜ x 8½.   65067-7

PROTECTION OF ELECTRONIC CIRCUITS FROM OVERVOLTAGES, Ronald B. Standler. Five-part treatment presents practical rules and strategies for circuits designed to protect electronic systems from damage by transient overvoltages. 1989 ed. xxiv+434pp. 6⅛ x 9¼.   42552-5

ROTARY WING AERODYNAMICS, W. Z. Stepniewski. Clear, concise text covers aerodynamic phenomena of the rotor and offers guidelines for helicopter performance evaluation. Originally prepared for NASA. 537 figures. 640pp. 6⅛ x 9¼.   64647-5

INTRODUCTION TO SPACE DYNAMICS, William Tyrrell Thomson. Comprehensive, classic introduction to space-flight engineering for advanced undergraduate and graduate students. Includes vector algebra, kinematics, transformation of coordinates. Bibliography. Index. 352pp. 5⅜ x 8½.   65113-4

HISTORY OF STRENGTH OF MATERIALS, Stephen P. Timoshenko. Excellent historical survey of the strength of materials with many references to the theories of elasticity and structure. 245 figures. 452pp. 5⅜ x 8½.   61187-6

ANALYTICAL FRACTURE MECHANICS, David J. Unger. Self-contained text supplements standard fracture mechanics texts by focusing on analytical methods for determining crack-tip stress and strain fields. 336pp. 6⅛ x 9¼.   41737-9

STATISTICAL MECHANICS OF ELASTICITY, J. H. Weiner. Advanced, self-contained treatment illustrates general principles and elastic behavior of solids. Part 1, based on classical mechanics, studies thermoelastic behavior of crystalline and polymeric solids. Part 2, based on quantum mechanics, focuses on interatomic force laws, behavior of solids, and thermally activated processes. For students of physics and chemistry and for polymer physicists. 1983 ed. 96 figures. 496pp. 5⅜ x 8½.   42260-7

Mathematics

FUNCTIONAL ANALYSIS (Second Corrected Edition), George Bachman and Lawrence Narici. Excellent treatment of subject geared toward students with background in linear algebra, advanced calculus, physics, and engineering. Text covers introduction to inner-product spaces, normed, metric spaces, and topological spaces; complete orthonormal sets, the Hahn-Banach Theorem and its consequences, and many other related subjects. 1966 ed. 544pp. 6⅛ x 9¼. 40251-7

ASYMPTOTIC EXPANSIONS OF INTEGRALS, Norman Bleistein & Richard A. Handelsman. Best introduction to important field with applications in a variety of scientific disciplines. New preface. Problems. Diagrams. Tables. Bibliography. Index. 448pp. 5⅜ x 8½. 65082-0

VECTOR AND TENSOR ANALYSIS WITH APPLICATIONS, A. I. Borisenko and I. E. Tarapov. Concise introduction. Worked-out problems, solutions, exercises. 257pp. 5⅜ x 8¼. 63833-2

THE ABSOLUTE DIFFERENTIAL CALCULUS (CALCULUS OF TENSORS), Tullio Levi-Civita. Great 20th-century mathematician's classic work on material necessary for mathematical grasp of theory of relativity. 452pp. 5⅜ x 8¼. 63401-9

AN INTRODUCTION TO ORDINARY DIFFERENTIAL EQUATIONS, Earl A. Coddington. A thorough and systematic first course in elementary differential equations for undergraduates in mathematics and science, with many exercises and problems (with answers). Index. 304pp. 5⅜ x 8½. 65942-9

FOURIER SERIES AND ORTHOGONAL FUNCTIONS, Harry F. Davis. An incisive text combining theory and practical example to introduce Fourier series, orthogonal functions and applications of the Fourier method to boundary-value problems. 570 exercises. Answers and notes. 416pp. 5⅜ x 8½. 65973-9

COMPUTABILITY AND UNSOLVABILITY, Martin Davis. Classic graduate-level introduction to theory of computability, usually referred to as theory of recurrent functions. New preface and appendix. 288pp. 5⅜ x 8½. 61471-9

ASYMPTOTIC METHODS IN ANALYSIS, N. G. de Bruijn. An inexpensive, comprehensive guide to asymptotic methods—the pioneering work that teaches by explaining worked examples in detail. Index. 224pp. 5⅜ x 8½ 64221-6

APPLIED COMPLEX VARIABLES, John W. Dettman. Step-by-step coverage of fundamentals of analytic function theory—plus lucid exposition of five important applications: Potential Theory; Ordinary Differential Equations; Fourier Transforms; Laplace Transforms; Asymptotic Expansions. 66 figures. Exercises at chapter ends. 512pp. 5⅜ x 8½. 64670-X

INTRODUCTION TO LINEAR ALGEBRA AND DIFFERENTIAL EQUATIONS, John W. Dettman. Excellent text covers complex numbers, determinants, orthonormal bases, Laplace transforms, much more. Exercises with solutions. Undergraduate level. 416pp. 5⅜ x 8½. 65191-6

CALCULUS OF VARIATIONS WITH APPLICATIONS, George M. Ewing. Applications-oriented introduction to variational theory develops insight and promotes understanding of specialized books, research papers. Suitable for advanced undergraduate/graduate students as primary, supplementary text. 352pp. 5⅜ x 8½.
64856-7

COMPLEX VARIABLES, Francis J. Flanigan. Unusual approach, delaying complex algebra till harmonic functions have been analyzed from real variable viewpoint. Includes problems with answers. 364pp. 5⅜ x 8½.
61388-7

AN INTRODUCTION TO THE CALCULUS OF VARIATIONS, Charles Fox. Graduate-level text covers variations of an integral, isoperimetrical problems, least action, special relativity, approximations, more. References. 279pp. 5⅜ x 8½.
65499-0

COUNTEREXAMPLES IN ANALYSIS, Bernard R. Gelbaum and John M. H. Olmsted. These counterexamples deal mostly with the part of analysis known as "real variables." The first half covers the real number system, and the second half encompasses higher dimensions. 1962 edition. xxiv+198pp. 5⅜ x 8½. 42875-3

CATASTROPHE THEORY FOR SCIENTISTS AND ENGINEERS, Robert Gilmore. Advanced-level treatment describes mathematics of theory grounded in the work of Poincaré, R. Thom, other mathematicians. Also important applications to problems in mathematics, physics, chemistry, and engineering. 1981 edition. References. 28 tables. 397 black-and-white illustrations. xvii+666pp. 6⅛ x 9¼.
67539-4

INTRODUCTION TO DIFFERENCE EQUATIONS, Samuel Goldberg. Exceptionally clear exposition of important discipline with applications to sociology, psychology, economics. Many illustrative examples; over 250 problems. 260pp. 5⅜ x 8½.
65084-7

NUMERICAL METHODS FOR SCIENTISTS AND ENGINEERS, Richard Hamming. Classic text stresses frequency approach in coverage of algorithms, polynomial approximation, Fourier approximation, exponential approximation, other topics. Revised and enlarged 2nd edition. 721pp. 5⅜ x 8½. 65241-6

INTRODUCTION TO NUMERICAL ANALYSIS (2nd Edition), F. B. Hildebrand. Classic, fundamental treatment covers computation, approximation, interpolation, numerical differentiation and integration, other topics. 150 new problems. 669pp. 5⅜ x 8½. 65363-3

THREE PEARLS OF NUMBER THEORY, A. Y. Khinchin. Three compelling puzzles require proof of a basic law governing the world of numbers. Challenges concern van der Waerden's theorem, the Landau-Schnirelmann hypothesis and Mann's theorem, and a solution to Waring's problem. Solutions included. 64pp. 5⅜ x 8½.
40026-3

THE PHILOSOPHY OF MATHEMATICS: An Introductory Essay, Stephan Körner. Surveys the views of Plato, Aristotle, Leibniz & Kant concerning propositions and theories of applied and pure mathematics. Introduction. Two appendices. Index. 198pp. 5⅜ x 8½. 25048-2

INTRODUCTORY REAL ANALYSIS, A.N. Kolmogorov, S. V. Fomin. Translated by Richard A. Silverman. Self-contained, evenly paced introduction to real and functional analysis. Some 350 problems. 403pp. 5⅜ x 8½.   61226-0

APPLIED ANALYSIS, Cornelius Lanczos. Classic work on analysis and design of finite processes for approximating solution of analytical problems. Algebraic equations, matrices, harmonic analysis, quadrature methods, more. 559pp. 5⅜ x 8½.   65656-X

AN INTRODUCTION TO ALGEBRAIC STRUCTURES, Joseph Landin. Superb self-contained text covers "abstract algebra": sets and numbers, theory of groups, theory of rings, much more. Numerous well-chosen examples, exercises. 247pp. 5⅜ x 8½.
65940-2

QUALITATIVE THEORY OF DIFFERENTIAL EQUATIONS, V. V. Nemytskii and V.V. Stepanov. Classic graduate-level text by two prominent Soviet mathematicians covers classical differential equations as well as topological dynamics and ergodic theory. Bibliographies. 523pp. 5⅜ x 8½.   65954-2

THEORY OF MATRICES, Sam Perlis. Outstanding text covering rank, nonsingularity and inverses in connection with the development of canonical matrices under the relation of equivalence, and without the intervention of determinants. Includes exercises. 237pp. 5⅜ x 8½.   66810-X

INTRODUCTION TO ANALYSIS, Maxwell Rosenlicht. Unusually clear, accessible coverage of set theory, real number system, metric spaces, continuous functions, Riemann integration, multiple integrals, more. Wide range of problems. Undergraduate level. Bibliography. 254pp. 5⅜ x 8½.   65038-3

MODERN NONLINEAR EQUATIONS, Thomas L. Saaty. Emphasizes practical solution of problems; covers seven types of equations. ". . . a welcome contribution to the existing literature. . . ."–*Math Reviews.* 490pp. 5⅜ x 8½.   64232-1

MATRICES AND LINEAR ALGEBRA, Hans Schneider and George Phillip Barker. Basic textbook covers theory of matrices and its applications to systems of linear equations and related topics such as determinants, eigenvalues, and differential equations. Numerous exercises. 432pp. 5⅜ x 8½.   66014-1

MATHEMATICS APPLIED TO CONTINUUM MECHANICS, Lee A. Segel. Analyzes models of fluid flow and solid deformation. For upper-level math, science, and engineering students. 608pp. 5⅜ x 8½.   65369-2

ELEMENTS OF REAL ANALYSIS, David A. Sprecher. Classic text covers fundamental concepts, real number system, point sets, functions of a real variable, Fourier series, much more. Over 500 exercises. 352pp. 5⅜ x 8½.   65385-4

SET THEORY AND LOGIC, Robert R. Stoll. Lucid introduction to unified theory of mathematical concepts. Set theory and logic seen as tools for conceptual understanding of real number system. 496pp. 5⅜ x 8¼.   63829-4

TENSOR CALCULUS, J.L. Synge and A. Schild. Widely used introductory text covers spaces and tensors, basic operations in Riemannian space, non-Riemannian spaces, etc. 324pp. 5⅜ x 8¼. 63612-7

ORDINARY DIFFERENTIAL EQUATIONS, Morris Tenenbaum and Harry Pollard. Exhaustive survey of ordinary differential equations for undergraduates in mathematics, engineering, science. Thorough analysis of theorems. Diagrams. Bibliography. Index. 818pp. 5⅜ x 8½. 64940-7

INTEGRAL EQUATIONS, F. G. Tricomi. Authoritative, well-written treatment of extremely useful mathematical tool with wide applications. Volterra Equations, Fredholm Equations, much more. Advanced undergraduate to graduate level. Exercises. Bibliography. 238pp. 5⅜ x 8½. 64828-1

FOURIER SERIES, Georgi P. Tolstov. Translated by Richard A. Silverman. A valuable addition to the literature on the subject, moving clearly from subject to subject and theorem to theorem. 107 problems, answers. 336pp. 5⅜ x 8½. 63317-9

INTRODUCTION TO MATHEMATICAL THINKING, Friedrich Waismann. Examinations of arithmetic, geometry, and theory of integers; rational and natural numbers; complete induction; limit and point of accumulation; remarkable curves; complex and hypercomplex numbers, more. 1959 ed. 27 figures. xii+260pp. 5⅜ x 8½. 42804-4

POPULAR LECTURES ON MATHEMATICAL LOGIC, Hao Wang. Noted logician's lucid treatment of historical developments, set theory, model theory, recursion theory and constructivism, proof theory, more. 3 appendixes. Bibliography. 1981 ed. ix+283pp. 5⅜ x 8½. 67632-3

CALCULUS OF VARIATIONS, Robert Weinstock. Basic introduction covering isoperimetric problems, theory of elasticity, quantum mechanics, electrostatics, etc. Exercises throughout. 326pp. 5⅜ x 8½. 63069-2

THE CONTINUUM: A Critical Examination of the Foundation of Analysis, Hermann Weyl. Classic of 20th-century foundational research deals with the conceptual problem posed by the continuum. 156pp. 5⅜ x 8½. 67982-9

CHALLENGING MATHEMATICAL PROBLEMS WITH ELEMENTARY SOLUTIONS, A. M. Yaglom and I. M. Yaglom. Over 170 challenging problems on probability theory, combinatorial analysis, points and lines, topology, convex polygons, many other topics. Solutions. Total of 445pp. 5⅜ x 8½. Two-vol. set.
Vol. I: 65536-9 Vol. II: 65537-7

INTRODUCTION TO PARTIAL DIFFERENTIAL EQUATIONS WITH APPLICATIONS, E. C. Zachmanoglou and Dale W. Thoe. Essentials of partial differential equations applied to common problems in engineering and the physical sciences. Problems and answers. 416pp. 5⅜ x 8½. 65251-3

THE THEORY OF GROUPS, Hans J. Zassenhaus. Well-written graduate-level text acquaints reader with group-theoretic methods and demonstrates their usefulness in mathematics. Axioms, the calculus of complexes, homomorphic mapping, p-group theory, more. 276pp. 5⅜ x 8½. 40922-8

Math–Decision Theory, Statistics, Probability

ELEMENTARY DECISION THEORY, Herman Chernoff and Lincoln E. Moses. Clear introduction to statistics and statistical theory covers data processing, probability and random variables, testing hypotheses, much more. Exercises. 364pp. 5⅜ x 8½. 65218-1

STATISTICS MANUAL, Edwin L. Crow et al. Comprehensive, practical collection of classical and modern methods prepared by U.S. Naval Ordnance Test Station. Stress on use. Basics of statistics assumed. 288pp. 5⅜ x 8½. 60599-X

SOME THEORY OF SAMPLING, William Edwards Deming. Analysis of the problems, theory, and design of sampling techniques for social scientists, industrial managers, and others who find statistics important at work. 61 tables. 90 figures. xvii +602pp. 5⅜ x 8½. 64684-X

LINEAR PROGRAMMING AND ECONOMIC ANALYSIS, Robert Dorfman, Paul A. Samuelson and Robert M. Solow. First comprehensive treatment of linear programming in standard economic analysis. Game theory, modern welfare economics, Leontief input-output, more. 525pp. 5⅜ x 8½. 65491-5

PROBABILITY: An Introduction, Samuel Goldberg. Excellent basic text covers set theory, probability theory for finite sample spaces, binomial theorem, much more. 360 problems. Bibliographies. 322pp. 5⅜ x 8½. 65252-1

GAMES AND DECISIONS: Introduction and Critical Survey, R. Duncan Luce and Howard Raiffa. Superb nontechnical introduction to game theory, primarily applied to social sciences. Utility theory, zero-sum games, n-person games, decision-making, much more. Bibliography. 509pp. 5⅜ x 8½. 65943-7

INTRODUCTION TO THE THEORY OF GAMES, J. C. C. McKinsey. This comprehensive overview of the mathematical theory of games illustrates applications to situations involving conflicts of interest, including economic, social, political, and military contexts. Appropriate for advanced undergraduate and graduate courses; advanced calculus a prerequisite. 1952 ed. x+372pp. 5⅜ x 8½. 42811-7

FIFTY CHALLENGING PROBLEMS IN PROBABILITY WITH SOLUTIONS, Frederick Mosteller. Remarkable puzzlers, graded in difficulty, illustrate elementary and advanced aspects of probability. Detailed solutions. 88pp. 5⅜ x 8½. 65355-2

PROBABILITY THEORY: A Concise Course, Y. A. Rozanov. Highly readable, self-contained introduction covers combination of events, dependent events, Bernoulli trials, etc. 148pp. 5⅜ x 8¼. 63544-9

STATISTICAL METHOD FROM THE VIEWPOINT OF QUALITY CONTROL, Walter A. Shewhart. Important text explains regulation of variables, uses of statistical control to achieve quality control in industry, agriculture, other areas. 192pp. 5⅜ x 8½. 65232-7

Math–Geometry and Topology

ELEMENTARY CONCEPTS OF TOPOLOGY, Paul Alexandroff. Elegant, intuitive approach to topology from set-theoretic topology to Betti groups; how concepts of topology are useful in math and physics. 25 figures. 57pp. 5⅜ x 8½. 60747-X

COMBINATORIAL TOPOLOGY, P. S. Alexandrov. Clearly written, well organized, three-part text begins by dealing with certain classic problems without using the formal techniques of homology theory and advances to the central concept, the Betti groups. Numerous detailed examples. 654pp. 5⅜ x 8½. 40179-0

EXPERIMENTS IN TOPOLOGY, Stephen Barr. Classic, lively explanation of one of the byways of mathematics. Klein bottles, Moebius strips, projective planes, map coloring, problem of the Koenigsberg bridges, much more, described with clarity and wit. 43 figures. 210pp. 5⅜ x 8½. 25933-1

CONFORMAL MAPPING ON RIEMANN SURFACES, Harvey Cohn. Lucid, insightful book presents ideal coverage of subject. 334 exercises make book perfect for self-study. 55 figures. 352pp. 5⅝ x 8¼. 64025-6

THE GEOMETRY OF RENÉ DESCARTES, René Descartes. The great work founded analytical geometry. Original French text, Descartes's own diagrams, together with definitive Smith-Latham translation. 244pp. 5⅜ x 8½. 60068-8

PRACTICAL CONIC SECTIONS: The Geometric Properties of Ellipses, Parabolas and Hyperbolas, J. W. Downs. This text shows how to create ellipses, parabolas, and hyperbolas. It also presents historical background on their ancient origins and describes the reflective properties and roles of curves in design applications. 1993 ed. 98 figures. xii+100pp. 6½ x 9¼. 42876-1

THE THIRTEEN BOOKS OF EUCLID'S ELEMENTS, translated with introduction and commentary by Thomas L. Heath. Definitive edition. Textual and linguistic notes, mathematical analysis. 2,500 years of critical commentary. Unabridged. 1,414pp. 5⅜ x 8½. Three-vol. set. Vol. I: 60088-2 Vol. II: 60089-0 Vol. III: 60090-4

GEOMETRY OF COMPLEX NUMBERS, Hans Schwerdtfeger. Illuminating, widely praised book on analytic geometry of circles, the Moebius transformation, and two-dimensional non-Euclidean geometries. 200pp. 5⅝ x 8¼. 63830-8

DIFFERENTIAL GEOMETRY, Heinrich W. Guggenheimer. Local differential geometry as an application of advanced calculus and linear algebra. Curvature, transformation groups, surfaces, more. Exercises. 62 figures. 378pp. 5⅜ x 8½. 63433-7

CURVATURE AND HOMOLOGY: Enlarged Edition, Samuel I. Goldberg. Revised edition examines topology of differentiable manifolds; curvature, homology of Riemannian manifolds; compact Lie groups; complex manifolds; curvature, homology of Kaehler manifolds. New Preface. Four new appendixes. 416pp. 5⅜ x 8½. 40207-X

History of Math

THE WORKS OF ARCHIMEDES, Archimedes (T. L. Heath, ed.). Topics include the famous problems of the ratio of the areas of a cylinder and an inscribed sphere; the measurement of a circle; the properties of conoids, spheroids, and spirals; and the quadrature of the parabola. Informative introduction. clxxxvi+326pp; supplement, 52pp. 5⅜ x 8½.   42084-1

A SHORT ACCOUNT OF THE HISTORY OF MATHEMATICS, W. W. Rouse Ball. One of clearest, most authoritative surveys from the Egyptians and Phoenicians through 19th-century figures such as Grassman, Galois, Riemann. Fourth edition. 522pp. 5⅜ x 8½.   20630-0

THE HISTORY OF THE CALCULUS AND ITS CONCEPTUAL DEVELOP-MENT, Carl B. Boyer. Origins in antiquity, medieval contributions, work of Newton, Leibniz, rigorous formulation. Treatment is verbal. 346pp. 5⅜ x 8½.   60509-4

THE HISTORICAL ROOTS OF ELEMENTARY MATHEMATICS, Lucas N. H. Bunt, Phillip S. Jones, and Jack D. Bedient. Fundamental underpinnings of modern arithmetic, algebra, geometry, and number systems derived from ancient civiliza-tions. 320pp. 5⅜ x 8½.   25563-8

A HISTORY OF MATHEMATICAL NOTATIONS, Florian Cajori. This classic study notes the first appearance of a mathematical symbol and its origin, the com-petition it encountered, its spread among writers in different countries, its rise to pop-ularity, its eventual decline or ultimate survival. Original 1929 two-volume edition presented here in one volume. xxviii+820pp. 5⅜ x 8½.   67766-4

GAMES, GODS & GAMBLING: A History of Probability and Statistical Ideas, F. N. David. Episodes from the lives of Galileo, Fermat, Pascal, and others illustrate this fascinating account of the roots of mathematics. Features thought-provoking refer-ences to classics, archaeology, biography, poetry. 1962 edition. 304pp. 5⅜ x 8½. (Available in U.S. only.)   40023-9

OF MEN AND NUMBERS: The Story of the Great Mathematicians, Jane Muir. Fascinating accounts of the lives and accomplishments of history's greatest mathe-matical minds–Pythagoras, Descartes, Euler, Pascal, Cantor, many more. Anecdotal, illuminating. 30 diagrams. Bibliography. 256pp. 5⅜ x 8½.   28973-7

HISTORY OF MATHEMATICS, David E. Smith. Nontechnical survey from ancient Greece and Orient to late 19th century; evolution of arithmetic, geometry, trigonometry, calculating devices, algebra, the calculus. 362 illustrations. 1,355pp. 5⅜ x 8½. Two-vol. set.   Vol. I: 20429-4   Vol. II: 20430-8

A CONCISE HISTORY OF MATHEMATICS, Dirk J. Struik. The best brief his-tory of mathematics. Stresses origins and covers every major figure from ancient Near East to 19th century. 41 illustrations. 195pp. 5⅜ x 8½.   60255-9

Physics

OPTICAL RESONANCE AND TWO-LEVEL ATOMS, L. Allen and J. H. Eberly. Clear, comprehensive introduction to basic principles behind all quantum optical resonance phenomena. 53 illustrations. Preface. Index. 256pp. 5⅜ x 8½.   65533-4

QUANTUM THEORY, David Bohm. This advanced undergraduate-level text presents the quantum theory in terms of qualitative and imaginative concepts, followed by specific applications worked out in mathematical detail. Preface. Index. 655pp. 5⅜ x 8½.   65969-0

ATOMIC PHYSICS: 8th edition, Max Born. Nobel laureate's lucid treatment of kinetic theory of gases, elementary particles, nuclear atom, wave-corpuscles, atomic structure and spectral lines, much more. Over 40 appendices, bibliography. 495pp. 5⅜ x 8½.   65984-4

A SOPHISTICATE'S PRIMER OF RELATIVITY, P. W. Bridgman. Geared toward readers already acquainted with special relativity, this book transcends the view of theory as a working tool to answer natural questions: What is a frame of reference? What is a "law of nature"? What is the role of the "observer"? Extensive treatment, written in terms accessible to those without a scientific background. 1983 ed. xlviii+172pp. 5⅜ x 8½.   42549-5

AN INTRODUCTION TO HAMILTONIAN OPTICS, H. A. Buchdahl. Detailed account of the Hamiltonian treatment of aberration theory in geometrical optics. Many classes of optical systems defined in terms of the symmetries they possess. Problems with detailed solutions. 1970 edition. xv+360pp. 5⅜ x 8½.   67597-1

PRIMER OF QUANTUM MECHANICS, Marvin Chester. Introductory text examines the classical quantum bead on a track: its state and representations; operator eigenvalues; harmonic oscillator and bound bead in a symmetric force field; and bead in a spherical shell. Other topics include spin, matrices, and the structure of quantum mechanics; the simplest atom; indistinguishable particles; and stationary-state perturbation theory. 1992 ed. xiv+314pp. 6⅛ x 9¼.   42878-8

LECTURES ON QUANTUM MECHANICS, Paul A. M. Dirac. Four concise, brilliant lectures on mathematical methods in quantum mechanics from Nobel Prize–winning quantum pioneer build on idea of visualizing quantum theory through the use of classical mechanics. 96pp. 5⅜ x 8½.   41713-1

THIRTY YEARS THAT SHOOK PHYSICS: The Story of Quantum Theory, George Gamow. Lucid, accessible introduction to influential theory of energy and matter. Careful explanations of Dirac's anti-particles, Bohr's model of the atom, much more. 12 plates. Numerous drawings. 240pp. 5⅜ x 8½.   24895-X

ELECTRONIC STRUCTURE AND THE PROPERTIES OF SOLIDS: The Physics of the Chemical Bond, Walter A. Harrison. Innovative text offers basic understanding of the electronic structure of covalent and ionic solids, simple metals, transition metals and their compounds. Problems. 1980 edition. 582pp. 6⅛ x 9¼.   66021-4

HYDRODYNAMIC AND HYDROMAGNETIC STABILITY, S. Chandrasekhar. Lucid examination of the Rayleigh-Benard problem; clear coverage of the theory of instabilities causing convection. 704pp. 5⅜ x 8¼. 64071-X

INVESTIGATIONS ON THE THEORY OF THE BROWNIAN MOVEMENT, Albert Einstein. Five papers (1905–8) investigating dynamics of Brownian motion and evolving elementary theory. Notes by R. Fürth. 122pp. 5⅜ x 8½. 60304-0

THE PHYSICS OF WAVES, William C. Elmore and Mark A. Heald. Unique overview of classical wave theory. Acoustics, optics, electromagnetic radiation, more. Ideal as classroom text or for self-study. Problems. 477pp. 5⅜ x 8½. 64926-1

PHYSICAL PRINCIPLES OF THE QUANTUM THEORY, Werner Heisenberg. Nobel Laureate discusses quantum theory, uncertainty, wave mechanics, work of Dirac, Schroedinger, Compton, Wilson, Einstein, etc. 184pp. 5⅜ x 8½. 60113-7

ATOMIC SPECTRA AND ATOMIC STRUCTURE, Gerhard Herzberg. One of best introductions; especially for specialist in other fields. Treatment is physical rather than mathematical. 80 illustrations. 257pp. 5⅜ x 8½. 60115-3

AN INTRODUCTION TO STATISTICAL THERMODYNAMICS, Terrell L. Hill. Excellent basic text offers wide-ranging coverage of quantum statistical mechanics, systems of interacting molecules, quantum statistics, more. 523pp. 5⅜ x 8½. 65242-4

THEORETICAL PHYSICS, Georg Joos, with Ira M. Freeman. Classic overview covers essential math, mechanics, electromagnetic theory, thermodynamics, quantum mechanics, nuclear physics, other topics. xxiii+885pp. 5⅜ x 8½. 65227-0

PROBLEMS AND SOLUTIONS IN QUANTUM CHEMISTRY AND PHYSICS, Charles S. Johnson, Jr. and Lee G. Pedersen. Unusually varied problems, detailed solutions in coverage of quantum mechanics, wave mechanics, angular momentum, molecular spectroscopy, more. 280 problems, 139 supplementary exercises. 430pp. 6½ x 9¼. 65236-X

THEORETICAL SOLID STATE PHYSICS, Vol. I: Perfect Lattices in Equilibrium; Vol. II: Non-Equilibrium and Disorder, William Jones and Norman H. March. Monumental reference work covers fundamental theory of equilibrium properties of perfect crystalline solids, non-equilibrium properties, defects and disordered systems. Total of 1,301pp. 5⅜ x 8½. Vol. I: 65015-4 Vol. II: 65016-2

WHAT IS RELATIVITY? L. D. Landau and G. B. Rumer. Written by a Nobel Prize physicist and his distinguished colleague, this compelling book explains the special theory of relativity to readers with no scientific background, using such familiar objects as trains, rulers, and clocks. 1960 ed. vi+72pp. 23 b/w illustrations. 5⅜ x 8½. 42806-0 $6.95

A TREATISE ON ELECTRICITY AND MAGNETISM, James Clerk Maxwell. Important foundation work of modern physics. Brings to final form Maxwell's theory of electromagnetism and rigorously derives his general equations of field theory. 1,084pp. 5⅜ x 8½. Two-vol. set. Vol. I: 60636-8 Vol. II: 60637-6

QUANTUM MECHANICS: Principles and Formalism, Roy McWeeny. Graduate student–oriented volume develops subject as fundamental discipline, opening with review of origins of Schrödinger's equations and vector spaces. Focusing on main principles of quantum mechanics and their immediate consequences, it concludes with final generalizations covering alternative "languages" or representations. 1972 ed. 15 figures. xi+155pp. $5\frac{3}{8}$ x $8\frac{1}{2}$. 42829-X

INTRODUCTION TO QUANTUM MECHANICS WITH APPLICATIONS TO CHEMISTRY, Linus Pauling & E. Bright Wilson, Jr. Classic undergraduate text by Nobel Prize winner applies quantum mechanics to chemical and physical problems. Numerous tables and figures enhance the text. Chapter bibliographies. Appendices. Index. 468pp. $5\frac{3}{8}$ x $8\frac{1}{2}$. 64871-0

METHODS OF THERMODYNAMICS, Howard Reiss. Outstanding text focuses on physical technique of thermodynamics, typical problem areas of understanding, and significance and use of thermodynamic potential. 1965 edition. 238pp. $5\frac{3}{8}$ x $8\frac{1}{2}$. 69445-3

TENSOR ANALYSIS FOR PHYSICISTS, J. A. Schouten. Concise exposition of the mathematical basis of tensor analysis, integrated with well-chosen physical examples of the theory. Exercises. Index. Bibliography. 289pp. $5\frac{3}{8}$ x $8\frac{1}{2}$. 65582-2

THE ELECTROMAGNETIC FIELD, Albert Shadowitz. Comprehensive undergraduate text covers basics of electric and magnetic fields, builds up to electromagnetic theory. Also related topics, including relativity. Over 900 problems. 768pp. $5\frac{5}{8}$ x $8\frac{1}{4}$. 65660-8

GREAT EXPERIMENTS IN PHYSICS: Firsthand Accounts from Galileo to Einstein, Morris H. Shamos (ed.). 25 crucial discoveries: Newton's laws of motion, Chadwick's study of the neutron, Hertz on electromagnetic waves, more. Original accounts clearly annotated. 370pp. $5\frac{3}{8}$ x $8\frac{1}{2}$. 25346-5

RELATIVITY, THERMODYNAMICS AND COSMOLOGY, Richard C. Tolman. Landmark study extends thermodynamics to special, general relativity; also applications of relativistic mechanics, thermodynamics to cosmological models. 501pp. $5\frac{3}{8}$ x $8\frac{1}{2}$. 65383-8

STATISTICAL PHYSICS, Gregory H. Wannier. Classic text combines thermodynamics, statistical mechanics, and kinetic theory in one unified presentation of thermal physics. Problems with solutions. Bibliography. 532pp. $5\frac{3}{8}$ x $8\frac{1}{2}$. 65401-X

Paperbound unless otherwise indicated. Available at your book dealer, online at **www.doverpublications.com**, or by writing to Dept. GI, Dover Publications, Inc., 31 East 2nd Street, Mineola, NY 11501. For current price information or for free catalogs (please indicate field of interest), write to Dover Publications or log on to **www.doverpublications.com** and see every Dover book in print. Dover publishes more than 500 books each year on science, elementary and advanced mathematics, biology, music, art, literary history, social sciences, and other areas.